INTRODUCTION TO NONLINEAR OPTICAL EFFECTS IN MOLECULES AND POLYMERS

INTRODUCTION TO NONLINEAR OPTICAL EFFECTS IN MOLECULES AND POLYMERS

PARAS N. PRASAD
Photonics Research Laboratory
Department of Chemistry
State University of New York
Buffalo, New York

and

DAVID J. WILLIAMS
Corporate Research Laboratories
Eastman Kodak Company
Rochester, New York

A Wiley-Interscience Publication

JOHN WILEY & SONS, Inc.

New York · Chichester · Brisbane · Toronto · Singapore

Library of Congress Cataloging in Publication Data:

Prasad, Paras N.
    Introduction to nonlinear optical effects in molecules and polymers / by Paras N. Prasad and David J. Williams.
        p.   cm.
    Includes bibliographical references.
    ISBN 0-471-51562-0
    1. Optical materials.   2. Polymers—Optical properties.
3. Nonlinear optics.   I. Williams, David J., 1943–   . II. Title.
QC374.P73   1990   90–37692
535—dc20   CIP

Printed in the United States of America

10 9 8 7 6 5 4 3 2 1

PREFACE

Nonlinear optics is a new frontier of science and technology that is to play a major role in the emerging technology of photonics. Photonics, which uses photons for information and image processing, has been labeled the technology of the 21st century, for which nonlinear optical processes provide the key functions of frequency conversion and optical switching. Molecular materials and polymeric systems have emerged in recent years as a new class of promising nonlinear optical materials because they offer the flexibility, both at the molecular and bulk levels, to optimize the nonlinearity and other required properties for device applications. This is a multidisciplinary area, the progress of which depends on active participation from chemists, physicists, material scientists, and engineers. Although several texts on nonlinear optics itself as well as several review books on nonlinear optical effects in organic and polymeric systems exist, there is no monograph or textbook dealing with nonlinear optics of molecules and polymers. Since the interest in this area is growing worldwide, a monograph dealing with this subject and addressing issues specific to molecular and polymeric materials will be of great value to newcomers to the field. It can also be used as a reference book by researchers of varied backgrounds. We hope that this book will be such a monograph for multidisciplinary readership. We have emphasized concepts and avoided mathematical rigor so as to provide useful conceptual information to readers outside the areas of their expertise.

We wish to express gratitude to our wives Nadia and Carol who, in spite of their own business and professional schedules, have provided valuable support and understanding for this project.

This book uses a fair amount of material from our research programs at

v

SUNY Buffalo and Eastman Kodak company. We thank the members of our research groups for their valuable research contributions and their comments pertaining to the book. The nonlinear optics research program at Buffalo has been supported by the Directorate of Chemical and Atmospheric Sciences of The Air Force Office of Scentific Research. We are indebted to Donald R. Ulrich (AFOSR) and Jack C. Chang (E.K.) for valuable support and encouragement. Finally we thank G. I. Stegeman, T. George, J. McIver, R. Boyd, and R. Fisher for taking time to read the manuscript and give their comments.

PARAS N. PRASAD
DAVID J. WILLIAMS

Buffalo, New York
Rochester, New York
October 1990

CONTENTS

INTRODUCTION TO NONLINEAR OPTICAL EFFECTS IN MOLECULES AND POLYMERS

1

INTRODUCTION

1.1 NONLINEAR OPTICS AND PHOTONICS

Nonlinear optics is expected to play a major role in the technology of photonics. Photonics is emerging as a multidiciplinary new frontier of science and technology that is capturing the imagination of scientists and engineers worldwide because of its potential applications to many areas of present and future information and image processing technologies. Photonics is the analog of electronics in that it describes the technology in which photons instead of electrons are used to acquire, store, transmit, and process information. Examples of nonlinear optical phenomena that are potentially useful in this context are the ability to alter the frequency or color of light and to amplify one source of light with another, switch it, or alter its transmission characteristics through a medium, depending on its intensity. It is the potential for providing these functions in suitable materials and devices that motivates much of the current fundamental and exploratory research in the field of nonlinear optics.

With the advent of lasers, which provide a source of high-intensity coherent light, much progress has been made in the field of nonlinear optics. The strong oscillating electric field of the laser beam creates a polarization response that is nonlinear in character and that can act as a source of new optical fields with altered properties. Among the nonlinear optical processes that have been studied, one of the most visually dramatic is frequency doubling. In the field of optical information storage this process can provide for the conversion of near-infrared laser light from diode lasers into deep blue light. Since the size of a focused spot of light is inversely proportional to its wavelength, second-harmonic generation can increase the capacity of stored information on optical disks immensely. Using related phenomena, one can also build devices, such as frequency mixers, that can act as new light sources or as amplification schemes, light modulators for controlling the phase or amplitude of a light beam, optical switches, optical logic, optical information storage, optical limiters, and numerous ways of processing the information content of data or images.

Optical processing of information and optical computing is one of the most appealing applications of photonics. Major potential advantages are a gain in speed in certain types of switching functions and the reconfigurable connectivity of light sources and detectors with little interference or cross-talk between adjacent channels. Photonic switching can take place with femtosecond speeds, thus providing a gain in speed that is many orders of magnitude over that of electronic processes. Working at optical frequencies provides a tremendous gain in the bandwidth of information processing. Optical processing functions are generally free from interference from electrical or magnetic sources. On the other hand, the weak interactions of optical fields with each other has necessitated the development of materials where in these interactions can be maximized, thus the need for considerable new fundamental understanding of the interaction of light with matter. Based on the prospect of three-dimensional interconnectivity between sources and receptors of light, concepts of optical neural networks that mimic the fuzzy algorithms by which learning takes place in the brain have been proposed and experimentation has begun. Integrated optical circuits, which are counterparts of electric circuits but where optical fibers or channels in waveguides conduct photons, can provide for various logic, memory, and multiplexing operations. Utilizing nonlinear optical effects, analogs of transistors or optical bistable devices with which light controls light have also been demonstrated.

Apart from the field of photonics there are other potentially interesting and important applications of nonlinear optics. Among these are the use of light intensity-dependent transmission properties of materials that might serve useful functions as eye or optoelectronic sensor protection from unwanted or stray sources of laser radiation. Here the medium would become nontransparent at some critical light intensity. Nonlinear optical processes are highly sensitive to many of the microstructural aspects or interfacial characteristics of the medium that exhibits the effect. They may, therefore, be suitable for sensor applications to probe structural relaxations in materials, chemical stimuli of various sorts, pressure, temperature, electric fields, and so on.

1.2   NONLINEAR OPTICAL MATERIALS

Basically, all materials exhibit nonlinear optical phenomena. This includes all forms of matter—gases, liquids, and solids. The power of the optical fields required to observe these effects varies over many orders of magnitude, depending on the detailed nature of the electronic structure of the atomic and molecular constituents of the medium, their dynamical behavior, as well as the symmetry and details of their geometrical arrangement in the medium. The important nonlinear optical materials from the device point of view are generally in solid formats and must meet a wide variety of ancillary material requirements for practical use. In general, they will require extraordinary stability with respect to ambient conditions and high-intensity light sources. They will have to meet many processing requirements for pattern or shape definition, and integration with additional dissimilar materials.

Before the major research issues in the development of new materials for nonlinear optics are discussed, several distinctions and features of materials for application in the field are commented on. The first type, molecular materials, consists of chemically bonded molecular units that interact in the bulk through weak van der Waals interactions. Many organic crystals and polymers typify this class of materials. For these materials the optical nonlinearity is primarily derived from the molecular structure. The expression of the nonlinearity is highly dependent on the geometrical arrangement of the molecules in the condensed medium in the case of second-order nonlinear processes, but much less so for third-order nonlinearities. One can define microscopic nonlinear coefficients (molecular hyperpolarizabilities) that are the molecular equivalents of bulk nonlinear optical susceptibilities. In fact, the bulk susceptibility can be readily related to the susceptibilities of the constituent molecules. The primary step in optimizing optical nonlinearities in this class of materials is at the molecular structural level, which then requires a detailed understanding of the relationship between molecular electronic structure and the nonlinear polarization that can be induced in a molecule. The molecular engineering issues of how to arrange molecules in solids encompass areas of chemistry, polymer science, and materials science.

The second major class of materials is bulk materials. Nonlinearities in these materials are thought of as arising from electrons not associated with individual nucleii, such as those in metals and semiconductors. The optical nonlinearity in this class is determined by the electronic characteristics of the bulk medium and thus requires different theoretical frameworks to account for the origins of nonlinear optical effects. Examples of materials in this category are quantum well structures derived from GaAs and II–VI semiconductors such as CdSe. Inorganic crystals, such as potassium dihydrogen phosphate (KTP) and potassium titanyl phosphate (KTP), are also regarded as bulk materials because no single molecular unit in the ionic lattice can be identified. However, in the systems the nonlinear responses are undoubtedly related to individual bond polarizabilities.

Compared to the more traditional inorganic nonlinear optical materials, the history of organic (or, in general, molecular) nonlinear optical materials is quite new. Although this book refers to one class of molecular materials, namely organic systems, most concepts and discussions are applicable to inorganic and organometallic molecular materials as well. Organic and other molecular materials are increasingly being recognized as the materials of the future because their molecular nature combined with the versatility of synthetic chemistry can be used to alter and optimize molecular structure to maximize nonlinear responses and other properties. Other benefits associated with molecular systems derive from the fabrication methods that are available or under development for building thin-film structures. These and other perceived benefits of organic materials are summarized in the following paragraphs.

Organic structures can be grown into thin crystalline layers, fabricated into structures that can be deposited into thin films layer by layer as in the Langmuir–Blodgett technique, or incorporated into polymers for deposition and processing as highly oriented thin-film structures. The resulting structures can exhibit optimized orientations for many types of nonlinear behavior, especially those of interest for waveguide formats.

Many organic materials, especially high-performance polymers, have high mechanical strength as well as excellent environmental and thermal stability. Recent developments have produced materials with thermal stabilities in excess of $350\,°C$. In contrast to misconceptions about the fraility of organic materials, the optical damage threshold for polymeric materials can easily be $>10\,GW/cm^2$ with picosecond pulses. In contrast, multiple-quantum well structures derived from GaAs will undergo optical damage at power densities many orders of magnitude lower.

Because of their unique chemical structures (π bonding), organic molecular materials exhibit the largest nonresonant (nonabsorptive) optical nonlinearities. For inorganic systems, important higher-order nonlinear optical effects (for example, third-order) are resonant (absorptive). Thus, heat dissipation tends to limit the cycle time of devices derived from these materials. For many device applications, such as in all-optical signal processing, the nonlinear optical response time is an important consideration. A nonresonant electronic optical nonlinearity, by its nature, would have the fastest response time, limited only by the width of the driving laser pulse. With current laser technology, femtosecond responses can be achieved and have, in fact, been demonstrated in organic polymers. In contrast, resonant optical nonlinearities have response times limited by the lifetime of the excitation. Other disadvantages associated with resonant optical nonlinearities are beam depletion due to absorption and thermal damage. Further complications arise from thermally induced nonlinearities associated with refractive index changes, which often can dominate the intrinsic electronic optical nonlinearity.

The dielectric constants of organic materials are considerably lower than those of inorganic crystals. This feature has important implications for electrooptic devices in which a low-frequency ac field is used to modulate the refractive index. The low dielectric constant yields a low RC time constant, thus permitting a large operating bandwidth ($>10\,GHz$) modulation. Furthermore, for organic materials the dielectric constants at low frequency are comparable to those at optical frequencies, which leads to minimization of phase mismatch between electrical and optical pulses in high-speed traveling wave devices.

From the above discussion of relative merits, it is clear that organic nonlinear optical materials are promising group of materials. This has provided the rationale for the present monograph which focuses on nonlinear optical effects in organic molecules and polymers.

1.3   BASIC RESEARCH OPPORTUNITIES

Although potential technological opportunities have provided the main impetus for the development of this field, the interest is not solely technological. This field offers challenging opportunities for fundamental research. The challenges are multidisciplinary, ranging from a basic understanding of physics of nonlinear optical interactions to molecular engineering and chemical synthesis of novel organic structures with enhanced optical nonlinearities. As we discuss in this monograph, a basic understanding of the relationship between molecular structure and microscopic optical nonlinearity is still very limited and must be sufficiently developed before we can take full advantage of the tailorability of molecular structures to enhance optical nonlinearity. Once the optical nonlinearity is maximized at the molecular structure level, the question that follows is how it transforms into bulk optical nonlinearity. Therefore, the relation between microscopic and bulk optical nonlinearities is also of fundamental importance. The relevant issues are the roles of (1) the intermolecular interaction and its importance in determining the local fields, (2) bulk excitations, (3) intermolecular charge transfer, and (4) molecular orientation effects. Although certain approximations (such as Lorentz approximation for the local electric field derived from an optical pulse) are widely used, their validity is far from conclusively established.

The effects of the dynamics of excitations and various excited-state resonances on optical nonlinearities of organic systems are not well explored. For inorganic semiconductors, it is well established that in many cases photogenerated or dopant-induced charge carriers have profound effects on nonlinear optical behavior. The role of excitation and the effect of various quantum confinements have been widely studied in inorganic semiconductors. In comparison, these types of studies are highly limited for organic systems. Organic polymers have shown that with appropriate structural features they generate novel types of excitations, such as solitons, polarons, and bipolarons. The role of these new excitations in influencing optical nonlinearities is another area of basic research opportunities.

1.4   MULTIDISCIPLINARY RESEARCH

The field of nonlinear optical effects in organic and polymer systems is truly a multidisciplinary one in which scientists and engineers of varied backgrounds from university and industrial environments can interface their expertise for its expeditious development.

The input from quantum theorists and physicists can significantly contribute to the understanding of nonlinear optical processes. Synthetic chemists can contribute significantly to the understanding of structure–property relationships by providing sequentially built and systematically derivatized structures. Measurements of optical nonlinearities on these varied structures provide a testing ground for theoretical models. The efforts of synthetic chemistry can lead to new molecular and polymeric structures to enrich the knowledge base, which currently is rather limited. The participation of polymer and material scientists and engineers is just as important. They can contribute by developing new molecular composite materials, designing processing schemes to improve optical quality, properly characterizing the bulk structures of the materials, and fabricating various device structures. Fabrication of optical quality films and fibers is an area of great need because organic systems tend to exhibit significant light scattering. Input from experimentalists and laser spectroscopists in measurement of nonlinear optical effects and study of excited-state dynamics is also needed. Important input for device processes involves the study of nonlinear optical processes in optical waveguides. Therefore, active participation of optical physicists and electrical engineers is essential. A strong input from device engineers is necessary to implement the device concepts in systems. Finally, it is clear that the most effective approach to help bring this field to maturity relatively quickly will require cross-talk and interactive feedback between participants of various backgrounds. This mode of interaction requires each group to be aware of the relevant issues outside its own area of expertise.

1.5 SCOPE OF THIS BOOK

This book is written with the objective of multidisciplinary appeal. Consequently, mathematical rigors are avoided and the focus is more on conceptual details. There are many excellent books that focus on the theory of nonlinear optical effects and we have left the detailed theoretical formulations to these works. Although several books containing expert reviews in the area of nonlinear optical effects in organic molecules and polymers exist by now, there is still a void as far as a monograph on this topic is concerned. The need for such a book became more apparent when one of the authors (PNP) was invited to offer a tutorial course on this topic at a meeting of the SPIE (the international society for optical engineers). The idea of writing a book in this field was conceived while teaching this course for several years at various SPIE meetings. The composition of the registrants for this tutorial course has been multi-disciplinary and over the years participants have constantly emphasized the need for a comprehensive and multidisciplinary monograph in this field. We have written this book with the objective that it would serve as a valuable guide to researchers working in this area in two ways: (1) by providing them with useful information in the areas outside their expertise and (2) by serving as a useful reference book. Since this area is experiencing a continuous rapid expansion, the book will be useful to newcomers who would like to build a working knowledge of this field in a relatively short time. It is our hope that the book can also be used for advanced level courses at universities and tutorial courses at various professional society meetings.

We begin by providing a brief background in the theory of nonlinear optical phenomena (Chapter 2), keeping in mind that the subject should also be readable to synthetic chemists, materials scientists, and engineers who may not be well versed in electromagnetic theory. Consequently, the concepts are developed using minimal mathematical detail. In Chapter 3 the theory of nonlinear optical effects in organic structures is developed, where the uniqueness of bonding in these structures and its manifestations in microscopic optical nonlinearities are discussed. The concepts of σ and π bonding in organic systems are briefly discussed to provide background for those who may not be familiar with them.

In Chapters 4 and 5 the various second-order nonlinear optical processes and the chemical and bulk structures required for observing these processes are discussed. These chapters should provide some synthetic guidance to chemists and suggest material issues to material scientists and engineers. Descriptions of various measurement techniques follow in Chapter 6. Experimental details are provided, as well as a discussion of the relative merits of each method and care needed in interpretation of the measurements. These discussions should be helpful to those interested in setting up a laboratory for experimental measurements of optical nonlinearities in organic structures. Chapter 7 provides a survey of data on second-order optical nonlinearities of various organic molecules and polymers. We list for various organic structures studied, the measurement technique, the nonlinear optical coefficients, and the nature of the bulk phase in which the measurement is made. In addition to providing the data in tabular forms, we discuss some specific molecules and polymers that have received increased attention. The same sequence of topics is followed in the subsequent chapters (Chapters 8–10) dealing with third-order nonlinearities. These surveys and discussions, we hope, will serve as a useful reference for researchers in this field.

The final topics discussed are related to device structures. A review of nonlinear optical processes in optical waveguides is provided in Chapter 11, as well as a discussion of relevant materials issues. Chapter 12 deals with some specific device structures to give a flavor of various applications of nonlinear optical phenomena in organic systems. Again, in addition to a discussion of device structure and device process, the material requirements are addressed. The objective is a multidisciplinary appeal so that these device-related chapters are useful not only for device scientists and engineers, but also for chemists and material scientists by giving them a qualitative feel for this topic and an awareness of materials requirement. We conclude with a discussion of the current status of this field and future directions of research (Chapter 13). This discussion is somewhat subjective and reflects our opinions based on participation in many international conferences, symposia, and workshops in this area.

2

BASIS AND FORMULATION OF NONLINEAR OPTICS

In this chapter a semiquantitative description of the interaction of light with matter is developed and used as the basis of understanding for nonlinear optics. The basic principles of electromagnetic theory that describe the properties of light and the constitutive equations that describe its interaction with matter are briefly reviewed. A rigorous and complete mathematical treatment of electromagnetic theory and linear and nonlinear optics is outside the scope of this book. These subjects have been thoroughly dealt with in excellent texts on the subject (Flytzanis et al. 1975, Shen 1984). Instead, we favor a descriptive approach to these subjects, bearing in mind the background and interests of the readers this book was written for. Some mathematical detail is presented where it is deemed necessary to illustrate key points, but the reader wishing to develop a deeper understanding and appreciation for the subject matter is encouraged to consult the references cited in this chapter.

2.1 INTERACTION OF LIGHT WITH A MEDIUM

The detailed microscopic description of the interaction of light with the molecular constituents of a medium is given in Chapter 3. In the present section a qualitative description of the interaction of light with a bulk medium based on the results of Maxwell's equations (Bloembergen 1965) is presented. A molecular medium, such as an organic crystalline or polymeric solid, is generally nonconducting and nonmagnetic and the electrons are regarded as being tightly bound to the nucleii. For such media the interaction with light can generally be regarded within the framework of a dielectric subjected to an electric field.

8

This approach is sometimes referred as the dipole approximation since the charge distribution induced in the molecule by the field is readily approximated by that of an induced dipole. The applied field polarizes the molecules in the medium, displacing them from their equilibrium positions and induces a dipole moment μ_{ind} given by

$$\mu_{ind} = -er \tag{2.1}$$

where e is the electronic charge and r is the field induced displacement. The bulk polarization P resulting from this induced dipole is given by

$$P = -Ner \tag{2.2}$$

where N is the electron density in the medium.

The electric field inside the material is lowered by the polarization that opposes the externally applied field. The reduction in field intensity in the volume element containing the molecule of interest is by the factor $1 + \varepsilon$, where ε is defined as the dielectric constant of the medium. If the field strength is relatively low, the polarization of the medium is linear in the applied field. The linear polarization is often expressed in terms of a susceptibility $\chi^{(1)}$ as

$$P = \chi^{(1)} \cdot E \tag{2.3}$$

where the proportionality constant is related to the dielectric constant by

$$\varepsilon = 1 + 4\pi\chi^{(1)} \tag{2.4}$$

The susceptibility is a second-rank tensor because it relates all of the components of the polarization vector to all of the components of the electric field vector. It contains all of the information about the medium needed to relate the polarization in a particular direction to the various Cartesian components of an electric field vector in an arbitrary direction. Some of the properties of tensors and the significance of the information they contain are reviewed later in this chapter.

Another quantity that is related to the polarization is the dielectric displacement, defined as

$$D = E + 4\pi P \tag{2.5}$$

The first term on the right side, E, gives the contribution to the electric flux density emanating from a distribution of charges, if those charges were in free space. The effect of a medium is usually to reduce the forces between charges by an amount proportional to the polarization of the medium. In a vacuum, D and E are equal. In an isotropic medium, D and E are parallel vectors and the polarization response is equal in all directions. From a comparison of equations 2.3–2.5, it is clear that the dielectric constant ε is equal to D/E. In anisotropic media, the vectors D and E are no longer parallel and ε is therefore a second-rank tensor with properties similar to the susceptibility tensor.

The wavelike properties of light are described by an oscillating electromagnetic field. The oscillating nature of the electric component can be described in the time domain as a propagating electric field $E(r,t)$ which is varying in space r and time t. Consequently, the material response P and its linear susceptibility are also time- and space-varying quantities. Therefore, equation 2.3 is modified by time and space convolutions of $\chi^{(1)}$ and E, as described by Shen (1984). Alternatively, one can use the frequency-domain representation where the electric field $E(\omega, k)$ is described by its oscillation frequency ω and its propagation vector k. Both of these representations are convenient for understanding various aspects of optical behavior. It is important to note that the principal physical phenomena are manifested in either representation and therefore both representations are equivalent.

The dielectric constant $\varepsilon(\omega)$ for a linear optical medium at optical frequencies is related to the linear optical susceptibility $\chi^{(1)}$ by an equation analogous to (2.4), except that $\chi^{(1)}$, specified to be the value appropriate for the optical frequency $\chi^{(1)}(\omega)$, and hence $\varepsilon(\omega)$, describes the linear optical response of the media. Such phenomena as absorption and refraction are accounted for by this relationship. The optical response of a medium is represented equivalently by its refractive index. For an isotropic medium

$$n_c^2(\omega) = \varepsilon(\omega) = 1 + 4\pi\chi^{(1)}(\omega) \tag{2.6}$$

Because of resonances in molecules and solids associated with electronic and nuclear motions, the refractive index and, hence, the dielectric constant are complex quantities. The optical properties associated with resonances in a harmonic oscillator are illustrated in Section 2.3. Near a resonance corresponding to an electronic absorption, for instance, the complex refractive index is given by

$$n_c = n + ik \tag{2.7}$$

The real part of $\varepsilon(\omega)$ and, hence, n accounts for refraction while the imaginary part (ik) describes the absorption of light in the dielectric medium. For a linear optical medium, the refractive index is independent of the electric field strength.

2.2   LIGHT PROPAGATION THROUGH AN OPTICAL MEDIUM

In the paragraphs above the response of a material to the oscillating electric field of radiation was discussed. However, the propagation characteristics of light need to be described to fully understand the interaction of light with the medium. While this is fully described by Maxwell's equation, a detailed discussion is outside the scope of this book. Here the wave equation, which follows from Maxwell's equation (Shen 1984) under the assumptions that the medium is insulating and nonmagnetic and that the interaction with the medium is electric dipolar, is simply stated and used to describe the propagation characteristics of light. The later assumption is valid for molecular entities where the wavelength of light is large compared to the dimensions site of the unit being polarized. The wave equation describing the propagation of an electromagnetic wave is

$$\nabla^2 E = -\frac{\varepsilon}{c^2}\frac{\partial^2 E}{\partial t^2} \tag{2.8}$$

where c is the speed of light, ε is the dielectric constant, and ∇^2 is the Laplace operator and is defined as

$$\nabla^2 = \frac{\partial^2}{\partial x^2} + \frac{\partial^2}{\partial y^2} + \frac{\partial^2}{\partial z^2} \tag{2.9}$$

The above equation relates the time and space variations of the electric field of light through the materials response, specified by the dielectric constant ε and is therefore of fundamental importance in understanding the interaction of the field with the medium. The usefulness of the wave equation can be illustrated conceptually by using an example in which light is propagating as a plane wave in the z direction. For such a case, equation 2.8 simplifies to

$$\frac{\partial^2 E}{\partial z^2} = -\frac{\varepsilon}{c^2}\frac{\partial^2 E}{\partial t^2} \tag{2.10}$$

One solution for the above equation is a traveling wave in the z direction with the electric field $E(z,t)$ given in the complex representation within an arbitrary phase factor as

$$E(z,t) = \tfrac{1}{2}(E^{(0)}e^{i(\omega t - kz)} + \text{complex conjugate}) \tag{2.11}$$

or, equivalently, as a sinusoidal oscillation

$$E(z,t) = E^{(0)}\cos(\omega t - kz) \tag{2.12}$$

with

$$k^2 = \frac{\varepsilon\omega^2}{c^2} \tag{2.13}$$

and with $E^{(0)}$ defining the amplitude of the field. The term k is the propagation constant in the material and is equal to 2π times the number of waves per unit length, that is,

$$k = \frac{2\pi}{\lambda} \tag{2.14}$$

The quantity k is a vector called the propagation vector. It characterizes the phase of the optical wave with respect to a reference point ($z = 0$, for instance) and kz describes the relative phase of the wave. The phase velocity v of a wave in a medium is given by $d\omega/dk$. Use of the relation $\varepsilon = n^2$ in equation 2.13 leads to

$$k = \frac{n\omega}{c} \tag{2.15}$$

and

$$v = \frac{c}{n} \tag{2.16}$$

Therefore, the propagation of an optical wave through a medium is slower than that in vacuum. The reduction factor is the refractive index n of the medium. As discussed above, the refractive index n of an optically linear medium is independent of the field strength E. Consequently, from equations 2.15 and 2.16, it is also apparent that the propagation constant k, the relative phase kz, and the light velocity v are all independent of the electric field strength E.

2.3 THE HARMONIC OSCILLATOR MODEL FOR LINEAR OPTICAL PROCESSES

An instructive way to visualize the optical properties of a medium is to consider it as an assembly of forced harmonic oscillators according to the model due to Lorentz (Zernike and Midwinter 1973). The bonding of electrons to the nucleii is approximated by that of charged particles attached to nucleii by springs. To simplify the discussion the interactions are assumed to be isotropic and vector notation will not be used. The force F_E exerted on an electron of charge e by the electric field E is then given by

$$F_E = eE \tag{2.17}$$

The bond, approximated as a spring, will exert a restoring force on the electron given by

$$F_R = -m\omega_0^2 x \tag{2.18}$$

where m is the mass of the electron, x is the displacement from the equilibrium position, and ω_0 is the natural frequency of the oscillator and is equal to the square root of the ratio of the elastic constant to m. Newton's second law states that the sum of the forces acting on a particle of mass m and charge e equals the mass times the acceleration. Hence, the equation of motion is

$$eE - m\omega_0^2 x = m\frac{d^2 x}{dt^2} \tag{2.19}$$

Providing for damping of the oscillator by adding a term proportional to the velocity and rearranging gives the familiar equation of motion

$$\frac{d^2 x}{dt^2} + 2\Gamma\frac{dx}{dt} + \omega_0^2 x = -\frac{e}{m}E \tag{2.20}$$

where Γ is the damping constant.

We now consider the action of an oscillating electric field of a plane wave described by equation 2.11 on the harmonic oscillator. The solution of differential equation, ignoring transient terms, then yields the following expression for displacement x where c.c. denotes the complex conjugate:

$$x = -\frac{e}{m}E\frac{e^{i\omega t}}{\omega_0^2 - 2i\Gamma\omega - \omega^2} + \text{c.c.} \tag{2.21}$$

Inspection of equation 2.21 suggests sinusoidal behavior in time with increasing displacement as the frequency of the field approaches the natural frequency of the oscillator. The use of equation 2.2 with 2.21 yields the following expression for the polarization:

$$P = \frac{Ne^2}{m}\frac{1}{\omega_0^2 - 2i\Gamma\omega - \omega^2}E(\omega)e^{i\omega t} + \text{c.c.} \tag{2.22}$$

This expression shows that the induced polarization is proportional to the amplitude of the electric field and has the same frequency dependence. Having obtained an expression for the polarization of a harmonic oscillator one can use it to describe the linear optical phenomena that it would exhibit.

As the electromagnetic wave propagates through the medium the electrons surrounding nucleii (here approximated as harmonic oscillators) are polarized and these oscillating dipoles act as new sources of radiation. The frequency of the radiated wave is identical to the incident wave but its phase lags behind the incident wave by a time determined by the natural frequency of the oscillator. If the wave encounters N oscillators as it passes through the medium, it will accumulate a phase delay proportional to N and will appear to have been delayed relative to a parallel wave that had traveled an identical distance but in a vacuum. As discussed above, the ratio between c, the velocity of light in a vacuum, and the apparent velocity in the medium or phase velocity v, is the refractive index of the material.

The use of the relationships between the refractive index, dielectric constant, and $\chi^{(1)}$ and the polarization leads to

$$n^2 = \varepsilon = 1 + 4\pi\chi^{(1)} = 1 + \frac{Ne^2}{m}\frac{4\pi}{\omega_0^2 - 2i\Gamma\omega - \omega^2} \tag{2.23}$$

When N is small enough, the absolute value of the second term on the right is small compared with unity. A good approximation to n then is the right side of (2.23) with the second term divided by 2. The real (n) and imaginary (k) parts are then

$$\mathrm{Re}\,n = 1 - \frac{Ne^2}{m}\frac{2\pi(\omega^2 - \omega_0^2)}{(\omega^2 - \omega_0^2)^2 + (2\Gamma\omega)^2} \tag{2.24}$$

and

$$\mathrm{Im}\,n = \frac{Ne^2}{m}\frac{4\pi\gamma\omega}{(\omega^2 - \omega_0^2)^2 + (2\Gamma\omega)^2} \tag{2.25}$$

The behavior of these expressions is shown schematically in Figure 2.1 where the real and imaginary parts of the refractive index corresponding to dispersion and absorption are plotted as a function of frequency in the region of an optical transition at ω_0. The familiar dispersion in the refractive index that occurs in the wavelength region below an optical transition is a manifestation of this behavior. Near the peak of the absorption the refractive index decreases rapidly and begins to increase again slowly on the high-energy side of the transition.

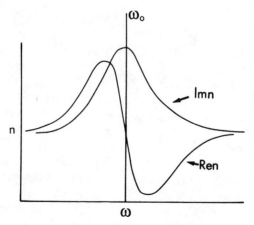

Figure 2.1 The real and imaginary parts of the refractive index for the Lorentz (harmonic) oscillator with frequency ω_0, which illustrate anomalous dispersion and absorption, respectively.

This relatively simplistic treatment has illustrated some of the important linear properties exhibited by typical organic materials. As will be shown later in this chapter and in subsequent chapters the understanding and control of linear optical properties are extremely important to the observation and practical utility of nonlinear optical properties.

2.4 NONLINEAR OPTICAL MEDIA

When a medium is subject to an intense electric field such as that due to an intense laser pulse, the polarization response of the material is not adequately described by equation 2.3. Assuming that the polarization of the medium is still weak compared to the binding forces between the electrons and nuclei, one can express the polarization in a power series of the field strength E

$$P = \chi^{(1)} \cdot E + \chi^{(2)} : EE + \chi^{(3)} : EEE + \tag{2.26}$$

In the above equation the term that is quadratic in the field strength E describes the first nonlinear effect. The coefficient $\chi^{(2)}$ relating the polarization to the square of the field strength E is called the second-order nonlinear susceptibility of the medium and is a third-rank tensor. The tensor and vector notation is explained more fully in Section 2.7. Its magnitude describes the strength of second-order processes. The $\chi^{(3)}$ term is referred to as the second nonlinear susceptibility describing third-order processes, and is a fourth-rank tensor. Similarly, higher-order terms describe the higher-order processes. For most materials, the higher-order effects are extremely difficult to observe. For this reason we limit our discussion up to and including third-order effects. Equation 2.26 can be recast as

$$P = \chi_{\text{eff}} E \tag{2.27}$$

which is analogous to equation 2.3 except that χ_{eff} is now dependent on the field strength. For optical waves this has important consequences. An inspection of equations 2.6, 2.15, and 2.16 reveals that with $\chi^{(1)}$ replaced by χ_{eff} the refractive index n, the phase of the wave kz, and the velocity v of the optical wave are now all dependent on E.

The manifestation of nonlinear optical behavior can be clearly seen by substituting a sinusoidal field equation (2.12) into the polarization expansion equation (2.26). This gives

$$P = \chi^{(1)} E_0 \cos(\omega t - kz) + \chi^{(2)} E_0^2 \cos^2(\omega t - kz) + \chi^{(3)} E_0^3 \cos^3(\omega t - kz) \tag{2.28}$$

and using appropriate trigonometric identities for $\cos^2\theta$ and $\cos^3\theta$ gives

$$\begin{aligned} P = \chi^{(1)} E_0 \cos(\omega t - kz) + \tfrac{1}{2}\chi^{(2)} E_0^2 [1 + \cos(2\omega t - 2kz)] \\ + \chi^{(3)} E_0^3 [\tfrac{3}{4}\cos(\omega t - kz) + \tfrac{1}{4}\cos(3\omega t - 3kz)] \end{aligned} \tag{2.29}$$

Equation 2.29 clearly shows the presence of new frequency components due to the nonlinear polarization. The second-order term gives a frequency independent contribution as well as one at 2ω. The former suggests that a dc polarization should appear in a second-order nonlinear material when it is appropriately irradiated. This phenomenon is referred as optical rectification. The latter term corresponds to second-harmonic generation. Fourier analysis of the nonlinear response leads to similar conclusions. This is illustrated graphically in Figure 2.2. The third term indicates a frequency response at the frequency of the optical field ω as well as a response at 3ω. The even- and odd-order terms in the expansion therefore lead to fundamentally different types of nonlinear responses. At this point, we simply note that contributions from the second- and third-order terms to the nonlinear polarization are predicated from different symmetry properties of the medium. A contribution from $\chi^{(2)}$ can come only from noncentrosymmetric media, whereas $\chi^{(3)}$ contributions can come from any medium regardless of symmetry.

Another manifestation of nonlinear optics is the so-called nonlinear index of refraction of the medium. In fact, all of the nonlinear effects discussed in

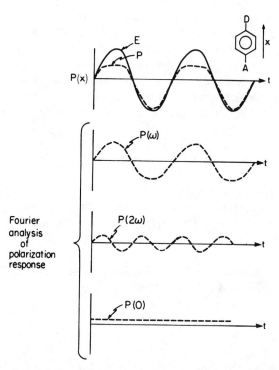

Figure 2.2 The polarization response $P(x)$ of a molecule in the x direction as a function of time due to an electric field oscillating at $E(\omega)$ (*top*), and Fourier components of $P(x)$ at frequencies $\omega, 2\omega$, and 0 (*bottom*). Adapted with permission from Yariv, 1975a.

the preceeding paragraph can be thought of as arising from field-induced modulation of the refractive index. In the case of second-harmonic generation, for example, one field component can be thought of as modulating the refractive index of the medium at frequency ω. This modulation can be quite weak or strong depending on the intensity of the beam. The second frequency component is phase modulated, giving rise to a sideband (or harmonic in this case because the frequency components are identical). Thus, second-harmonic generation can be explained in terms of modulation of the refractive index.

A more important distinction occurs when the medium is subjected to a dc (or low-frequency ac) electric field, which we designate as $E(0)$, and an optical field $E(\omega)$. The total field the medium is subjected to is, therefore,

$$E = E(0) + E(\omega)$$
$$= E(0) + E_0 \cos(\omega t - kz) \tag{2.30}$$

Substituting equation 2.30 into equation 2.26 gives

$$P = \chi^{(1)}[E(0) + E_0 \cos(\omega t - kz)] + \chi^{(2)}[E(0) + E_0 \cos(\omega t - kz)]^2$$
$$+ \chi^{(3)}[E(0) + E_0 \cos(\omega t - kz)]^3 + \cdots \tag{2.31}$$

Expanding these terms, applying the appropriate trigometric identities, and collecting terms that describe oscillation at ω (since we are interested in contributions to the refractive index at that frequency) gives

$$P(\omega) = \chi^{(1)}E_0 \cos(\omega t - kz) + 2\chi^{(2)}E(0)E_0 \cos(\omega t - kz) + 3\chi^{(3)}E_{(0)}^2 E_0 \cos(\omega t - kz)$$
$$+ \tfrac{3}{4}\chi^{(3)}E_0^3 \cos(\omega t - kz) = \chi_{\text{eff}} E(0) \cos(\omega t - kz) \tag{2.32}$$

In analogy with equation 2.6, one can write for the nonlinear index of refraction

$$n^2 = 1 + 4\pi\chi_{\text{eff}} = 1 + 4\pi[\chi^{(1)} + 2\chi^{(2)}E(0) + 3\chi^{(3)}E^2(0) + \tfrac{3}{4}\chi^{(3)}E_0^2] \tag{2.33}$$

Designating the linear refractive index as n_0, one obtains by substitution of equation 2.6 into 2.33

$$n^2 - n_0^2 = 8\pi\chi^{(2)}E(0) + 12\pi\chi^{(3)}E(0)^2 + 3\pi\chi^{(3)}E_0^2 \tag{2.34}$$

which leads to the relationship

$$n = n_0 + \frac{4\pi\chi^{(2)}}{n_0}E(0) + \frac{6\pi\chi^{(3)}}{n_0}E(0)^2 + \frac{3\pi}{2n_0}\chi^{(3)}E_0^2 \tag{2.35}$$

The definition of the light intensity in cgs units is

$$E_0^2 = \frac{8\pi}{cn} I(\omega) \tag{2.36}$$

which when substituted into (2.35) gives

$$n = n_0 + \frac{4\pi\chi^{(2)}}{n_0} E(0) + \frac{6\pi\chi^{(3)}}{n_0} E(0)^2 + \frac{12\pi^2}{cn_0^2} \chi^{(3)} I \tag{2.37}$$

The value of the nonlinear index of refraction at frequency ω can therefore be written as

$$n(\omega) = n_0(\omega) + n_1 E(0) + n_2(0)E(0)^2 + n_2(\omega)I(\omega) \tag{2.38}$$

where the terms $n_1, n_2(0)$, and $n_2(\omega)$ are defined as

$$n_1 = \frac{4\pi\chi^{(2)}}{n_0} \tag{2.39}$$

$$n_2(0) = \frac{6\pi\chi^{(3)}}{n_0} \tag{2.40}$$

$$n_2(\omega) = \frac{12\pi\chi^{(3)}}{cn_0^2} \tag{2.41}$$

and correspond to the linear electroptic effect, the quadratic electrooptic effect, and the optical Kerr effect, respectively. The linear electrooptic effect (or Pockels effect) is proportional to $\chi^{(2)}$ and is discussed fully in Chapter 6. The optical Kerr effect corresponds to the light intensity-dependent refractive index and is potentially important for high-speed all-optical switching functions and devices and is discussed in detail in Chapter 8.

Second-order nonlinear optical effects can be visualized as a three-wave mixing process where the waves exchange energy with one another through the intercession of the nonlinear medium. In general, the process is described by the interaction of waves at frequencies ω_1 and ω_2 to produce a new frequency ω_3. The susceptibility, in the frequency domain, is provided with an argument to describe this process, for example, $\chi^{(2)}(-\omega_3; \omega_1, \omega_2)$. For second-harmonic generation the term is $\chi^{(2)}(-2\omega; \omega, \omega)$. For the electrooptic effect it is $\chi^{(2)}(-\omega; \omega, 0)$. The negative sign is a convention to indicate that momentum must be conserved in the process. Similarly, third-order nonlinear optical effects can be described as four-wave mixing processes where waves at frequencies $\omega_1, \omega_2, \omega_3$ interact to produce ω_4. In general, the waves can be at any combination of frequencies that satisfies the momentum conservation requirement. This coefficient is usally represented as $\chi^{(3)}(-\omega_4; \omega_1, \omega_2, \omega_3)$. For third-harmonic generation for instance, it becomes $\chi^{(3)}(-3\omega; \omega, \omega, \omega)$.

In a nonlinear medium, the wave equation (2.8) must be modified to include the impact of the medium on the time and space variations of the electric field. The equation now becomes

$$\nabla^2 E(\omega) = -\frac{\varepsilon}{c^2} \frac{\partial^2 E(\omega)}{\partial t^2} - \frac{4\pi}{c^2} \frac{\partial^2 P}{\partial t^2} \qquad (2.42)$$

in which P is the nonlinear polarization given as

$$P = \chi^{(2)}(-\omega_3; \omega_1, \omega_2) E(\omega_1) E(\omega_2)$$
$$+ \chi^{(3)}(-\omega_4; \omega_1, \omega_2, \omega_3) E(\omega_1) E(\omega_2) E(\omega_3) + \cdots \qquad (2.43)$$

The linear polarization of the medium is included in ε.

Equation 2.42 describes the coupling of various input and output waves through the nonlinear polarization terms P. The wave equation describes the conditions for phase matching (momentum conservation) for the generation and propagation of new fields. This point is further illustrated in Chapter 4 where the wave equation is used to describe the phase-matching condition for second-harmonic generation.

2.5 THE ANHARMONIC OSCILLATOR MODEL FOR NONLINEAR OPTICAL EFFECTS

At the most fundamental level, the nonlinear optical response of a medium can be related to the deviation of electronic displacement from harmonic behavior. In a relatively straightforward and similar manner to the harmonic oscillator treatment for linear optics, the occurrence of nonlinear optical properties can be visualized by introducing anharmonic terms such as ax^2 for second-order effects and bx^3 for third-order effects into equation 2.20 to account for anharmonicity. In the following discussion we will consider only the lowest-order anharmonic term. With the first anharmonic term included, equation 2.20 becomes

$$\frac{d^2 x}{dt^2} + 2\Gamma \frac{dx}{dt} + \omega_0^2 x + ax^2 = -\frac{e}{m} E \qquad (2.44)$$

The addition of the anharmonic term prevents the straightforward solution of the equation in a manner similar to that for equation 2.20. The usual course of action is to assume that the anharmonic contribution to the polarization is small compared to the harmonic or linear term and to approximate the solution as a power series in the displacement x. Considering the first two terms in the expansion we can write

$$x = x_1 + x_2 \tag{2.45}$$

The solution for the first term x_1 is exactly as obtained in equation 2.21 for the linear case. The solution for the second term is obtained by approximating ax^2 in equation 2.44 by ax_1^2, which linearizes the equation and leads to a straightforward solution in terms of a component at frequency 2ω and another at zero frequency or dc. This solution is of the form

$$x_2 = x_2(2\omega) + x_2(0) + \text{c.c.} \tag{2.46}$$

where

$$x_2(2\omega) = -a\left(\frac{e}{m}\right)^2 E^2(\omega)\frac{e^{i2\omega t}}{(\omega_0^2 - 2i\Gamma\omega - \omega^2)(\omega_0^2 - 4i\Gamma\omega - 4\omega^2)} + \text{c.c.} \tag{2.47}$$

and

$$x_2(0) = -2a\left(\frac{e}{m}\right)^2\frac{E(\omega)}{\omega_0^2}\frac{1}{(\omega_0^2 - 2i\Gamma\omega - \omega^2)} + \text{c.c.} \tag{2.48}$$

The first term in equation 2.46 shows the response of the oscillator at 2ω and leads to the phenomenon of second-harmonic generation. The second term shows that the nonlinear response leads to a dc polarization or displacement in the equilibrium position of the oscillator. Substituting these results into equation 2.2 leads directly to expressions for the nonlinear polarizations at the corresponding frequencies and to the susceptibility functions commonly used to describe the response to the field.

$$P_{NL} = -Ne[x_2(2\omega) + x_2(0)] \tag{2.49}$$

$$\chi^{(2)}(2\omega) = \frac{Nex_2(2\omega)}{E(\omega)^2} \tag{2.50}$$

$$\chi^{(2)}(0) = \frac{Nex_2(0)}{E(\omega)^2} \tag{2.51}$$

If we were to generalize the analysis by considering that the polarizing field contained frequency components at arbitrary frequencies ω_1 and ω_2, terms would appear in equation 2.46 with frequency arguments of the type $x_2(\omega_1 + \omega_2)$ and $x_2(\omega_1 - \omega_2)$ at the sum and difference frequencies that through Maxwell's equations are the source of new electromagnetic fields at those frequencies.

By adding the next higher-order anharmonic term to equation 2.44 and engaging in a similar iterative process to obtain solutions we would find terms containing up to three frequency components, leading to a variety of third-order processes including third-harmonic generation and four-wave mixing processes $\chi^{(3)}(-\omega; \omega, -\omega, \omega)$, which are discussed in detail in later chapters.

2.6 ANISOTROPIC MEDIA

In the preceding discussion of the harmonic oscillator and anharmonic oscillator models we have assumed that the interactions have been isotropic, which causes the induced polarization to be parallel to the field and related to it by a scalar proportionality factor. In general, however, materials that are likely to be of interest in nonlinear optics are not likely to be isotropic. Organic crystals, for instance, are composed of molecules having definite orientations within the unit cell and the molecules themselves often have highly anisotropic electronic structures. We might, therefore, expect polarization responses in directions other than that of the applied electric fields. Similarly, polymer chains in a solid are often oriented other than randomly and the exact nature of their orientational distribution will determine their response. For this reason, the proportionality factors, often referred to as susceptibilities, are tensor quantities relating the polarization response in one direction to field components in three directions. The relevant properties of tensors are reviewed in Section 2.7 and for those who are unfamiliar with them, it might be profitable to read that section before proceeding.

One manifestation of anisotropy in a medium is the dielectric constant ε_{ij}, which relates the electric field and dielectric displacement and is a second-rank tensor quantity. As was pointed out in equation 2.4, ε_{ij} is also related to the linear susceptibility, which must therefore also be a second-rank tensor. For the sake of illustration we write out all of the components of the dielectric displacement vector in terms of the tensor elements ε_{ij} and the components of the electric field vector in an arbitrary coordinate system:

$$D_x = \varepsilon_{11}E_x + \varepsilon_{12}E_y + \varepsilon_{13}E_z$$
$$D_y = \varepsilon_{21}E_x + \varepsilon_{22}E_y + \varepsilon_{23}E_z \qquad (2.52)$$
$$D_z = \varepsilon_{31}E_x + \varepsilon_{32}E_y + \varepsilon_{33}E_z$$

A suitable rotation of the coordinate system can always be found that causes $\varepsilon_{12}, \varepsilon_{13}$, and so on, and all other off-diagonal terms to go to zero. Under these circumstances we have

$$D_x = \varepsilon_x E_x$$
$$D_y = \varepsilon_y E_y \qquad (2.53)$$
$$D_z = \varepsilon_z E_z$$

where ε_x, ε_y, and ε_z are referred to as the principal dielectric axes of the medium. These axes determine the allowed polarization directions and, through Maxwell's equations, the flow of energy through the medium.

One interesting consequence of the dielectric anisotropy in crystals is the phenomenon of birefringence. Referring to Figure 2.3, visualize a uniaxial crystal with $n_x = n_y \neq n_z$ and a light wave propagating from right to left along the x axis with its electric field polarized in the z direction. The propagation constant k_z will be determined by n_z from

$$k_z = \frac{\omega}{c} n_z \qquad (2.54)$$

and $n_z = \sqrt{\varepsilon_z}$. Another wave propagating along the x direction but with y polarization would have a different propagation constant k_y. The waves would accumulate a phase difference traveling through the crystal. If these waves were the projections of a plane wave incident from the x direction with a 45° inclination from the z axis then the cumulative phase lag would cause the resultant electric vector to rotate as it propagated, resulting in elliptically polarized light.

A further consequence of birefringence is that for light incident in any direction other than along a principal axis the flow of energy does not occur normal to the wavefront. In a vacuum or isotropic medium the wavefront perpendicular to the propagation direction is always parallel to energy flow. In anisotropic media this is not the case except for the condition stated above.

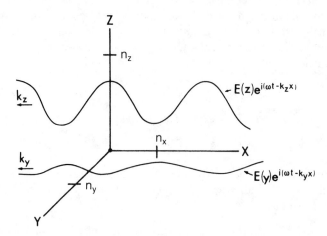

Figure 2.3 The projection of electric field components along principal directions of a birefringent medium. The phase velocity determined by $n = kc/\omega$ is different for the two components so that the resultant polarization vector rotates in the direction of propagation resulting in elliptically polarized light.

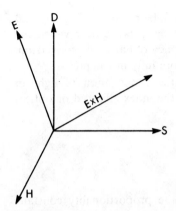

Figure 2.4 E, D, S, H, and Poynting $(E \times H)$ vectors in a uniaxial medium. See text for explanation.

The proof of this statement requires a relatively complex analysis of Maxwell's equations. Here we will try to illustrate the point in a more conceptual way. Referring to Figure 2.4, the relevant vectors needed to describe the propagating electromagnetic wave and its interactions with the medium are illustrated. The electric E and magnetic H vectors of the indicent wave are always mutually orthogonal. The electric vector produces a dielectric displacement D which is in general not parallel to E because of the anisotropic polarizability tensor. The magnetic vector H produces a magnetic induction B (not shown) but which is for all practical purposes parallel to H because of the weakness of the interaction of H with a nonmagnetic medium. The propagation of the wavefront in the crystal is perpendicular to D and H but the flow of power S, as required by Maxwell's equations, is proportional to $E \times H$. S is often referred to as the Poynting vector. In other words, the propagation of the polarization wave and the flow of optical power are not parallel. These points are illustrated in Figure 2.5 where E, D, $E \times H$, k, and S are all in the same plane perpendicular to H. In the nonlinear process of second-harmonic generation where the flow of power from the fundamental to the harmonic beam is dependent on the spatial overlap of the polarization wave propagating with wave vector k_ω and

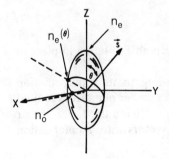

Figure 2.5 The index ellipsoid in a uniaxial medium. The dotted lines represent the axes along which polarization propagates. n_e and n_o are the extraordinary and ordinary components of the refractive index and $n_e(0)$ is the extraordinary value of the refractive index for nonnormal incidence of light with respect to the principal (optic) axis.

the electric field propagating with $k_{2\omega}$, the nonparallel nature of $E_{2\omega} \times H$ and S, in general, puts a fundamental limitation on the distance over which the interaction can be maintained and thus the efficiency of harmonic conversion. This phenomenon, called *walkoff*, is described more fully in Chapter 4.

Another very useful method for analyzing the propagation of light in anisotropic media is by a method referred to as the index ellipsoid or optical indicatrix (Yariv 1975b).

2.7 TENSORS

In the previous section, the need to represent the proportionality constant between a vector stimulus and its response in anisotropic media as a tensor was mentioned. In this section, some of the tensor quantities encountered in nonlinear optics, notation, and assumptions regarding their properties are introduced. Since tensor notation is used throughout nonlinear optics it is important to define tensor quantities precisely and discuss the notations and simplifying assumptions frequently used. A tensor is defined as the relationship between two vectors. For example, two vectors A and B in a Cartesian coordinate might be related by (Dence 1975)

$$\begin{pmatrix} A_x \\ A_y \\ A_z \end{pmatrix} = \begin{pmatrix} \lambda_{11} & \lambda_{12} & \lambda_{13} \\ \lambda_{21} & \lambda_{22} & \lambda_{23} \\ \lambda_{31} & \lambda_{32} & \lambda_{33} \end{pmatrix} \begin{pmatrix} B_x \\ B_y \\ B_z \end{pmatrix} \tag{2.55}$$

or, equivalently,

$$A_x = \lambda_{11} B_x + \lambda_{12} B_y + \lambda_{13} B_z$$
$$A_y = \lambda_{21} B_x + \lambda_{22} B_y + \lambda_{23} B_z \tag{2.56}$$
$$A_z = \lambda_{31} B_x + \lambda_{32} B_y + \lambda_{33} B_z$$

or, alternatively, in vector notation by

$$A = \lambda : B \tag{2.57}$$

It should also be apparent that $\lambda_{xx}, \lambda_{xy}, \ldots$ is acceptable notation in place of $\lambda_{11}, \lambda_{12}, \ldots$.

The quantity λ is a second-rank tensor according to its transformation properties in a Cartesian coordinate system and the number of indices required to specify each element. The quantity $\chi^{(2)}$ in equation 2.26 is a third-rank tensor transforming the action of the two electric field vectors into a polarization vector and can be written as

$$
\begin{pmatrix} P_x^{(2)} \\ P_y^{(2)} \\ P_z^{(2)} \end{pmatrix} = \begin{pmatrix} \chi_{xxx}^{(2)} & \chi_{xyy}^{(2)} & \chi_{xzz}^{(2)} & \chi_{xyz}^{(2)} & \chi_{xxz}^{(2)} & \chi_{xxy}^{(2)} \\ \chi_{yxx}^{(2)} & \chi_{yyy}^{(2)} & \chi_{yzz}^{(2)} & \chi_{yyz}^{(2)} & \chi_{yxz}^{(2)} & \chi_{yxy}^{(2)} \\ \chi_{zxx}^{(2)} & \chi_{zyy}^{(2)} & \chi_{zzz}^{(2)} & \chi_{zyz}^{(2)} & \chi_{zxz}^{(2)} & \chi_{zxy}^{(2)} \end{pmatrix} \begin{pmatrix} E_x^2 \\ E_y^2 \\ E_z^2 \\ 2E_yE_z \\ 2E_xE_z \\ 2E_xE_y \end{pmatrix}
\tag{2.58}
$$

where the $\chi^{(2)}$ tensor contains 18 elements and according to its transformation properties and number of indices is termed a third-rank tensor. Here the ith component of the nonlinear polarization vector $P_i^{(2)}$ is given by

$$
P_I^{(2)} = \sum_{J,K} \chi_{IJK}^{(2)} E_J E_K
\tag{2.59}
$$

or equivalently in vector notation by

$$
P^{(2)} = \chi^{(2)} : EE
\tag{2.60}
$$

Commenting further on nonlinear optical notations for vector and tensor quantities, the dispersion in the susceptibility function (as illustrated in the harmonic and anharmonic oscillator examples) requires that the frequencies of the interacting field components must be specified in order to fully specify the polarization response. This was mentioned in Section 2.4 and is reiterated here. Because of this, frequency arguments are often included with the susceptibility tensor, for example, $\chi^{(2)}(-\omega_3; \omega_1, \omega_2)$ or $\chi^{(3)}(-\omega_4; \omega_1, \omega_2, \omega_3)$. The first frequency in the argument is usually the frequency component of the resultant polarization. In the case of second-harmonic generation the tensor is written $\chi^{(2)}(-2\omega; \omega, \omega)$

$$
P^{(2)} = \chi^{(2)}(-2\omega; \omega, \omega) : E(\omega)E(\omega)
\tag{2.61}
$$

with Cartesian components of the polarization vectors given by

$$
P_I^{(2)} = \sum_{J,K} \chi_{IJK}^{(2)}(-2\omega; \omega, \omega) E_J(\omega) E_K(\omega)
\tag{2.62}
$$

Similar and consistent notation applies to third-order nonlinear optical processes.

Historically, much of the experimental information that has been obtained on second-order nonlinear properties has been from second-harmonic generation (SHG) measurements and a SHG nonlinear coefficient d is often cited. The coefficient d is related to $\chi^{(2)}$ by

$$\chi^{(2)}_{IJK} = 2d_{IJK} \tag{2.63}$$

A contracted form of notation for the coefficient has arisen from the observation that the contribution to $P^{(2)}_I$ in equation 2.62 from $E_J(\omega)$ should be indistinguishable from that of $E_K(\omega)$ for the process of second-harmonic generation so that interchange of the last two indices of the coefficient

$$d_{IJK} = d_{IKJ} \tag{2.64}$$

should not effect its value. Under these circumstances the last two indices can be contracted as illustrated below:

$$
\begin{aligned}
d_{1\underline{11}} &= d_{1\underline{1}} \\
d_{1\underline{22}} &= d_{1\underline{2}} \\
d_{1\underline{33}} &= d_{1\underline{3}} \\
d_{1\underline{23}} &= d_{1\underline{32}} = d_{1\underline{4}} \\
d_{1\underline{31}} &= d_{1\underline{13}} = d_{1\underline{5}} \\
d_{1\underline{12}} &= d_{1\underline{21}} = d_{1\underline{6}}
\end{aligned}
\tag{2.65}
$$

Similar contractions can be made for $d_{211}, d_{311}, \ldots, d_{333}$ and the 27 element tensor reduces to 18 elements.

Another significant and simplifying symmetry consideration is attributed to Kleinman (1962) who recognized that in many nonlinear processes where all the interacting frequencies are far away from resonances, energy is simply exchanged between fields and is not dissipated in the medium. This leads to the simplification that for a particular combination of interacting frequencies, $\chi^{(2)}_{IJK}(-2\omega; \omega, \omega)$, for instance, the value of the coefficient is independent of interchange of its Cartesian indices:

$$\chi^{(2)}_{IJK} = \chi^{(2)}_{KIJ} = \chi^{(2)}_{JKI} = \chi^{(2)}_{JIK} = \chi^{(2)}_{KJI} = \chi^{(2)}_{IKJ} \tag{2.66}$$

This reduces the number of independent $\chi^{(2)}$ tensor elements from 27 to 10 and for $\chi^{(3)}$ processes from 81 to 15. The susceptibility tensors for the various polar point groups can be found in a number of texts and reviews on nonlinear optics such as Byer (1977) and are based on the standards developed by the IRE (1949). In the appendix to this chapter we include the SHG tensors for the 21 out of 32 crystal classes that lack a center of inversion, which is a requirement for nonzero nonlinear coefficients in the dipolar approximation.

2.8 SYMMETRY

Our discussion of nonlinear optics has generally referred to a medium consisting of oscillators either isotropically or anistotropically distributed throughout a medium. Organic molecules themselves consist of a system of electrons and nucleii moving under the influence of the molecular quantum field. The electrons are bound to the nucleii very strongly compared with their attraction to nucleii on neighboring molecules. The π electrons generally have relatively high kinetic and low potential energy relative to the electrons in σ bonds. For the benefit of readers not familiar with these terminologies, the concepts of σ and π bonds are discussed in Chapter 3. The molecule can therefore be considered as a polarizable entity itself in contrast to the medium which consists of an assembly of such molecules. The reason for making this distinction is that molecular quantum field theory can account for the way that the many subtle features of molecular structure account for its polarization response. Understanding the molecular nonlinear response and solid-state structure allows a connection to be established between the response of the molecule and that of the medium. These theories and methods are discussed in Chapters 3 and 4.

For now, we would like to point out that a series of molecular coefficients $a_{ij}, \beta_{ijk}, \gamma_{ijkl}$ corresponding to $\chi_{IJ}^{(1)}, \chi_{IJK}^{(2)}, \chi_{IJKL}^{(3)}$ describe the polarization responses of molecules and have similar tensor properties and symmetry considerations as the macroscopic coefficients. Here the indices are distinguished from the macroscopic coefficients since they refer to the molecular coordinate system. The relationship between the macroscopic and microscopic coefficients are discussed in some detail in Chapter 4. Here we wish to discuss some important considerations regarding the symmetry of the molecule and the medium.

In Figure 2.6 we consider a molecule with a center of symmetry, such as benzene. When a field is applied in the positive x direction the linear polarization is $P^{(1)} = aE$. For a field in the minus x direction it is $-aE$. Since the molecule is symmetric with respect to the plus or minus x direction, the potential energy must be as well. Considering the polarization in one direction only, we can write

$$P_x = \alpha E + \gamma E^3 + \cdots \tag{2.67}$$

and the potential energy associated with that polarization is

$$V_x(E) = - \left[\tfrac{1}{2}\alpha E^2 + \tfrac{1}{4}\gamma E^4 + \cdots \right] \tag{2.68}$$

The inclusion of only odd-order terms in equation 2.67 results in a symmetric potential and $V_x(E)$ and $V_x(-E)$ are equal. If we had included an even-order term, say βE^2 in equation 2.67, the potential energy would be different for the plus and minus x direction, but this is inconsistent with the symmetry of the molecule. Thus, we draw the conclusion that in molecules with a center of symmetry all the tensor elements β_{ijk} are equal to zero and no second-order

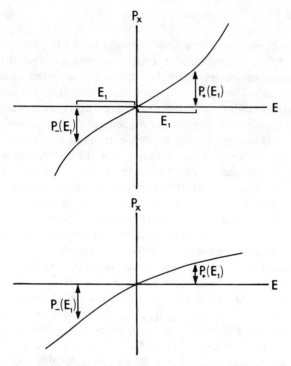

Figure 2.6   Schematic illustration of polarization versus electric field in a molecule with (top) and without (bottom) inversion symmetry.

nonlinear response exists (in the approximation that only dipolar contributions to the nonlinear polarization are significant).

On the other hand, a molecule without a center of symmetry, as is schematically shown in Figure 2.7, will exhibit an asymmetric potential in the plus and minus x directions. Here the polarization is given by

$$P_x = \alpha E + \beta E^2 + \gamma E^3 + \cdots \qquad (2.69)$$

and the potential energy is

$$V_x(E) = -\left[\tfrac{1}{2}\alpha E^2 + \tfrac{1}{3}\beta E^3 + \tfrac{1}{4}\gamma E^4 + \cdots\right] \qquad (2.70)$$

and $V_x(E)$ is obviously different for E than for $-E$. The point is further illustrated by the schematic plots of polarization versus electric field in Figure 2.7 where the nonlinear polarization response is symmetrical in a molecule with inversion symmetry and asymmetrical when the molecule lacks it.

The arguments regarding symmetry of the medium extend to the macroscopic coefficients $\chi^{(2)}_{IJK}$ and are characteristic of the medium. Therefore, even though a molecule may lack inversion symmetry, if it forms a crystal possessing a center

SYMMETRIC Π SYSTEM

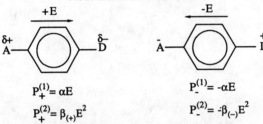

$$P_+^{(1)} = \alpha E$$
$$P_+^{(2)} = \beta E^2$$

$$P_-^{(1)} = -\alpha E$$
$$P_-^{(2)} = \beta(-E)^2 = \beta E^2$$

ASYMMETRIC Π SYSTEM

$$P_+^{(1)} = \alpha E$$
$$P_+^{(2)} = \beta_{(+)} E^2$$

$$P_-^{(1)} = -\alpha E$$
$$P_-^{(2)} = -\beta_{(-)} E^2$$

Figure 2.7 *Upper*: Nonlinear polarization response to a system having a center of symmetry. *Lower*: Nonlinear polarization response to a system without a center of symmetry.

of inversion, all of the tensor elements will be zero in the dipolar approximation. Similar considerations pertain to liquids and isotropic solids such as glasses. For this reason only the polar point groups referred to in the previous section and illustrated in the appendix have nonzero second-order tensor elements.

On the other hand, all molecules and media exhibit nonzero third-order nonlinear responses. Although polar symmetry is not required, third-order nonlinear coefficients can be highly anisotropic and enhanced in oriented media. These points are discussed in detail in Chapter 5.

APPENDIX

Second-Order Nonlinear Susceptibility Tensors for 21 of the 32 Crystal Classes Lacking Inversion Symmetry

TRICLINIC SYSTEM

Class 1—C₁

$$\begin{pmatrix} d_{11} & d_{12} & d_{13} & d_{14} & d_{15} & d_{16} \\ d_{21} & d_{22} & d_{23} & d_{24} & d_{25} & d_{26} \\ d_{31} & d_{32} & d_{33} & d_{34} & d_{35} & d_{36} \end{pmatrix}$$

MONOCLINIC SYSTEM

Class m—C_{1h}

$$\begin{pmatrix} d_{11} & d_{12} & d_{13} & 0 & 0 & d_{16} \\ d_{21} & d_{22} & d_{23} & 0 & 0 & d_{26} \\ 0 & 0 & 0 & d_{34} & d_{35} & 0 \end{pmatrix} m \perp Z$$

Class m—C_{1h}

$$\begin{pmatrix} d_{11} & d_{12} & d_{13} & 0 & d_{15} & 0 \\ 0 & 0 & 0 & d_{24} & 0 & d_{26} \\ d_{31} & d_{32} & d_{33} & 0 & d_{35} & 0 \end{pmatrix} \begin{matrix} m \perp Y \\ \text{(IRE-convention)} \end{matrix}$$

Class 2—C_2

$$\begin{pmatrix} 0 & 0 & 0 & d_{14} & d_{15} & 0 \\ 0 & 0 & 0 & d_{24} & d_{25} & 0 \\ d_{31} & d_{32} & d_{33} & 0 & 0 & d_{36} \end{pmatrix} 2 \parallel Z$$

Class 2—C_2

$$\begin{pmatrix} 0 & 0 & 0 & d_{14} & 0 & d_{16} \\ d_{21} & d_{22} & d_{23} & 0 & d_{25} & 0 \\ 0 & 0 & 0 & d_{34} & 0 & d_{36} \end{pmatrix} \begin{matrix} 2 \parallel Y \\ \text{(IRE-convention)} \end{matrix}$$

ORTHORHOMBIC SYSTEM

Class mm 2—C_{2v}

$$\begin{pmatrix} 0 & 0 & 0 & 0 & d_{15} & 0 \\ 0 & 0 & 0 & d_{24} & 0 & 0 \\ d_{31} & d_{32} & d_{33} & 0 & 0 & 0 \end{pmatrix}$$

Class 222—D_2

$$\begin{pmatrix} 0 & 0 & 0 & d_{14} & 0 & 0 \\ 0 & 0 & 0 & 0 & d_{25} & 0 \\ 0 & 0 & 0 & 0 & 0 & d_{36} \end{pmatrix}$$

TETRAGONAL SYSTEM

Class 4—C_4

$$\begin{pmatrix} 0 & 0 & 0 & d_{14} & d_{15} & 0 \\ 0 & 0 & 0 & d_{15} & -d_{14} & 0 \\ d_{31} & d_{31} & d_{33} & 0 & 0 & 0 \end{pmatrix}$$

Class $\bar{4}$—S_4

$$\begin{pmatrix} 0 & 0 & 0 & d_{14} & d_{15} & 0 \\ 0 & 0 & 0 & -d_{15} & d_{14} & 0 \\ d_{31} & -d_{31} & 0 & 0 & 0 & d_{36} \end{pmatrix}$$

Class 4mm—C_{4v}

$$\begin{pmatrix} 0 & 0 & 0 & 0 & d_{15} & 0 \\ 0 & 0 & 0 & d_{15} & 0 & 0 \\ d_{15} & d_{15} & d_{33} & 0 & 0 & d_{36} \end{pmatrix}$$

Class $\bar{4}2m$—D_{2d}

$$\begin{pmatrix} 0 & 0 & 0 & d_{14} & 0 & 0 \\ 0 & 0 & 0 & 0 & d_{14} & 0 \\ 0 & 0 & 0 & 0 & 0 & d_{36} \end{pmatrix}$$

Class 422—D_4

$$\begin{pmatrix} 0 & 0 & 0 & d_{14} & 0 & 0 \\ 0 & 0 & 0 & 0 & -d_{14} & 0 \\ 0 & 0 & 0 & 0 & 0 & 0 \end{pmatrix}$$

TRIGONAL SYSTEM

Class 3—C_3

$$\begin{pmatrix} d_{11} & -d_{11} & 0 & d_{14} & d_{15} & -d_{22} \\ -d_{22} & d_{22} & 0 & d_{15} & -d_{14} & -d_{11} \\ d_{31} & d_{31} & d_{33} & 0 & 0 & 0 \end{pmatrix}$$

Class $3m$—C_{3v}

$$\begin{pmatrix} 0 & 0 & 0 & 0 & d_{15} & -d_{22} \\ -d_{22} & d_{22} & 0 & d_{15} & 0 & 0 \\ d_{31} & d_{31} & d_{33} & 0 & 0 & 0 \end{pmatrix} \begin{matrix} m \perp X \\ \text{(IRE-convention)} \end{matrix}$$

Class $3m$—C_{3v}

$$\begin{pmatrix} d_{11} & -d_{11} & 0 & 0 & d_{15} & 0 \\ 0 & 0 & 0 & d_{15} & 0 & -d_{11} \\ d_{31} & d_{31} & d_{33} & 0 & 0 & 0 \end{pmatrix} m \perp Y$$

Class 32—D_3

$$\begin{pmatrix} d_{11} & -d_{11} & 0 & d_{14} & 0 & 0 \\ 0 & 0 & 0 & 0 & -d_{14} & -d_{11} \\ 0 & 0 & 0 & 0 & 0 & 0 \end{pmatrix}$$

HEXAGONAL SYSTEM

Class $\bar{6}$—C_{3h}

$$\begin{pmatrix} d_{11} & -d_{11} & 0 & 0 & 0 & -d_{22} \\ -d_{22} & d_{22} & 0 & 0 & 0 & -d_{11} \\ 0 & 0 & 0 & 0 & 0 & 0 \end{pmatrix}$$

Same as Class 4—C_4

Class $\bar{6}m2$—D_{3h}

$$\begin{pmatrix} 0 & 0 & 0 & 0 & 0 & -d_{22} \\ -d_{22} & d_{22} & 0 & 0 & 0 & 0 \\ 0 & 0 & 0 & 0 & 0 & 0 \end{pmatrix} \begin{matrix} m \perp X \\ \text{(IRE-convention)} \end{matrix}$$

Class $\bar{6}m2$—D_{3h}

$$\begin{pmatrix} d_{11} & -d_{11} & 0 & 0 & 0 & 0 \\ 0 & 0 & 0 & 0 & 0 & -d_{11} \\ 0 & 0 & 0 & 0 & 0 & 0 \end{pmatrix} m \perp Y$$

Class 6mm—C$_{6v}$

$$\begin{pmatrix} 0 & 0 & 0 & 0 & d_{15} & 0 \\ 0 & 0 & 0 & d_{15} & 0 & 0 \\ d_{31} & d_{31} & d_{33} & 0 & 0 & 0 \end{pmatrix}$$

Same as Class 4mm—C$_{4v}$

Class 622—D$_6$

$$\begin{pmatrix} 0 & 0 & 0 & d_{14} & 0 & 0 \\ 0 & 0 & 0 & 0 & -d_{14} & 0 \\ 0 & 0 & 0 & 0 & 0 & 0 \end{pmatrix}$$

Same as Class 422—D$_4$

CUBIC SYSTEM

Class 23—T

$$\begin{pmatrix} 0 & 0 & 0 & d_{14} & 0 & 0 \\ 0 & 0 & 0 & 0 & d_{14} & 0 \\ 0 & 0 & 0 & 0 & 0 & d_{14} \end{pmatrix}$$

Class $\bar{4}$3m—T$_4$

$$\begin{pmatrix} 0 & 0 & 0 & d_{14} & 0 & 0 \\ 0 & 0 & 0 & 0 & d_{14} & 0 \\ 0 & 0 & 0 & 0 & 0 & d_{14} \end{pmatrix}$$

Class 432—O

All elements vanish

REFERENCES

Bloembergen, N., *Nonlinear Optics*, Benjamin, New York, 1965.

Byer, R. L., unpublished doctoral dissertation, Standard University, Palo Alto, CA, 1968.

Byer, R. L., P. G. Harper, and B. S. Wherret (Eds.), *Nonlinear Optics*, Academic, London, 1977, p. 55ff.

Dence, J. B., *Mathematical Techniques in Chemistry*, Wiley-Interscience, New York, 1975, p. 345ff.

Flytzanis, C., H. Rabin, and C. L. Tang (Eds.), *Quantum Electronics: A Treatise*, Vol. 1, Academic, New York, 1975.

IRE, Standards on piezoelectric crystals, *Proc. IRE* **37**, 1378 (1949).

Kleinman, D. A., *Phys. Rev.* **126**, 1977 (1962).

Shen, Y. R., *The Principles of Nonlinear Optics*, Wiley, New York, 1984, Chap. 2.

Yariv, A., *Quantum Electronics*, Wiley, New York, 1975a, p. 419.

Yariv, A., *Quantum Electronics*, Wiley, New York, 1975b, p. 450ff.

Zernike, F., and J.E. Midwinter, *Applied Nonlinear Optics*, Wiley, New York, 1973, p. 3ff.

3

ORIGIN OF MICROSCOPIC NONLINEARITY IN ORGANIC SYSTEMS

3.1 MICROSCOPIC NONLINEARITY

Organic systems consist of molecular units (molecules or polymers) that in the absence of net charge or intermoleclar charge transfer interact only weakly. An oriented-gas model is often used to relate the molecular properties to the corresponding bulk properties. This oriented-gas model has also been used to relate the molecular optical nonlinearity to the bulk nonlinearity. In this model the optical nonlinearity of the organic medium is determined primarily by the nonlinear optical properties of the molecular unit. It is therefore of fundamental importance to create an understanding of the electronic origins of the microscopic nonlinearity and its dependence on molecular structure. Only through elucidation of fundamental properties can one use molecular engineering principles to optimize the nonlinear optical behavior of bulk systems. In this chapter the molecular original of nonlinear optics is discussed and various theoretical approaches for computing nonlinear optical inter-actions are described. The relative merits of the various approaches are also addressed.

In discussing nonlinear optical properties, the polarization of the molecule by the radiation field is often approximated as the creation of an induced dipole by an electric field. Under the weak polarization limit, one can use a power expansion in the electric field components (discussed in Chapter 2) to describe the dipolar interaction with the radiation field. The two equivalent approaches are using either the Stark energy or dipole moment, which is the negative derivative of the energy with respect to the field as follows:

$$U(E) = U^0 - \sum_i \mu^0 E_i - \frac{1}{2} \sum_{ij} \alpha_{ij} E_i E_j$$

$$- \frac{1}{3} \sum_{ijk} \beta_{ijk} E_i E_j E_k - \frac{1}{4} \sum_{ijkl} \gamma_{ijkl} E_i E_j E_k E_l \tag{3.1}$$

$$\mu_i(E) = \mu_i^0 + \sum_j \alpha_{ij} E_j + \sum_{jk} \beta_{ijk} E_j E_k + \sum_{jkl} \gamma_{ijkl} E_j E_k E_l \tag{3.2}$$

In equation 3.1, U^0 is the energy in the absence of field and μ^0 is the permanent dipole moment. The coefficient α is the polarizability of the molecule and is related to the second derivative of the energy and first derivative of the dipole moment with respect to the optical field E. The polarizability term describes the linear interaction with the optical field and accounts for linear absorption and refraction behavior of the molecule. The higher-order terms involving β and γ describe the microscopic nonlinear optical interactions and likewise are sensitive to and manifest the electronic structure of the molecule. The coefficients β and γ, being third- and fourth-rank tensors, respectively, are called first and second hyperpolarizabilities and constitute the molecular origin of the second- and third-order nonlinear optical interactions. Symmetry requirements, as discussed in Chapter 2, Section 2.8, impose the condition that in the power series expansion (equation 3.2), the even terms with respect to the applied field vanish for a centrosymmetric structure. In other words, only for non-centrosymmetric molecules are μ_i and β_{ijk} nonzero. The odd terms α and γ are nonzero for all molecules and for all media, including air.

3.2   BRIEF REVIEW OF σ AND π ELECTRONS AND BONDING

This section reviews some elementary concepts of σ and π electrons and bonding and is intended for readers who are not familiar with these concepts. Many unique features exhibited by organic structures are easily understood by invoking the conceptual separation of σ and π electrons and the differences in their behaviors. A carbon atom, which occupies a position in group IVA, of the periodic table exhibits maximum valence and can form four covalent bonds involving 2s and 2p orbitals. These orbitals when mixed as four sp^3 hybrid orbitals can form four single directed bonds in a tetrahedral geometry. Each bond involves the overlap of an sp^3 hybrid orbital of carbon and an orbital of the bonded atom (carbon or other) along the internuclear axis. Such bonds are called σ bonds by virtue of their axial symmetry and the electrons involved in the σ bonds are referred to as σ electrons. If a carbon atom forms four σ bonds involving all the four sp^3 hybridized orbitals, its valence is completely satisfied and it yields a saturated structure.

One very important reason nature has chosen to involve carbon chemistry for evolution of life is its versatility to form multiple bonds as well as bonds between two carbon atoms. Let us consider, specifically, a compound such as ethylene

which involves a double bond between two carbon atoms. In this case each carbon exhibits sp^2 hybridization and yields three sp^2 hybridized orbitals which can form three σ bonds, two with hydrogen atoms and one between the carbon atoms themselves. The orbitals involved in bonding overlap along the internuclear axis to give bonds with axial symmetry. However, the four valence requirements for each carbon atom are not satisfied. There is one electron left in the 2p orbitals of each of the carbon atom. Due to geometrical restrictions, the unhybridized 2p orbitals on these two carbon atoms (usually designated as $2p_z$) cannot overlap along the internuclear axis, but can overlap laterally. This mode of overlap of orbitals leads to a π bond which has a mirror plane. The electrons in a π bond are called π electrons. Similarly, in the case of the acetylene molecule

$$H-C\equiv C-H$$

each carbon atom involves two sp-hybridized orbitals for two σ bonds (four σ electrons per carbon atom) and two p orbitals for two π bonds between the carbon atoms (a total of four π electrons). The rule is that when a carbon atom forms a multiple bond with another atom, only one bond is of σ type. The remaining bond(s) is (are) of the π type. The π bonds are weaker, giving the carbon atom the tendency to form σ bonds through chemical reaction whenever possible. For this reason the compounds containing π bonds are called unsaturated.

A special case occurs when a compound involves a chain or ring of bonded carbon atoms with alternate single and multiple bonds. The two examples of this situation are hexatriene (**I**) and benzene (**II**):

In both these structures, the π electrons can move over the entire length of the molecule and the structure is said to be conjugated. In the case of hexateriene, all of the carbon bonds are not equivalent and one finds experimentally bond-length alternation corresponding to alternate double and single bonds. In the case of benzene, however, there are two resonance structures which are energetically equivalent and indistinguishable:

These systems are referred to as aromatic molecules. In structures of this class the electrons have considerably more kinetic energy and less potential energy than their nonaromatic counterparts. In aromatic molecules all bonds are of equal length and the structure is generally written as

It is the delocalization behavior of the π electrons that makes the π-electron distribution highly deformable in conjugated electronic systems, which also gives rise to large optical nonlinearities. The nonlinear polarization of the π electrons is large even for frequencies of radiation far away from electronic resonances (where the molecule absorbs the incident photon). It is the π electrons that make the organic systems so different from inorganic materials for nonlinear optical applications.

Because of the different nature of orbital overlap in π bonding and the mobility of the π electrons, it is often useful to treat σ and π electrons separately and apply semiempirical approaches using a simplified Hamiltonian that takes into account only the π electrons. At the simplest level, the π electrons are treated by the free-electron model in which the π electrons are approximated as moving in a box of some effective length determined by the size of the molecule. This model has been used to describe both the linear absorption spectra (band gap) as well as the nonlinear optical properties of conjugated linear structures (such as hexateriene discussed above). Chemists have widely used another semiempirical approach called Hückel's method to treat the π boding in a conjugated system. In this treatment, the Hamiltonian interaction terms $I_{mn} = \langle \phi_m / H / \phi_n \rangle$ are replaced by a diagonal interaction parameter α and a nearest neighbor ($m = n \pm 1$) transfer integral parameter β ($= I_{n,n\pm1}$). In the case of benzene or a conjugated ring system, only one transfer integral (β) is required, since all carbon–carbon bonds are equivalent. For systems like hexatriene, which is a representative of the series that in the infinite limit would yield the polyacetylene polymer, the bond alternation would suggest that two different transfer integrals β_1 and β_2, corresponding to two different bonds, should be used. In the semiempirical method of calculation of microscopic nonlinearities, one often neglects the contribution due to the σ electrons and considers only the π-electron contribution.

3.3   EQUIVALENT INTERNAL FIELD MODEL

The equivalent internal field model was developed (Oudar and Chemla 1975) to explain observed trends in β for some of the first organic substances for which it was determined. The premise of the model is simple. The delocalized π-electronic cloud of a centrosymmetric molecule such as benzene is polarized by an electropositive or electronegative substituent at one of the hydrogen positions, thus introducing a dipole moment into the π-electron cloud and rendering it noncentrosymmetric. This induced dipole will be either positive or negative, depending on whether the substituent is more electronegative or electropositive than hydrogen. The magnitude of the dipole created in the symmetric π system is given by

$$\Delta\mu = \alpha E_0 + \gamma E_0^3 \sim \alpha E_0 \tag{3.3}$$

where E_0 is the internal electric field arising from the substituent. The tensorial nature of the interaction is ignored for the sake of simplicity in this discussion. The induced dipole is parallel to the molecular axis containing the substituent. When an external field E is applied to the system the total E_T is $E + E_0$ and the polarization p (μ_{ind}) is given by

$$\mu_{ind} = \alpha E_T + \gamma E_T^3 \tag{3.4}$$

Substituting for E_T and collecting terms gives

$$\mu_{ind} = \alpha E_0 + \gamma E_0^3 + (\alpha + 3\gamma E_0^2)E + 3\gamma E_0 E^2 + \gamma E^3 \tag{3.5}$$

Chemical substitution, therefore, slightly modifies the polarizability (third term) and introduces a term quadratic in E, which is designated as β where

$$\beta = 3\gamma E_0 \tag{3.6}$$

and substituting for E_0 from equation 3.3 gives

$$\beta = \frac{3\gamma\Delta\mu}{\alpha} \tag{3.7}$$

In the equivalent Internal Field model, the derived value of β is that associated with a DC field. The relationship between $\beta(2\omega)$, the value for the second harmonic generation process, and β as defined in this model is $\beta(2\omega) = (1/2)\beta$ (Ward and New 1969).

This correlation has been verified by Oudar and Chemla (1975) for both monosubstituted benzenes and stilbenes and is exhibited in Figure 7.2 (Chapter 7) for monosubstituted benzenes. Using the mesomeric moments obtained by Everard and Sutton (1951) a reasonable correlation was established, although there is some disagreement regarding the validity of the procedures for obtaining $\Delta\mu$ (Levine and Bethea 1975). It is clear from inspection of the figure that highly electropositive andelectronegative substituents produce the largest values of β. On the other hand, the procedure used for obtaining $\Delta\mu$, which is to subtract the dipole moment of a substituted alkane from that of a similarly substituted phenyl ring to obtain the π contribution to the dipole moment, would appear to be a gross oversimplification. There is undoubtedly a combination of electrostatic and resonance effects that produces the true value of the hyperpolarizability, and sophisticated calculations are required to account for this type of effect.

3.4   ADDITIVITY MODEL FOR MOLECULAR HYPERPOLARIZABILITIES

The approach of treating a molecular property as a sum of contributions coming from structural elements of a molecule (atoms, bonds, or functional groups, depending on the details of the approach) provides in many cases a useful estimate of this property. Additivity is known to provide good estimates for molecular polarizabilities, or, at least, for their scalar parts. However, α is a second-rank tensor, and the tensor components of α are, generally, not easily predicted by additivity. An alternative to a simple additivity is a model of interacting entities. Such models are generally shaped after the Silberstein's model of polarizability (Applequist et al. 1972).

In the field of molecular nonlinear optics there were several attempts at compiling values of additive increments to the first and second hyperpolarizabilities that could be used in predicting values of β and γ for more complicated molecules. Bond additivity rather than atom additivity has been preferred. The vector part of β for substituted hydrocarbons has been calculated by Levine et al. (1975) by using vector components of β oriented along the bonds. Levine et al. also attempted to find increments for the scalar part of γ. The specific bond values of β and γ have been determined by measurements on a large umber of derivatized substituted saturated hydrocarbons (Levine and Bethea 1975, Meredith et al. 1983, Kajzar and Messier 1985, 1987). Some of the extracted values for γ reported by Kajzar and Messier are listed in Table 3.1.

Several other approaches have been described to improve on the simple bond additivity model. One approach (Kajzar and Messier 1985, Miller et al. 1981, Sundberg 1977) assigns full tensorial β and γ to each bond and calculates the molecular β and γ tensors by conversion to molecular axes and using additivity. In another approach (Kajzar and Messier 1987), the bond hyperpolarizability for each bond (or chemical group) is calculated by assigning an appropriate local field factor for each group. These local field factors relate the microscopic hyperpolarizabilities to the bulk nonlinear susceptibilities as discussed in detail in Chapter 4. Miller et al. (1981) proposed the interacting segment model in which the molecular properties are considered to be fitted in terms of a set of "bare" electric tensor parameters for each bond, which are modified by

TABLE 3.1   Bond Values of Gamma[a]

Bond Substitution	$\gamma(C-X) \times 10^{-36}$
C–H	0.2
C–C	0.118
C–Cl	0.772
C–Br	1.359
C–I	3.061

[a]From measurement of third-harmonic generation using a fundamental wavelength of 1.064 μm.
Source: Data from Kajzar and Messier (1987).

intramolecular electrostatic interactions. In other words allowance is made for electrostatic interaction between adjacent segments of the molecule.

It has become evident that while the simple additivity is a relatively good approximation for σ-bonded molecules of, for example, substituted hydrocarbons, it fails misreably for π-electron molecules. The delocalization of π electrons over the molecular structure makes the π-electron contribution a molecular property and not localized in a specific bond. This failure is especially important for substituted benzene with high β in which the deviation from additivity has usually been discussed in terms of an internal charge-transfer contribution to the first hyperpolarizability β.

Another additivity model proposed recently to interpret the nonlinear optical behavior of conjugated oligomeric structures (several monomeric repeat structures chemically bonded) is based on an anharmonic oscillator model introduced by Bloembergen (Prasad et al. 1989). The anharmonic oscillator model discussed in Chapter 2, Section 2.5, has been extended by Prasad et al. to the case of a chain of anharmonic oscillators coupled together linearly, each oscillator representing a molecular repeat unit in an oligomeric or polymeric structure. The main features of this coupled anharmonic oscillator model are as follows:

1. Each monomeric unit is assumed to be a single electronic oscillator ignoring the details of structure and bonding. The local anharmonicity is treated by a cubic term a and a quartic term b, which represent the first and second hyperpolarizabilities of the monomeric unit, respectively. The local anharmonicity depends on the monomeric unit functionality.

2. A single coupling constant k is used to represent the π-electron delocalization effect throughout the oligomeric series. This assumption implies that each monomeric unit bears the same structural correlation with its neighboring units.

3. The model assumes a single resonance frequency ω_0, implying a two-level model. A two-level model has been successfully used for explaining the second-order nonlinearity (as discussed below). The first step of this model

is to solve the equation of motion to get the electronic displacement r of the oligomeric chain:

$$\frac{d^2 r_i}{dt^2} + \Gamma \frac{dr_i}{dt} + \omega_0^2 r_i + a r_i^2 + b r_i^3 + k(r_{i-1} + r_{i+1}) = -\left(\frac{e}{m}\right) E \quad (3.8)$$

where r_i represents the displacement of the ith single unit, Γ is the damping factor, e and m are the charge and the mass of an electron, and E is the applied electric field. The nonlinearities are then calculated using the expansion of the dipole moment in field strength:

$$\mu = -er = \alpha E + \beta E^2 + \gamma E^3 \quad (3.9)$$

To account for the four physical characteristics, that is, the band gap, the polarizability, and the first and second hyperpolarizabilities, only three parameters need to be determined: the coupling constant k, and the anharmonic coefficients a and b. These can be calculated by fitting the experimental values of band gap, polarizabilities and first and second hyperpolarizabilities using a simple least-squares method.

A simple classical model of coupled locally anharmonic oscillators can give a good description of nonlinearities of oligomeric series, such as benzene and thiophene series. Calculated with this model, the dependence of the polarizability and the second hyperpolarizability on the number of repeat units for the oligomers of thiophene and benzene is in agreement with the power laws obtained by fitting the experimental results (Prasad et al. 1989). In addition, this model shows that the π-electron delocalization along the oligomeric chain axis is more extended for the thiophenes than for the benzene series. This classical model has the merit that one can get a chemical intuition about (1) monomeric structures with large local anharmonicity and (2) effective delocalization in going from monomer to oligomers to polymers.

3.5   QUANTUM-CHEMICAL APPROACHES

3.5.1   Derivative versus Sum-over-State Method

For the calculation of microscopic nonlinearities, there are two conceptually different approaches which are summarized in Table 3.2. The derivative method relates different derivatives of the energy and dipole moment to various terms of the power series expansions (expansions 3.1 and 3.2). As listed in the table, the third-order nonlinear optical coefficient (the second hyperpolarizability γ_{ijkl}), for example, is simply given by the fourthe derivative of the energy or the third derivative of the dipole moment with respect to the applied field. The computation, therefore, involves quantum calculations of energy or dipole

TABLE 3.2   Microscopic Theory of Optical Nonlinearity

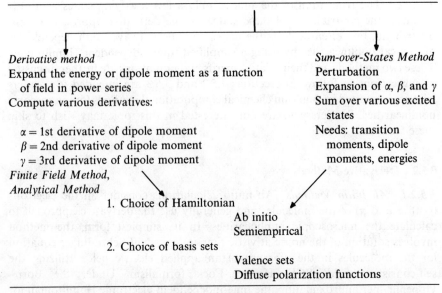

Derivative method
Expand the energy or dipole moment as a function
of field in power series
Compute various derivatives:

α = 1st derivative of dipole moment
β = 2nd derivative of dipole moment
γ = 3rd derivative of dipole moment
Finite Field Method,
Analytical Method

Sum-over-States Method
Perturbation
Expansion of α, β, and γ
Sum over various excited
states
Needs: transition
moments, dipole
moments, energies

1. Choice of Hamiltonian

Ab initio
Semiempirical

2. Choice of basis sets

Valence sets
Diffuse polarization functions

moment followed by obtaining different derivatives either numerically or
analytically. The sum-over-states method is based on the perturbation expansion
of the Stark energy. Different-order Stark energy terms are related to optical
nonlinearities based on their order in the field strength. In this approach,
therefore, the optical nonlinearities are introduced as a result of Stark mixing
with various excited states. The mixing terms are summed over various excited
states to obtain α, β, and γ.

After the selection of the method of approach, one has to make choice as to
the nature of the Hamiltonian. One can take an ab initio approach to start
with a complete Hamiltonian. Alternatively, a semiempirical approach can be
adopted whereby a simplified Hamiltonian involving empirical parameters is
used. The nature of the approximation used to simplify the Hamiltonian and
various interaction terms defines various semiempirical approaches. The next
step of calculation involves the choice of atomic orbital basis functions, which
are used to construct molecular orbitals by well-known LCAO–MO methods.
As shall be discussed below, the choice of an appropriate basis set is very
important for the calculation of optical nonlinearities. The basis sets that may
adequately describe the geometry, energies of various states, and other valence
parameters may not be adequate to describe the nonlinear optical processes.

In principle, both the derivative and the sum-over-states methods should
yield the same result, if exact calculations are done. However, exact calculations are not possible. Consequently, the two approaches use different sets of
approximations. The sum-over-states calculations, to be exact, should involve
sums over not only the low-lying molecular states, as is usually done, but also

atomic Rydberg states and continuum (unbound) states, since completeness is assumed. The truncation of the sum to only a few low-lying states artificially enhances the contribution of these states. In the derivative approach, on the other hand, one often neglects the dispersion effects (wavelength dependence) on optical nonlinearities by using a simplified time-independent Hamiltonian. These two methods and their relative merits and limitations are further discussed in subsections to follow. Subsections 3.5.2 and 3.5.3 are intended for readers who are interested in quantum chemical computational methods for microscopic nonlinearities. Readers who are not interested in this topic may wish to skip these sections.

3.5.2 Derivative Methods

3.5.2.1 Ab Initio Methods   Ab initio calculations consider all the electrons (both σ and π) of the molecules and generally use the derivative approach to calculate the microscopic nonlinearities. In its simplest form, the method involves solution of the nonrelativistic time-independent Schrödinger equation for the molecules in the presence of an applied electric field, utilizing the self-consistent field (SCF) Hartree–Fock formalism. Under the Born–Oppenheimer approximation, the time-independent electronic Hamiltonian for a molecular system in the presence of an electric field E can be written as

$$H(E) = \sum_{I>J} \frac{Z_I Z_J}{R_{IJ}} - \sum_I Z_I R_I \cdot E - \sum_i \left(\tfrac{1}{2} \nabla_i^2 + \sum_I \frac{Z_I}{r_{iI}} - E \cdot r_i \right) + \sum_{i>j} \frac{1}{r_{ij}} \quad (3.10)$$

where i and j label electrons and I and J label nuclei; Z_I are the nuclear charge; R_I and r_i are the positions of nucleus I and electron i, respectively; R_{IJ} is the distance between nuclei I and J; r_{iI} is the distance between electron i and nucleus I; r_{ij} is the distance between electrons i and j.

In equation 3.10, the terms defining the Hamiltonian on the right side are as follows: The first term describes the nuclear–nuclear repulsion; the second term represents the interactions of nuclei with the electric field; the third term represents the kinetic energy of electrons; the fourth term is due to the attraction between electrons and nuclei; the fifth term represents the interactions of electrons and the electric field; the sixth term describes the interelectron repulsion.

The electronic wavefunction, ψ, representing a given electronic configuration (electronic state) must be antisymmetric with respect to interchange of the coordinates of any pair of electrons to satisfy Pauli's exclusion principle. To ensure antisymmetrization, the wave functions are often represented by a determinant called the Slater determinant (Hehre et al. 1986). A single Slater determinant represents the wave function describing a closed-shell molecular electronic state (Hehre et al. 1986), if one ignores configuration interaction (mixing between various electronic configurations). A Slater determinant is an antisymmetrized product of spin-molecular orbitals $\phi_k(E)$. For a detailed discussion of the quantum chemical methods discussed here the readers are referred to the book by Hehre et al. (1986). The molecular orbitals $\phi_k(E)$ are often expressed as a linear combination of nuclear-centered, atomic-orbital basis functions f_s as follows

$$\phi_k(E) = \sum f_s C_{sk}(E) \qquad (3.11)$$

where the expansion coefficients (C) above depend on E. They are chosen to minimize the energy. These atomic-orbital basis functions constitute the basis set. The two types of atomic basis functions that have received widespread usage are (1) Slater-type orbitals (STO), which have exponential radial parts, and (2) Gaussian-type atomic functions, which have Gaussian radial parts. A series of minimal basis sets called STO-3G consists of expansions of Slater-type atomic orbitals in terms of the Gaussian functions. An example of an improved basis set is the split-valence basis set such as 321-G in which two basis functions, instead of one, are used to describe each atomic orbital. A detailed discussion of the various bases sets is outside the scope of this book. The interested reader is referred to the book by Hehre et al. (1986).

The self-consistent field method is now used to optimize these molecular orbitals to obtain the coefficients $C_{sk}(E)$. One optimizes the geometry without the field and then includes the field-dependent terms in the Hamiltonian to calculate the energy and the dipole moment. The molecular energy and the dipole moment in the presence of the field are given by

$$\varepsilon(E) = \langle \psi(E) | H(E) | \psi(E) \rangle \qquad (3.12)$$

$$\mu(E) = \langle \psi(E) | \sum_i q_i \cdot r_i | \psi(E) \rangle \qquad (3.13)$$

In equation 3.13, q_i and r_i are the charge the position vector, respectively, of the particles i and the summation runs over all electrons and nuclei. Then various derivatives are calculated using either a numerical method (finite field method) or an analytical method. The simplest numerical method involves fitting of the computed energy dipole moment into a polynomial of the applied field E. With the numerical method, care must be exercised in selecting the range of field strengths to avoid numerical instability. The analytical approach involves mathematical relations widely known as the generalized coupled perturbed Hartree–Fock (CPHF) equations, which are derived from the variational properties of the wavefunction $\psi(E)$. Aside from the issues of numerical stability and computational efficiency, the finite field methods and the analytical derivative method (CPHF) are equivalent approaches and should yield identical results. A fundamental question is whether one should use the derivative of energy or that of the dipole moment. Differentiation of equation 3.12 with respect to E_z yields

$$\frac{\partial \varepsilon}{\partial E_z} = 2\left\langle \frac{\partial \psi(E)}{\partial E_z} | H(E) | \psi(E) \right\rangle + \left\langle \psi(E) | \frac{\partial H}{\partial E_z}(E) | \psi(E) \right\rangle \qquad (3.14)$$

The second term on the right side of equation 3.14 is $\langle \mu_z(E) \rangle$. If the first term on the right side of equation 3.14 is zero, that is, if ψ satisfies the Hellmann–Feynman theorem, the two approaches (Eqs. 3.1 and 3.2) using energy and dipole moment derivatives are equivalent. The SCF ab initio formalisms using atomic-orbital basis functions whose centers are either fixed on nuclei or allowed to float (optimized in the presence of the field) satisfy the Hellmann–Feynman theorem. Therefore, the energy derivative and the dipole moment derivative methods are equivalent. However, the dipole moment derivative approach may be easier if a finite field method is used for calculation of derivatives because a lower order of numerical differentiation is required.

To get different components of the nonlinear hyperpolarizability tensor β and γ along various molecular axes, one can change the field (E) direction and compute the nonlinearity. The ab initio method yields the total nonlinearity due to both σ and π electrons. A quantum mechanical technique called corresponding orbital transformation is one approach that can be used to analyze contribution due to individual occupied molecular orbitals and thus estimate the relative contributions of the σ and π electrons (Chopra et al., 1989).

Linear and nonlinear polarizabilities can also be interpreted in terms of the spatial regions that contribute to the field dependence of the electron charge density. This type of analysis can be of considerable value in understanding the substituent effect on optical nonlinearities. The charge density can be expanded in a Taylor series as a power of field strength in the same way that energy and dipole moments are expanded. Similar to equations 3.1 and 3.2, one can write

$$\rho(r, E) = \rho^{(0)}(r) + \rho^{(1)}(r)E + \frac{1}{2!}\rho^{(2)}(r)EE + \frac{1}{3!}\rho^{(3)}(r)EEE + \cdots \qquad (3.15)$$

The dipole moment $\mu_z(E)$ can be obtained by integration of equation 3.15 as

$$\mu_z(E) = \int z\rho(r, E)\, dr^3 \qquad (3.16)$$

Using Eqs. 3.2, 3.15, and 3.16, one can see that for a field directed along the z axis

$$\beta_{zzz} = \frac{1}{2}\int z\rho^{(2)}(r)\, dr^3 \qquad (3.17)$$

$$\gamma_{zzzz} = \frac{1}{6}\int z\rho^{(3)}(r)\, dr^3 \qquad (3.18)$$

Therefore, plots of r, $\rho^{(2)}(r)$, and $r\rho^{(3)}(r)$ can yield information about spatial regional contributions to β and γ. As was stated earlier, the choice of basis set plays a key role in the calculation of optical nonlinearity. For ab initio calculations, the minimum basis set is STO-3G. This basis set is totally inadequate for calculation of even the linear polarizability α. The split valence set 3-21G, which generally is adequate to describe bonding, is not sufficient for computation of nonlinear terms. The reason is simple: The nonlinear optical behavior is derived mainly from the tail portion (farther away from the nucleus) of the wavefunction. To describe this tail portion, one must include diffuse polarization functions. A more detailed discussion of the basis set effect is beyond the scope of this book. For details, the reader is referred to the paper by Chopra et al. (1989).

In its simplest form, ab initio calculations at the SCF level uses a time-independent Hamiltonian and ignores the electron correlation, as implied by the uses of a single Slater determinant. The use of time-independent Hamiltonian yields only static hyper-polarizabilities. Therefore, no information is obtained on the frequency dependence (dispersion) of the hyperpolarizabilities. In an actual experiment, if one measures the hyperpolarizabilities at different wavelengths, one is likely to get variations in the values of α, β, and γ. Even more pronounced are the changes near an electronic resonance (when the frequency of the light used approaches the energy gap for an excited state). The time-independent ab initio calculation does not provide any information on these resonance effects. However, for creating an understanding of the structure–property relationship it provides a highly useful approach whereby one can investigate the contribution of a specific functional group to hyperpolarizabilities.

Because of the computational complexity, very few calculations of optical nonlinearities have been done in the past using the ab initio method. However, as the availability of supercomputers increases and improved computational methods are developed, this method is going to receive considerable attention. The existing calculations have computed α and γ in π-bonded structures such as

as a function of the number of the repeat unit N. The calculations yield a positive sign for γ and a large contribution due to π electrons, with only a small contribution (of opposite sign) derived from the σ electrons. It suggests that the value of γ increases rapidly (to the power of $N^{3.2}$ to $N^{3.5}$) as the number of the repeat unit increases. Therefore, the π-electron delocalization (which increases as N increases) enhances the optical nonlinearity (γ). The largest component of γ is along the chain direction (the direction of the π-electron delocalization). For the linear polarizability α, the dependence on the repeat unit is much milder ($N^{1.3}$ to $N^{1.6}$).

3.5.2.2 *Semiempirical Methods*

In contrast with the ab initio methods where analytical or numerical solutions to integrals appearing in the quantum mechanical problem are sought, semiempirical methods use a combination of approximations and experimental data to simplify the mathematical problem of solving the integrals. In the following paragraphs the main approximations and most important semiempirical methods are discussed.

Pariser–Parr–Pople (PPP) Method Instead of attempting a rigorous SCF calculation, the PPP method (Pariser and Parr 1953, Pople 1953) makes certain approximations for some integrals and uses experimental data to evaluate other integrals that are encountered in the computations. The overlap between atomic wavefunctions on adjacent atoms is assumed to be zero so that the overlap integral

$$S_{rs} = \langle f_r^{(1)} | f_s^{(1)} \rangle = \delta_{rs} \tag{3.19}$$

Where δ_{rs} is the Kronecker delta and the $f_g^{(n)}$ is the atomic orbital occupied by electron n.

In the PPP method, only the π-electron Hamiltonian including electron repulsions is used. In the π-electron approximation, the $n_\pi \pi$ electrons are treated separately by incorporating the effects of the σ electrons and the nuclei into an effective π-electron Hamiltonian H_π:

$$H_\pi = \sum_{i=1}^{n_\pi} H_{\text{core}}(i) + \sum_{i=1}^{n\pi} \sum_{j>i} \frac{1}{r_{ij}}$$

where

$$H_{\text{core}}(i) = -\tfrac{1}{2}\nabla_i^2 + V_i \tag{3.20}$$

where V_i is the potential energy of the ith π electron in the field produced by the nuclei and the σ electrons.

In addition, the π-electron wavefunction is written as an antisymmetrized product of spin orbitals. The π MOs are taken as LCAOs, with $2p\pi$ carbon orbitals used for hydrocarbons. The bond integral β_{rs}^{core} where

$$\beta_{rs}^{\text{core}} = \langle f_r^{(1)} | H_{\text{core}} | f_s^{(1)} \rangle \tag{3.21}$$

is taken as an empirical parameter for bonded atoms and vanishes for nonbonded atoms. The integral corresponding to the energy of the core electrons α_r^{core} is given by

$$\alpha_r^{\text{core}} = \langle f_r | H_{\text{core}} | f_r \rangle \tag{3.22}$$

and has different values for different atoms. The PPP method assumes, as stated above, that the differential overlap is zero (ZDO). From this it follows that

$$\langle rs | tu \rangle = \left\langle f_r^{(1)} f_s^{(1)} \left| \frac{e^2}{r_{12}} \right| f_t^{(2)} f_u^{(2)} \right\rangle = \delta_{rs}\delta_{tu}\langle rr | tt \rangle \tag{3.22}$$

The physical interpretation of this integral is that it gives the repulsion between an electron distributed in the space described by $f_r f_s$ and another described by $f_t f_u$. Since the overlap integral is zero between adjacent atoms according to the approximation, the expression accounts for repulsion of electron density on two centers. All three and four center integrals are assumed to be zero. Thus, this method ignores many (but not all) of the electron repulsion integrals, thereby greatly simplifying the calculation. The integrals $\langle rr|tt \rangle$ are often treated as semiempirical parameters, rather than being evaluated theoretically.

The Extended Hückel Method (EH)   The extended Hückel (EH) method, developed and widely applied by Hoffman (1963, 1964), begins with the approximation of treating the valence electrons separately from the core by taking the molecular energy as the sum of core- and valence-electron energies. The core electrons are treated as a charge distribution that provides an effective repulsive potential for the valence electrons. This leads to an effective Hamiltonian for the valence electrons which is taken as the sum of one electron Hamiltonians:

$$H_{val} = \sum H_{eff}(1) \qquad (3.24)$$

The molecular orbitals are approximated as linear combinations of valence atomic orbitals of the atoms:

$$\phi_i = \sum c_{ij} F_j \qquad (3.25)$$

In the simple Hückel theory of planar hydrocarbons, each π MO contains contributions from one $2p\pi$ AO on each carbon atoms; in the extended Hückel treatment of planar hydrocarbons, each valence MO contains contributions from four AOs on each carbon atom (one 2s and three 2p's) and one 1s AO on each hydrogen atom. The AOs are usually the familiar Slater type with fixed orbital exponents.

Unlike the Hückel theory, the EH theory does not neglect overlap. Rather, all overlap integrals are explicitly evaluated using the forms chosen for the AOs and the internuclear distances at which the calculation is being done.

Complete Neglect of Differential Overlap (CNDO)   The CNDO method (Pople et al. 1965) treats only the valence electrons explicitly. The valence-electron Hamiltonian has the following form:

$$H_{val} = \sum_{i=1}^{n} (-\tfrac{1}{2}\nabla_i^2 + V_i) + \sum_{i<j} \frac{1}{r_{ij}} \qquad (3.26)$$

where there are n valence electrons and where V_i is the potential energy of valence electron i in the field of the nuclei and inner-shell electrons; the quantity in parentheses is $H_{cor}(i)$. The CNDO method generally uses a minimal basis set of Slater AOs with fixed orbital exponents on each atoms. The CNDO method makes the zero differential overlap approximation for all pairs of AOs in two-electron integrals. Thus, as in the case of the PPP method, the nonvanishing repulsion integrals in CNDO are of the form $\langle rr|tt \rangle$. Since only two AOs occur in these integrals, all three- and four-center integrals are neglected. In the PPP treatment of conjugated hydrocarbons, there is only one basis AO per atom—the $2p\pi$ AO. In the CNDO method, there are several basis valence AOs on every atom (except hydrogens), and the ZDO approximation leads to neglect of electron-repulsion integrals containing the product $f_r(1)f_s(1)$ where f_r and f_s are different AOs centered on the same atom. The CNDO method also neglects overlap integrals of equation 3.19. The core-Hamiltonian integrals of equations 3.21 and 3.22 are approximated by using data such as atomic ionization potentials and electron affinities and by taking some of them as semiempirical parameters adjusted so that CNDO calculations will give a good overall fit with the results of minimal basis ab initio SCF calculations.

Intermediate Neglect Differential Overlap (INDO)   As in the CNDO method, the INDO method (Pople et al. 1967) treats only the valence electrons explicitly. However, it is an improvement on CNDO. Here, differential overlap between AOs on the same atom is not neglected in one-center electron-repulsion integrals, but is still neglected in two-center electron-repulsion integrals. Otherwise, the two methods are similar.

Modified Neglect of Diatomic Overlap (MNDO)   The MNDO method (Dewar and Thiel 1977) is based on the neglect only of differential overlap between AOs centered on different atoms. The MNDO method is parameterized to reproduce gas-phase $\Delta H^0_{f,298}$ values.

Comparisons of Quantum Mechanical Methods   In this section, we compare briefly the accuracies of various semiempirical and ab initio methods in calculating molecular properties. The discussion is mainly focused on optical nonlinear properties that represent the molecular distortion to the external electric field.

We first consider the calculation of the equilibrium geometry for molecules, which is the first step when calculating nonlinearities. The more efficient procedure is to calculate the energy and the first derivatives of the energy (energy gradient) with respect to the nuclear coordinates at a relatively small number of points and then use these data to find the equilibrium geometry. Calculations of the energy gradient from SCF wavefunctions is relatively straightforward. For semiempirical MO methods based on neglect of differential overlap (CNDO, INDO, and MNDO), calculations of the energy gradient is much faster than for the ab initio SCF method. The CNDO, INDO, and MNDO all require the same amount of computing time. Now we consider dipole moments. The accuracy of ab initio SCF dipole moments calculated with the STO-3G and 4-31G basis set is fair; the 6-31G* basis set gives substantially more accurate results. Although sometimes accurate, CNDO and INDO dipole moments often show very substantial errors. MNDO is reasonably reliable for dipole moments. Now we consider linear polarizabilities calculated in the static case ($w = 0$). Ab initio orientationally average linear polarizabilities calculated using the STO-3G and 4-31G basis set are approximately 50% smaller than experimental linear polarizabilities (Hurst et al. 1988). Calculated using 6-31G augmented with several s, p, and d functions of diffuse and polarization character, they differ by less than 15% (Perrin et al. 1989). However, the anisotropy of the linear polarizability is slightly overestimated due to an underestimation of the out-of-plane component. CNDO and INDO orientationally averaged linear polarizabilities differ very often from experimental values by 40% (Waite et al. 1982) and 50% (Zyss 1978), respectively. For dipoles, MNDO is reasonably reliable for linear polarizabilities and gives a relative error of 30% for the in-plane components. However, the anisotropy of the linear polarizability is always grossly overestimated.

Finally, we consider together first and second hyperpolarizabilities calculated in the static case ($w = 0$). Davidson and Feller (1986) have shown that the choice of the basis set is an important consideration for ab initio calculations of second and third optical nonlinear effects. 4-31G and 6-31G basis sets have to be augmented with several s, p, and d diffuse functions of diffuse and polarization character on the atoms to give a relatively good description of the optical nonlinearities. One study carried out on the haloform series (Karna et al. in Press) shows that ab initio and experimental first hyperpolarizabilities differ only by 50%. In contrast, the ab initio and experimental second hyperpolarizabilities differ by a factor of 4. The PPP and EH methods are not very accurate. They overestimate second- and third-order nonlinearities due to the neglect of σ electrons, which contribute to the nonlinearities with a sign opposite that of the π electrons (Chopra et al., 1989) CNDO third-order optical nonlinear effects differ by factors of 3 and 5, respectively, from the corresponding experimental data. However, White et al. (1982) have improved the CNDO calculations of optical nonlinear properties. For a model compound, they calibrated the semiempirical CNDO parameters to recover the experimental values. Then, they calculated the optical nonlinearities for other compounds. With such a procedure their calculated values are in relatively reasonable agreement with the experimental results (30% for the second hyperpolarizability). INDO first and second hyperpolarizabilities, reported by Zyss (1978), are four and ten times smaller, respectively, than the corresponding ab initio results. The accuracy of MNDO first and second hyperpolarizabilities is very poor, especially for the out-of-plane components, which can even differ in sign from the experimental results. The in-plane components of the first and second hyperpolarizabilities are approximately 2 and 3 times smaller, respectively, than the corresponding ab initio results (Dewar et al. 1978).

In conclusion, semiempirical methods give substantial savings in computation time over ab initio methods, and optical nonlinearities of larger molecules can be calculated. However, the overall reliability of the ab initio SCF MO method is generally good when one uses a basis set of suitable size (the size required depends on the property being calculated). Of course, one is quite restricted in the size of the molecules for which ab initio calculations can be done.

3.5.3 Sum-over-States Approach

The sum-over-states (SOS) approach has its basis in a perturbation theory method developed by Ward (1965) to account for the effects of an externally applied electromagnetic field on the motions of electrons associated with the molecule of interest. Under the influence of the oscillating field the electrons will be perturbed and oscillating currents produced in the molecule. The dipolar contribution to the current (which polarizes the molecule) can be obtained by inclusion of the field as a perturbation

$$H' = -e(E^{\omega} \cdot r)\sin \omega t \tag{3.27}$$

to the Hamiltonian and collecting terms of appropriate orders in the electric field. In this expression E^{ω} is the amplitude of the field component at frequency ω, and r is a coordinate associated with the position of the electrons and is calculated from

$$r = \sum_{a} r_a \tag{3.28}$$

in which a is summed over all of the electrons and

$$-er = \mu \tag{3.29}$$

The result is expressions for the polarizability and hyperpolarizabilities expressed as infinite sums over various excited states in which the numerators contain dipolar integrals of the type $er_{nn'} = \langle n|er|n' \rangle$. When $n = n'$ this quantity corresponds to the dipole moment of state n' and when $n \neq n'$ it corresponds to a transition dipole moment between the two states.

The expression for the polarizability is given by

$$\alpha(\omega) = \sum_{n} \frac{e^2}{\hbar} \left(\frac{r_{gn} r_{ng}}{\omega - \omega_{ng}} + \frac{r_{gn} r_{ng}}{\omega + \omega_{ng}} \right) \tag{3.30}$$

In this expression g refers to the ground state and n are the various excited states with $\omega_{ng} = \omega_n - \omega_g$. The first hyperpolarizability term $\beta(-2\omega; \omega, \omega)$, which is responsible for second-harmonic generation, is given by

$$\beta(-2\omega; \omega, \omega) = p \sum_{n_1, n_2} \left(\frac{e^3}{2\hbar^2} \right) \frac{r_{gn_2} r_{n_2 n_1} r_{n_1 g}}{(\omega - \omega_{n_1 g})(2\omega - \omega_{n_2 g})} \tag{3.31}$$

The symbol p indicates that the summations must be performed over all permutations of the Cartesian indices in the molecular framework i, j, and k, with the electric field frequencies ω and 2ω. This summation generates terms that are products of transition dipole moment matrix elements and also sums and differences of dipole moments between ground and excited states as well as between various excited states. This expression simplifies into an intuitively appealing expression if we consider the properties of the molecule to be approximated by a simple two-level model.

The second hyperpolarizability term $\gamma(-3\omega; \omega, \omega, \omega)$ responsible for third-harmonic generation is expressed as

$$\gamma(-3\omega; \omega, \omega, \omega) = p \sum_{n_1, n_2, n_3} \left(\frac{e^4}{4\hbar^3} \right) \frac{r_{gn_3} r_{n_3 n_2} r_{n_2 n_1} r_{n_1 g}}{(\omega - \omega_{n_3 g})(2\omega - \omega_{n_2 g})(3\omega - \omega_{n_1 g})} \tag{3.32}$$

The computation of the polarizabilities thus involves the evaluation of various dipole moment operators r_{nn} and the energies, then summing all the terms. These sums are infinite sums over various states including the atomic Rydberg states. However, one often utilizes the sum over only low lying π-electron states. For example, Garito and coworkers (Lalama and Garito 1979) have used a SCF-MO method with the CNDO approximation including configuration interaction for the computation of the dipole moments and excitation energies. The limitation of this approach is the arbitrary truncation of the summation for the sake of computational efficiency where contributions of only a limited number of excited states are considered.

The SOS expression serves a useful purpose in that it allows one to examine the effect of resonance enhancement on optical nonlinearities. For example, the expression for $\beta(-2\omega; \omega, \omega)$ shows that the β value is considerably resonantly enhanced when the fundamental frequency (ω) and/or the second-harmonic frequency (2ω) is near an electronic resonance (i.e., ω, $2\omega = \omega_{n_1 g}$ or $\omega_{n_2 g}$). At resonance, one has to consider a complex energy $\omega_{n_1 g} + i\Gamma_{n_1 g}$ with the damping term $\Gamma_{n_1 g}$. Therefore, β is a complex quantity that is enhanced by one and/or two photon resonances. Similarly, an inspection of equation 3.32 shows that $\gamma(-3\omega; \omega, \omega, \omega)$, responsible for third-harmonic generation, is resonance enhanced by one-photon resonance ($\omega = \omega_{n_1 g}$), two-photon resonance ($2\omega = \omega_{n_1 g}$), and three-photon resonance ($3\omega = \omega_{n_1 g}$). Again, γ would be complex near a resonance.

3.5.4   Two-Level Model

The simplest model to take into account the contribution of charge-transfer resonances within a molecule to the first hyperpolarizability was the two-level model (Oudar and Chemla 1977). This model assumes that the electronic properties of the molecule are determined by a ground state and a low-lying charge-transfer excited state. Polarization results primarily from the admixing

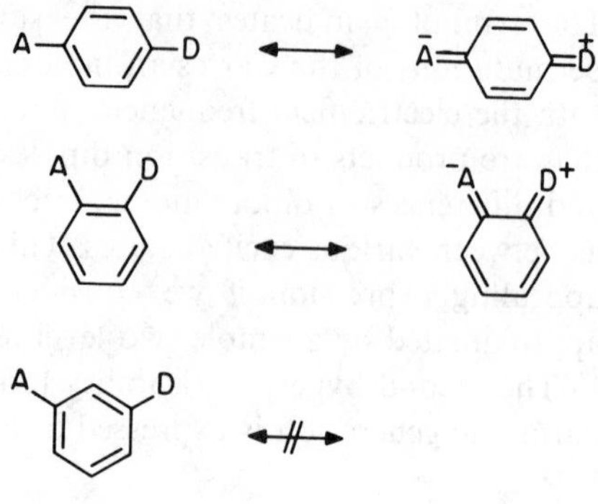

Figure 3.1 Ground-state and lowest energy polar resonance forms for p and 0 substitution. Charge transfer resonance is forbidden for the m substituent.

of the charge-transfer resonance state with the ground state through the action of the electric field $E(w)$ (Fig. 3.1).

In the two-level model, equation 3.31 reduces to

$$\beta(-2\omega;\omega,\omega) \cong \frac{3e^2}{2\hbar m} \frac{\omega_{eg} f \Delta\mu}{(\omega_{eg}^2 - \omega^2)(\omega_{eg}^2 - 4\omega^2)} \tag{3.33}$$

where ω_{eg} is the frequency of the optical transition, f is the oscillator strength and is related to the transition moment between ground and charge-transfer excited state $\langle g|er|n\rangle^2$, and $\Delta\mu$ is the difference between ground- and excited-state dipole moment. With this expression it is possible to establish trends in the relationship between β and the molecular structure in terms of relatively simple physical organic notions. As the oscillator strength or extinction coefficient increases, β is expected to increase. As the difference in polarity between ground and excited state increases, an increase in β is also anticipated. Polarity differences can be estimated from solvent-induced spectral shifts (see Chapter 5 for a more detailed discussion). The model also shows that when ω or 2ω is close to the absorption band, β is enhanced. The applicability of this model in deriving structure property relationships is discussed further in Chapters 5 and 7.

3.5.5   The Free-Electron Model

There have been several semiempirical approaches to calculate the nonlinear optical property of organic molecules, starting from very early work by Rustagi and Ducuing in 1974. They used the free-electron model for conjugated linear polymers in which the π electrons are confined in a one-dimensional box. Since then there have been several other semiempirical models. Among them is the pioneer work by Hameka and coworkers using Hückel and PPP Hamiltonians. The more recent calculations are those of de Melo and Silbey (1987).

For a conceptual understanding of the third-order nonlinearity γ in conjugated linear polymers, the one-dimensional box model of Rustagi and Ducuing (1974) is still used because of its simplicity. A brief description of this model is presented here. The Hamiltonian for the π electrons in a one-dimensional box, in the presence of an electric field along the chain direction, can be written

$$H = H^{(0)} + H^{(1)} = -\left(\frac{h^2}{2m}\right)\left(\frac{\partial^2}{\partial Z^2}\right) - eEZ \qquad (3.34)$$

In the above equation the first term is the kinetic energy of the electron in the box of length $2L$ (the dimension of effective π conjugation), and the second term is the electric dipole interaction with the field. The field direction is along the chain axis Z. Rayleigh–Schrödinger perturbations are used to obtain the infinite sum over the intermediate states, to evaluate the Stark energies in higher orders, and to calculate γ. The expresson for γ obtained by this method is

$$\gamma = \frac{128L^{10}}{a_0^3 e^2} \sum_{n=1}^{N} \left(\frac{-2}{9\pi^6 n^6} + \frac{140}{3\pi^8 n^8} - \frac{440}{\pi^{10} n^{10}}\right) \qquad (3.35)$$

In the above equation $a_0 = \hbar^2/me^2$ is the atomic Bohr radius and the sum n runs over all the occupied levels N ($2N$ is the total number of π electrons in an alternately single- and double-bonded structure; N is also the number of repeat units). For a highly conjugated structure (large conjugation length and, consequently, large N), equation 3.35 to the lowest order in $(1/N)$ yields

$$\gamma = \frac{256L^{10}}{45 a_0^3 e^2 \pi^6 N^5} \qquad (3.36)$$

Assuming that $N \propto L$, equation 3.36 suggests that $\gamma \propto N^5$. This dependence on the conjugation length (or the number of the repeat unit) is stronger than what is predicted by a simple ab initio calculation as discussed above. The above model has been extended to calculate the dependence of $\chi^{(3)}$ on the band gap of the conjugated polymer. The band gap for this purpose is the gap between the highest occupied π orbital and the lowest unoccupied π^* orbital of the molecule (π–π^* transition). The derived expression is

$$\chi^{(3)}_{ZZZZ} = \frac{2^5}{45}\pi^2 \frac{e^{10}}{\sigma}\left(\frac{a_0}{d}\right)^3 \frac{1}{E_g^6} \qquad (3.37)$$

In this equation, d is the average C–C distance, σ the cross-sectional area per chain, and E_g is the band-gap energy. Equation 3.37 predicts that the $\chi^{(3)}$ component along the conjugated polymer chain direction shows an inverse dependence on the sixth power of the $\pi\pi^*$ band gap. Therefore, an estimation of the band gap by linear spectroscopy can yield a rough estimate of third-order nonlinearity in conjugated polymeric structures. The more colored the molecule is, the smaller is the band gap (band gap shifted in the visible spectral range), and, consequently, the higher is its $\chi^{(3)}$ value.

3.6   ORIENTATIONAL NONLINEARITY

Anisotropic molecules, especially in a fluid phase, can also exhibit orientational optical nonlinearity derived from the molecular alignment either in an electric field or in an optical field. The orientational alignment in an electric field gives rise to important nonlinear optical effects such as electric field-induced second-harmonic generation (EFISH). This subject is discussed in detail in a later chapter. In an optical field, the molecular alignment occurs for molecules whose polarizability tensor α is not fully symmetric. Consequently, different components of the polarizability tensor produce different induced dipole moments resulting in reorientation of the molecule to align the long axis of the polarizability along the polarization (field) direction of the optical field. As a result, the medium becomes anisotropic (birefringent) and contributes to the γ value, which describes the intensity dependence of refractive index (but not to third-harmonic generation). This contribution also describes optical Kerr effect.

A simple fundamental approach to describe this optimally induced birefringence in relation to the Kerr effect was presented by Buckingham and Pople (1955) and subsequently by Kielich (1967). The details of the derivation of expressions appropriate for the molecular Kerr constant are presented in numerous papers. One should be aware, however, that the expressions derived for the Kerr effect usually refer to the birefringence induced by the electric field, that is, the difference between two components of the refractive index indicatrix: that along the electric field and that perpendicular to it (i.e., $\Delta n^{\text{Kerr}} = (n^{\parallel}(E) - n^{\perp}(E))$. Following the derivation of Buckingham and remembering that γ defined by the power expansion equations (3.1) and (3.2) is sixfold smaller than that used by Buckingham, one can obtain the expression for the orientational contribution $\gamma_{\text{orientation}}$ which relates to the difference between the refractive indices in the presence and in the absence of the electric field. The technique of degenerate four-wave mixing (DFWM), discussed in Chapter 9, measures such orientational γ. The expression for γ measured by DFWM with all parallel polarization is given by (Samoc et al., to be published)

$$\gamma_{\text{orientation}} = \tfrac{1}{135}kT[(\alpha_{11} - \alpha_{22})^2 + (\alpha_{22} - \alpha_{33})^2 + (\alpha_{11} - \alpha_{33})^2] \qquad (3.38)$$

where α_{ii} are the components of the linear polarizability tensor.

The derivation of equation 3.38 assumes that the electric field causing the molecular reorientation is present long enough for the new equilibrium orientational distribution of molecules to be established. This is true only if the laser pulse duration is much longer than the involved reorientational times. Femtosecond optical Kerr studies (McMorrow et al. 1988) of several simple molecular liquids (such as CS_2) show that the kinetics of molecular reorientation in liquids can involve many different time constants, ranging from a few hundred femtoseconds to several picoseconds. These processes, therefore, can be extremely fast for small molecules.

REFERENCES

Applequist, J., J. R. Carl, and K. K. Fung, *J. Am. Chem. Soc.* **94**, 2952 (1972).

Bair, N. C. and M. J. S. Dewar, *J. Chem. Phys.* **50**, 1262 (1969).

Buckingham, A. D., and A. J. Pople, *Proc. Phys. Soc.* (*London*) **68**, 905 (1955).

Chopra, P., L. Carlacci, H. F. King, and P. N. Prasad, *J. Phys. Chem.* **93**, 7120 (1989).

Davidson, E. R., and D. Feller, *Chem. Rev.* **86**, 681 (1986).

de Melo, C. P., and R. Silbey, *Chem. Phys. Lett.* **140**, 537 (1987).

Dewar, M. J. S., and W. Thiel, *J. Am. Chem. Soc.* **99**, 4899, 4907 (1977).

Dewar, M. J. S., Y. Yamagushi, and S. H. Such, *Chem. Phys.* **59**, 541 (1978).

Ditchfield, R., in D. R. Lide and M. Paul (Eds.), *Critical Evaluation of Chemical and Physical Structural Informations*, National Academy of Sciences, Washington, DC, 1974.

Everard, K. B., and L. E. Sutton, *J. Chem. Soc.* **25**, 402 (1951).

Gordon, M. S., *J. Am. Chem. Soc.* **100**, 2670 (1978).

Halgren, T. A., *J. Am. Chem. Soc.* **100**, 6595 (1978).

Hehre, W. J., L. Radom, P. R. Schleyer, and J. A. Pople, *Ab Initio Molecular Orbital Theory*, Wiley Interscience, New York 1986.

Hoffmann, R., *J. Chem. Phys.* **39**, 1397 (1963); **40**, 2445, 2474, 2480, (1964).

Hurst, G. J. B., M. Dupuis, and E. Clementi, *J. Chem. Phys.* **89**, 385 (1988).

Hush, N. S., and M. L. Williams, *Chem. Phys. Lett.* **6**, 163 (1970).

Jug, K., and D. N. Nanda, *Theor. Chim. Acta* **57**, 107, 131 (1980).

Kajzar, F., and J. Messier, *Phys. Rev. A* **32**, 2352 (1985).

Kajzar, F., and J. Messier, *J. Opt. Soc. Am. B* **4**, 1040 (1987).

Karna, S. P., M. Dupuis, E. Perrin, and P. N. Prasad, *J. Chem. Phys.,* **90**, 7418 (1990).

Kielich, S., *Acta Phys. Polon.* **31**, 689 (1967).

Lalama, S. J., and A. F. Garito, *Phys. Rev.* **A20**, 1179 (1979).

Levine, B. F., and C. G. Bethea, *J. Chem. Phys.* **63**, 2666 (1975).

McMorrow, D., W. T. Lotshaw, and G. A. Kenney-Wallace, *Chem. Phys. Lett.* **145**, 309 (1988).

Meredith, G. R., B. Buchalter, and C. Hanzlik, *J. Chem. Pys.* **78**, 1543 (1983).

Miller, C. K., B. J. Orr, and J. F. Ward, *J. Chem. Phys.* **74**, 4858 (1981).

Nanda, D. N., and K. Jug, *Theor. Chim. Acta* **57**, 95 (1980).

Oudar, J. L., and D. S. Chemla, *Opt. Commun.* **13**, 164 (1975).

Oudar, J. L., and D. S. Chemla, *J. Chem. Phys.* **66**, 2664 (1977).

Pariser, R., and R. G. Parr, *J. Chem. Phys.* **21**, 466, 711 (1953).

Perrin, E., P. N. Prasad, P. Mougenot, and M. Dupuis, *J. Chem. Phys.* **91**, 4728 (1989).

Pople, J. A., *Trans. Faraday Soc.* **49**, 1375 (1953).

Pople, J. A., D. P. Santry, and G. A. Segal, *J. Chem. Phys.* **43**, S129 (1965).

Pople, J. A., D. L. Beveridge, and P. A. Dobosh, *J. Chem. Phys.* **47**, 2026 (1967).

Prasad, P. N., E. Perrin, and M. Samoc, *J. Chem. Phys.* **91**, 2360 (1989).

Roothaan, C. C. J. *Rev. Mod. Phys.* **23**, 69 (1951).

Rustagi, K. C., and J. Ducuing, *Opt. Commun.* **10**, 258 (1974).

Samoc, A., M. Samoc, P. N. Prasad, C. S. Willard, and D. J. Williams, to be published.

Sundberg, K. R., *J. Chem. Phys.* **66**, 1475 (1977); **67**, 4314 (1977).

Waite, J., M. G. Papadopoulos, and C. A. Nicolaidos, *J. Chem. Phys.* **63**, 2666 (1982).

Ward, J., *Rev. Mod. Phys.* **37**, 1 (1965).

Ward, J. and G. H. C. New, *Phys. Rev.* **185**, 57, (1969, Appendix A).

Zyss, J., *J. Chem. Phys.* **70**, 3333 (1978).

4

BULK NONLINEAR OPTICAL SUSCEPTIBILITY

4.1 INDUCED POLARIZATION

In the previous chapter the interaction of the radiation field with a molecular unit was described in terms of a power series expansion of the Stark energy of the molecule, or equivalently the dipole moment, with the electric field components. This expansion is valid as long as the total induced polarization is small and dipole approximation is valid. From equation 3.2 the ith Cartesian component of the induced molecular polarization, p_i (the same as μ_i), is given by

$$p_i = \alpha_{ij}E_j + \beta_{ijk}E_jE_k + \gamma_{ijkl}E_jE_kE_l \tag{4.1}$$

where the subscripts i, j, k, and l refer to the molecular coordinate system and E_j, E_k, and so on, denote components of the applied field. Similarly, the nonlinear polarization induced in bulk media can be expressed by the expansion

$$P_I = \chi_{IJ}^{(1)}E_J + \chi_{IJK}^{(2)}E_JE_K + \chi_{IJKL}^{(3)}E_JE_KE_L + \cdots \tag{4.2}$$

The validity of the dipolar approximation requires that in addition to the induced polarization being small, the size of the polarizable unit must be small compared to the wavelength of light, and the material or medium nonmagnetic. A question naturally arises as to the relationship between the properties of the constituent molecules or atoms and the properties of the bulk medium.

To a first approximation it is possible to infer the nonlinear optical properties of the bulk medium from those of the constituent molecules and vice versa, although great care must be taken in doing so. The factors that need to be considered are numerous and different for $\chi^{(2)}$ and $\chi^{(3)}$ processes. For $\chi^{(2)}$ processes the symmetry of the medium and orientation of the molecular constituents in the bulk medium is of paramount importance. The contribution to the macroscopic process from the molecular oscillators is intimately dependent on their orientation. Randomness at the molecular level leading to isotropy in the medium results in a virtual complete cancellation of the second-order nonlinear effect in the bulk. The nonlinear optical properties of the molecules themselves are anisotropic so that their projection onto the macroscopic frame of reference must also be considered. Also, the effects of local fields arising from any free charges as well as the mutual polarization from electrons and nuclei of neighboring molecules can greatly affect the magnitude of the electric field in the bulk. Any orientation relaxation at the molecular site as a result of its dipolar field must also be considered. Local field effects can change the response of individual molecules to externally applied electric field components by significant amounts relative to that of an isolated molecule, as in a vacuum.

In third-order processes orientation is less important since the macroscopic response is an average of all orientations of the polarizable constituents and cancellation effects do not occur as a result of orientational averaging. In cases where the nonlinear response is extremely anisotropic and large in one direction within the polarizable unit, axial orientation to enhance the interaction of that entity with the field components can lead to an increased nonlinear response up to a factor of five. Subtleties of the electronic structure, such as charged or uncharged mobile defects in conjugated chains, excitonic effects, and multiphoton processes, can also have enormous effects on third-order nonlinear processes and the polarizable unit can consist of multiple repeat units in polymeric nonlinear optical media.

In this chapter we address the various issues relating molecular structure and orientation to the observed second- and third-order nonlinear coefficients that are obtained in the bulk.

4.2   RELATIONSHIP BETWEEN MICROSCOPIC AND MACROSCOPIC NONLINEARITIES IN CRYSTALS

A particularly incisive analysis of the relationship between microscopic and macroscopic second-order nonlinearities in noncentrosymmetric crystals was developed by Oudar and Zyss (1982). The analysis was extended to the case of poled polymer films by Meredith et al. (1982, 1984), Williams (1987), and Singer et al. (1987) and can be extended to third-order nonlinearities as well.

The analysis of Oudar and Zyss (1982) begins by assuming that the collection of molecules in the crystal can be treated as an oriented gas with well-defined relationships between the hyperpolarizabilities of the molecular constituents and the macroscopic nonlinear coefficients. The macroscopic second-order coefficient $\chi^{(2)}_{IJK}$ is related to the microscopic one β_{ijk} by

$$\chi^{(2)}_{IJK} = N f_I(\omega_3) f_J(\omega_1) f_K(\omega_2) b_{IJK} \tag{4.3}$$

where

$$b_{IJK} = \frac{1}{N_g} \sum_{ijk} \left(\sum_{s=1}^{N_g} \cos\theta_{Ii}^{(s)} \cos\theta_{Jj}^{(s)} \cos\theta_{Kk}^{(s)} \right) \beta_{ijk} \qquad (4.4)$$

In this equation N is the number of molecules per unit of volume in the crystal. It is related to the unit cell volume by $N = N_g/V$, where N_g is the number of equivalent sites, s, in the unit cell and V is unit cell volume. For disordered or partially ordered materials such as poled polymers or Langmuir–Blodgett films where no unit cell can be defined, N is simply the concentration of the active hyperpolarizable unit and $N_g = 1$.

The tensor b_{IJK} has the characteristics of the β_{ijk} tensor but is expressed in the crystal frame of reference and refers to the contribution of a single molecule to the macroscopic nonlinear coefficients. The angle $\theta_{Ii}^{(s)}$ expresses the relationship between the crystallographic axis I and the molecular axis i_s for a particular constituent of the unit cell. The $\cos\theta_{Ii}^{(s)}$ and so on are the scaler products of unit vectors I and i and give the projection of the molecular coefficients into the crystal frame of reference. The local field factors $f_I(\omega)$ and so on are associated with the appropriate crystal axis and frequency components. For the sake of simplicity in the notation the usual frequency arguments were not included in the susceptibilities in equation 4.3.

Although a number of complexities can arise for certain combinations of unit cell sizes and redundancies within the unit cell, the application of (4.3) is relatively straightforward when crystal symmetry and the inherent symmetries associated with nonlinear processes under certain conditions are taken into account. A hypothetical case in point that illustrates these relationships is the monoclinic point group 2 where two molecules in the unit cell are interchanged by a twofold rotation about a particular axis.

Consider the molecule in the unit cell illustrated in Figure 4.1. In this example Z is chosen as the two-fold rotation axis that produces the other unit cell constituent. The molecular z axis is chosen as the one that is parallel to the

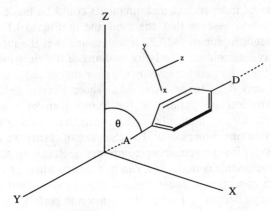

Figure 4.1   The relationship between molecular and crystal frame of reference.

direction of the dipole moment and is at some angle θ with respect to the Z axis. For planar aromatic molecules the molecular nonlinear optical properties are conveniently described if we choose x perpendicular to z and in the plane of the molecule. The macroscopic X direction is chosen to be perpendicular to Z and to also contain the molecular plane x. Before further discussion of the example several comments and observations regarding b_{IJK} would be helpful.

The b_{IJK} matrix is quite useful in determining what molecular coefficients will contribute to crystal nonlinearity through observations on crystal symmetry. In this particular case the Z axis is the polar axis in the crystal. Applying a twofold rotation about Z, only tensor components b_{IJK} containing a Z index will be nonzero. Otherwise, the average of that value over the two molecules in the unit cell will cancel. Another way of saying this is that the X and Y crystal axes are symmetric with respect to a twofold rotation about Z. With the further assumption of Kleinman symmetry (discussed in Chapter 2), which equates elements of the tensors related by permutation of the indices, there are four independent elements for the b tensor for this particular case:

$$b_{ZXX} = \frac{1}{2}\sum_{ijk} \cos\theta_{Zi}\cos\theta_{Xj}\cos\theta_{Xk}\beta_{ijk} \tag{4.5}$$

$$b_{ZZZ} = \frac{1}{2}\sum_{ijk} \cos\theta_{Zi}\cos\theta_{Zj}\cos\theta_{Zk}\beta_{ijk} \tag{4.6}$$

$$b_{ZYY} = \frac{1}{2}\sum_{ijk} \cos\theta_{Zi}\cos\theta_{Yj}\cos\theta_{Yk}\beta_{ijk} \tag{4.7}$$

$$b_{ZYX} = \frac{1}{2}\sum_{ijk} \cos\theta_{Zi}\cos\theta_{Yj}\cos\theta_{Xk}\beta_{ijk} \tag{4.8}$$

Now, with a knowledge of the molecular coefficients β_{ijk} it may be possible to determine the nonlinear coefficients of the crystal or vice versa, assuming a sufficient number of independent measurements could be made. For the sake of discussion we will assume that the molecule in Figure 4.1 is methyl(2,4-dinitrophenyl)aminopropionate (MAP). If we assume that the molecular hyperpolarizability β is determined entirely by motions of the electrons in the plane of the π system then the number of nonzero elements of the β tensor are four: β_{xxx}, β_{xzz}, β_{zzz}, and β_{zxx}. In practice, β_{zzz}, where electric field components producing polarization are parallel to the charge transfer z axis, would be expected to be the largest molecular component, thus making the molecular nonlinearity pseudo one dimensional. For the sake of clarity we calculate b_{ZXX} and state the results for the remaining components. Recalling an earlier statement that only coefficients containing Z can be nonzero when a twofold rotation is performed, we can make a more general observation. That is, coefficients $\cos\theta_{Li}$ will be multiplied by $+1$ when the rotation is performed only if $L = Z$;

otherwise it will be multiplied by -1. With this is mind, the coefficient b_{ZXX}, as defined in (4.4) and (4.5), can be expressed as

$$b_{ZXX} = \tfrac{1}{2}[\cos\theta_{Zx}\cos\theta_{Xx}\cos\theta_{Xx}\beta_{xxx} + \cos\theta_{Zx}\cos\theta_{Xz}\cos\theta_{Xz}\beta_{xzz}$$
$$+ \cos\theta_{Zz}\cos\theta_{Xz}\cos\theta_{Xz}\beta_{zzz} + \cos\theta_{Zz}\cos\theta_{Xx}\cos\theta_{Xx}\beta_{zxx}$$
$$+ (\text{permutations for two fold } Z \text{ axis})] \qquad (4.9)$$

Referring to Figure 4.1 and the discussion of the relationship between the various axis, a table of terms can be generated (Table 4.1) for the two molecules in the unit cell.

Evaluating the terms in equation 4.9 yields

$$b_{ZXX} = \cos\alpha\beta_{zxx} \qquad (4.10)$$

where α is the angle between the molecular z axis and the macroscopic Z axis. Carrying out a similar analysis for the remaining terms in β gives

$$b_{ZZZ} = \cos^3\alpha\beta_{zzz} \qquad (4.11)$$

$$b_{ZYY} = \cos\alpha\sin^2\alpha\beta_{zzz} \qquad (4.12)$$

$$b_{XYZ} = -\sin\alpha\cos\alpha\beta_{zzz} \qquad (4.13)$$

An interesting observation can be made at this point: The second-order nonlinear optical properties of the crystal are independent of the value β_{xxx} due to the nature of the symmetry of the medium. It would therefore not be possible to determine that coefficient from measurements on the crystal. We also see that the angle α is given by the ratio

$$\frac{b_{ZYY}}{b_{ZZZ}} = \tan^2\alpha \qquad (4.14)$$

TABLE 4.1   Values of Dot Products $Z \cdot x$,
$X \cdot x$, $X \cdot z$, and $Z \cdot z$ for two Molecules in a Unit,
All Related by a Twofold Rotation Axis and
Subject to Other Assumption in the Text

$s = 1$	$s = 2$
$\cos\theta_{Zx} = \cos 90° = 0$	$\cos\theta_{Zx} = 0$
$\cos\theta_{Xx} = \cos 0 = 1$	$\cos\theta_{Xx} = -1$
$\cos\theta_{Xz} = \cos(90 - \alpha) = \sin\alpha$	$\cos\theta_{Xz} = \sin\alpha$
$\cos\theta_{Zz} = \cos\alpha$	$\cos\theta_{Zz} = \cos\alpha$

Finally we note that transformations to other convenient frames of reference can be accomplished by suitable rotations in the XY plane. Typically the d_{IJK} tensor is expressed in a reference frame $1, 2, 3$ where the numbers refer to the principal dielectric axes of the crystal. In our case $3 = Z$ and $1, 2$ are related to X and Y by rotation.

This analysis is quite useful is relating the microscopic and macroscopic coefficients in solids and has been extended to all crystal point groups and numbers of molecules within the unit cell (Zyss and Oudar 1982). By extending this analysis to the other polar point groups, the crystal point groups giving the largest phase-matchable second-order nonlinear (NL) coefficients as well as the optimum molecular orientation within that point group were identified. A summary of those results was tabulated by Zyss and Oudar (1982) and is partially reproduced here in Table 4.2.

The angles Θ and Ψ are Euler angles between the crystal Y and molecular x direction (as x was defined in Figure 4.1) and between the projection of z in the xy plane and the Y axis, respectively. Under the NL coefficient column the $\frac{1}{4}$ and so on after d_{22} signifies the $\frac{1}{4}$ of the maximum molecular nonlinearity is available for phase-matched second-harmonic generation due to the projections of the molecular coefficients onto b as discussed earlier in the chapter. The reader is referred to Oudar and Zyss (1982) for a complete discussion of this approach. Here it is sufficient to note that 38% of the maximum molecular nonlinearity is the best that can be achieved for phase-matched second-harmonic generation in a crystal.

At this point some comments on the local field factors in equation 4.3 are warranted. Local field factors relate the externally applied field $E(\omega)$ to the field $e(\omega)$ actually felt by a molecule in the condensed phase as a result of charges on neighboring molecules and any orientational polarization that occurs on neighboring molecules in reaction to the charge distribution on the molecule of interest. This relationship, subject to assumptions outlined by Zyss and

TABLE 4.2 Maximum Values of Effective Nonlinear Coefficients for the Specified Molecular Orientations in the Unit Cell for the Polar Crystal Point Groups (See text for definitions for Θ and Ψ.)

Group	NL Coefficient	Θ	Ψ	$b_{\text{eff}}^{\max}$
$1, 2mm, 2$	—	—	—	$\frac{2}{3}\sqrt{3} \sim 0.38$
$\bar{6}2m$	$d_{22}, \frac{1}{4}$	$90°$	$60°$	$\frac{1}{4}$
$\bar{6}, 3$	$d_{22}, \frac{1}{4}$	$90°$	—	$\frac{1}{4}$
$3m$	$d_{22}, \frac{1}{4}$	$90°$	$30°$	$\frac{1}{4}$
32	$d_{11}, \frac{1}{4}$	$90°$	$60°$	$\frac{1}{4}$
222	—	—	—	$\frac{1}{3}\sqrt{3}$
$\bar{6}mm, 6, 4mm, 4$	$d_{15}, \frac{1}{3}\sqrt{3}$	$54.74°$	—	$\frac{1}{3}\sqrt{3}$
$\bar{4}2m, 23, \bar{4}3m$	$d_{14}, \frac{1}{3}\sqrt{3}$	$54.74°$	$45°$	

Source: Data from Zyss and Oudar (1982).

Chemla (1987), can be written as

$$e(\omega) = E(\omega) + 4\pi L P(\omega) \tag{4.15}$$

where L is a tensor whose value is related to the geometry of the local environment surrounding the molecule of interest. If polarization of the surrounding medium is considered to be uniform then the components of L are related to the semiaxes of an ellipsoidal cavity. For a spherical cavity $L_1 = L_2 = L_3 = \frac{1}{3}$. For a long cylinder $L_1 = L_2 = \frac{1}{2}$ and $L_3 = 0$; for a flat one $L_1 = L_2 = 0$ and $L_3 = 1$ (Böttcher 1952). In the case of a linear dielectric, a local field factor $f(\omega)$ can be defined as

$$f(\omega) = \frac{e(\omega)}{E(\omega)} \tag{4.16}$$

or can be expressed alternatively in terms of the dielectric constant ε_ω as

$$f(\omega) = 1 + (\varepsilon_\omega - 1)L \tag{4.17}$$

For media with high symmetry, such as liquids or cubic crystals, equation 4.17 reduces to

$$f(\omega) = \frac{\varepsilon_\omega + 2}{3} \tag{4.18}$$

which is the well-known Lorentz correction factor. This correction pertains to effects of induced dipoles in the medium through electronic polarization and is probably the most appropriate one to use in crystals using refractive indices $n^2 = \varepsilon_\omega$ appropriate for the principal directions in the crystal. It is also commonly used in determining the molecular hyperpolarizability β of solutions and liquids, as is discussed in Section 6.1.

The Lorentz field factor pertains only to dipoles induced through electronic motions. However, external fields and fields associated with neighboring dipoles can induce orientation of permanent dipoles, which requires further correction if the internal field is to be properly accounted for. The local field factors ascribed to this process are known as the Onsager local field factors and given by (Onsager 1936):

$$f(0) = \frac{\varepsilon_0(\varepsilon_\omega + 2)}{(2\varepsilon_0 + \varepsilon_\omega)} \tag{4.19}$$

where ε_0 is the static dielectric constant and ε_ω is the value at frequency ω. The Lorentz and Onsager field factors are used extensively in the interpretation of a variety of nonlinear optical experiments.

4.3 POLED POLYMERS

A macroscopic second-order optical nonlinearity can be induced in certain polymeric materials containing dipolar (intramolecular donor–acceptor) chromophores (Meredith et al. 1982). The chromophore can be a molecular species dissolved in a polymeric host where the role of the host is to provide a rigid matrix and other suitable properties for guest alignment. Alternatively, the chromophore could be incorporated into the polymeric structure in which case a range of physical properties would have to be incorporated into the polymer. The chromophore might also be incorporated into the polymer main chain, leading to further potential advantages and complexities. To achieve non-centrosymmetric ordering of the chromophores, electric field poling of the bulk medium can be used. The poling process is conceptually simple but in practice can be quite complex. It involves raising the temperature of a film of the polymer to near its glass transition temperature and applying a dc electric field to the film. The permanent dipoles experience a force tending to align them in the direction of the field. The net orientation is determined by the potential energy of the molecular dipoles in the applied field relative to the energy associated with thermal fluctuations of the system, as well as the dielectric and electric properties of medium and interfaces. The system is then cooled under the influence of the field and the alignment retained in the rigid state. Here we are mainly concerned with the nature of the macroscopic $\chi^{(2)}$ tensor and its relationship to the microscopic coefficient as well as the structural features required for good poling.

The poling process induces a polar axis in the polymer film. The axis is essentially an infinite-fold rotational axis with an infinite number of mirror planes. Its symmetry is designated as ∞mm or $C_{\infty v}$. The essential symmetry relationships are illustrated in Figure 4.2.

The symmetry is similar to but higher than that described for the monoclinic system depicted in Figure 4.1. In this case the molecules are distributed cylindrically about the Z axis and the angle α varies from molecule to molecule. In the weak poling limit the distribution of α will be extremely broad, but with each molecule exhibiting a slight tendency to tip toward the Z axis relative to the unpoled state.

Before pursuing the relationship between the macroscopic and microscopic coefficients the statistical thermodynamics of electric field-induced alignment are discussed. The problem begins with determining the probability of finding a molecule at an angle α with respect to the polar axis induced by the poling field (Williams 1987). The probability of finding a molecule in a small angular segment $d\alpha$ is given by

$$F(\alpha) \sin \alpha d\alpha \tag{4.20}$$

where $F(\alpha)$ is the orientational distribution function defined as

$$F(\alpha) = e^{-U(\alpha)/kT} \tag{4.21}$$

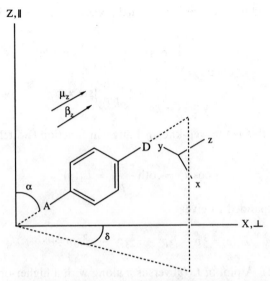

Figure 4.2 The relationship between the molecular dipole moment μ_z, the vector component of the hyperpolarizability tensor β_z, and the molecular coordinates and the macroscopic frame of reference characterized by Z (the direction parallel to the external field) and X (the direction perpendicular to it).

and $U(\alpha)$ is the potential energy of the molecule in the medium in the presence of the poling field. There are contributions to $U(\alpha)$ from potentials associated with local structure or anisotropy in the medium, $U_1(\alpha)$, and from the interaction of the dipole μ with the applied field E. We therefore have for $U(\alpha)$

$$U(\alpha) = U_1(\alpha) - \mu \cdot E \tag{4.22}$$

$$E = E_{ext} f(0) \tag{4.23}$$

and $f(0)$ and E_{ext} are the local field factor and applied external field, respectively.

Rather than attempting to specify $F(\alpha)$ to describe the state of alignment of the system we will characterize it with a related quantity, the average orientation of a dipole in the system. According to the Boltzman distribution law this is given by

$$\langle \cos \alpha \rangle = \frac{\displaystyle\int_0^\pi F(\alpha) \cos \alpha \sin \alpha \, d\alpha}{\displaystyle\int_0^\pi F(\alpha) \sin \alpha \, d\alpha} \tag{4.24}$$

In an isotropic medium the extent to which dipoles can be directed is given by

$$\langle \cos \alpha \rangle = \frac{\int_0^\pi \cos \alpha \exp\left(\frac{\mu \cdot E}{kT}\right) \sin \alpha \, d\alpha}{\int_0^\pi \exp\left(\frac{\mu \cdot E}{kT}\right) \sin \alpha \, d\alpha} \tag{4.25}$$

The solution to this is the well-known Langevin function (Böttcher 1973)

$$\langle \cos \alpha \rangle = \coth \frac{p-1}{p} = L_1(p) \tag{4.26}$$

which can be expanded to give

$$L_1(p) = \tfrac{1}{3}p - \tfrac{1}{45}p^3 + \tfrac{2}{945}p^5 - \tfrac{2}{9450}p^7 + \cdots \tag{4.27}$$

where $p = \mu \cdot E/kT$. A plot of $L_1(p)$ versus p along with a higher-order Langevin function (discussed below) is given in Figure 4.3. The main point to note is that for low p (low field or dipole moment) $L_1(p)$ is essentially linear in p but becomes sublinear and saturates for high values.

At this point we consider the nonlinear coefficient appropriate for ∞mm symmetry. At the outset it is clear that we will have to deal with averages since the unpoled polymer is isotropic and only partial orientation is induced by poling. The tensor elements can therefore be written as

$$\chi^{(2)}_{IJK} = N f_I(\omega_1) f_J(\omega_2) f_K(\omega_3) \langle b_{IJK} \rangle \tag{4.28}$$

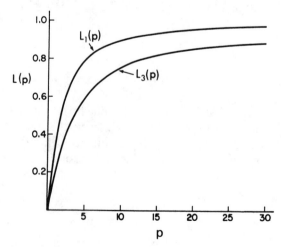

Figure 4.3 A plot of $L_1(P)$ and $L_3(p)$ which is proportional to $\chi^{(2)}_{zzz}$ versus p, where $p = \mu E/kt$.

where the quantity in brackets denotes an average, the nature of which is described below. In Figure 4.2 and the associated discussion it was pointed out that this system exhibits cylindrical (polar) symmetry. Intuitively, we would expect two independent nonlinear coefficients; one perpendicular and the other parallel to the polar axis. The tensor describing this situation is similar to that for classes 6mm and 4mm. The elements of the tensor are

$$
\begin{pmatrix}
0 & 0 & 0 & 0 & \chi_{15}^{(2)} & 0 \\
0 & 0 & 0 & \chi_{24}^{(2)} & 0 & 0 \\
\chi_{31}^{(2)} & \chi_{32}^{(2)} & \chi_{33}^{(2)} & 0 & 0 & 0
\end{pmatrix}
$$

where $\chi_{15}^{(2)} = \chi_{XXZ}^{(2)} = \chi_{24}^{(2)} = \chi_{YYZ}^{(2)}$ and $\chi_{31}^{(2)} = \chi_{ZXX}^{(2)} = \chi_{32}^{(2)} = \chi_{ZYY}^{(2)}$ due to the symmetry of the medium. Taking Kleinman symmetry into account we are left with two independent coefficients, $\chi_{p(ZXX)}^{(2)}$ (p indicating all the permutations of the indices) and $\chi_{32}^{(2)} = \chi_{ZZZ}^{(2)}$. In reality these quantities are not independent, but are related through the statistical relationships imposed by the Boltzman distribution law. This is discussed in more detail below.

In pursuing the relationship between the macroscopic and microscopic coefficients, the following analysis emerges. For the cylindrical symmetry associated with poled polymers and consistent with the discussion of the $\chi^{(2)}$ tensor components in the previous paragraph there are two b tensor elements. The first is

$$\langle b_{zzz} \rangle = \langle \cos^3 \alpha \rangle \beta_z \tag{4.29}$$

where $\beta_z = \beta_{zxx} + \beta_{zyy} + \beta_{zzz} \sim \beta_{zzz}$. Note that β_{zzz} is independent of the azimuthal angle λ. The average indicated by the brackets is given by the Boltzman distribution law as

$$\langle \cos^3 \alpha \rangle = \frac{\displaystyle\int_0^\pi F(\alpha) \cos^3 \alpha \sin \alpha \, d\alpha}{\displaystyle\int_0^\pi F(\alpha) \sin \alpha \, d\alpha} \tag{4.30}$$

The solution to this expression is a higher-order Langevine function

$$L_3(p) = \left(1 + \frac{6}{p^2}\right) L_1(p) - \frac{2}{p} \cdots \tag{4.31}$$

This function is plotted in Figure 4.3 and rises less steeply than $L_1(p)$, as might be anticipated, and also eventually saturates. For second-harmonic generation the expression for the macroscopic coefficient associated with β_{zzz} can now be written as

$$\chi^{(2)}_{ZZZ} = Nf^2_Z(\omega)f_Z(2\omega)\beta_z L_3(p) \tag{4.32}$$

For the second β coefficient, β_{zxx}, the situation is somewhat different. The X and Y axes are orthogonal to each other and to the Z axis but are not strictly defined with respect to a principal direction because of the cylindrical symmetry of the medium. It is therefore convenient to define the X axis as the one that is coincident with the component of incident light in the plane perpendicular to the Z axis. In this case the polarization induced in the Z direction by electric field components in the X direction will depend on the angle α with respect to the molecular z axis and the angle δ between the projection of the z axis onto the xy plane and the x axis. For $b_{p(ZXX)}$ we therefore have

$$\langle b_{p(ZXX)} \rangle = \langle (\cos \alpha)(\sin^2 \alpha)(\cos^2 \delta) \rangle \beta_z \tag{4.33}$$

Before writing the explicit expression for the Boltzmann thermal average, the relationship between alignment of the molecular z axis along the macroscopic polar Z axis and net alignment in other directions should be considered. In the case of a poled isotropic solid, the plane perpendicular to the poling direction should remain isotropic for all practical purposes and the term $(\cos^2 \delta)$ should not appear in the Boltzman thermal average. We therefore have

$$\langle (\cos \alpha)(\sin^2 \alpha)(\cos^2 \delta) \rangle = \frac{\displaystyle\int_0^\pi F(\alpha)(\cos \alpha)(\sin^3 \alpha)\, d\alpha}{\displaystyle\int_0^\pi F(\alpha)\sin^3 \alpha\, d\alpha} \int_0^{2\pi} \cos^2 \delta\, d\delta \tag{4.34}$$

$$= \frac{\langle \cos \alpha - \cos^3 \alpha \rangle}{2}$$

$$= \tfrac{1}{2}(L_1(p) - L_3(p))$$

The expression for $\chi^{(2)}_{p(ZXX)}$ is therefore given by

$$\chi^{(2)}_{p(ZXX)} = \frac{Nf^2_X(\omega)f_Z(2\omega)\beta_z[L_1(p) - L_3(p)]}{2} \tag{4.35}$$

In the limit of low poling fields and low dipole moments equations (4.32) and 4.35 take the relatively simple form

$$\chi^{(2)}_{ZZZ} = \frac{Nf^2_Z(\omega)f_Z(2\omega)\mu E\beta_z}{5kT} \tag{4.36}$$

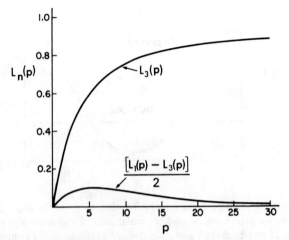

Figure 4.4 A plot of $L_3(p)$ which is proportional to $\chi^{(2)}_{zzz}$ and $[L_1(p) - L_3(p)]/2$ which is proportional to $\chi^{(2)}_{zxx}$ versus p, where $p = \mu E/kT$.

and

$$\chi^2_{zxx} = \frac{N f_z^2(\omega) f_z(2\omega) \mu E \beta_z}{15kT} \tag{4.37}$$

At high field the expression for $\chi^{(2)}_{zzz}$ approaches a saturation value corresponding to $L_3(p) = 1$ and $\chi^{(2)}_{zxx}$ approaches zero. This is illustrated in Figure 4.4.

Thus far we have assumed the medium to be isotropic in character prior to poling. Anisotropy in the medium can have a significant impact on the equilibrium degree of poling in a system and could come form a variety of sources. To illustrate this more fully we consider the case of a long rod-shaped chromophore dissolved in or attached as a side-chain pendent unit of a liquid-crystalline polymer. We assume that the chromophore is pleochroic, in that it has a tendency to align with the liquid-crystalline director S. The potential energy of the chromophore of interest will be lowered when it is parallel to the director and somewhat higher in the perpendicular direction. This is illustrated schematically in Figure 4.5 where the local potential energy is plotted against the angle ξ with respect to the director.

Bearing in mind that the liquid-crystalline director is a nonpolar axis resulting from gross alignment due to highly directional (nondipolar) van der Waals interactions at the molecular level, the interaction with an external electric field produces an axially aligned state. Permanent dipoles in such a medium interacting with an external field will feel the potential associated with the director as well as the field components arising from the external source. Referring back to equation 4.22, the potential of the molecular dipole of the nonlinear chromo-

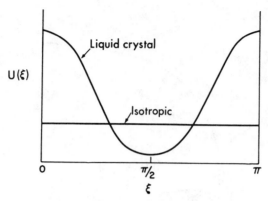

Figure 4.5 A possible local potential energy function $U(\xi)$ vs. the angle ξ which is the angle between the nematic director and the polar axis of the nonlinear chromophore. For the purposes of this diagram 0 and π are defined as the angles of minimum potential energy. Reprinted with permission from G. R. Meredith, J. C. VanDusen, and D. J. Williams, *Nonlinear Optical Properties of Organic and Polymeric Materials*, D. J. Williams, Ed., ACS Symposium Series 223, American Chemical Society, Washington, DC, 1983, p. 115.

phore in a liquid-crystalline environment is (DeGennes 1974)

$$U(\alpha) = U_s(\alpha) - \mu \cdot E \tag{4.38}$$

Since ξ and the induced polar axis are codirectional we can substitute α for ξ in Figure 4.5 for the sake of simplicity. The contributions of $U_S(\alpha)$ to the orientational distribution function obviously impact the degree of alignment one can achieve in such a system and, fortunately, in a positive way. Substitution of equation 4.38 into 4.9 and the result into the expressions for $\langle \cos \alpha \rangle$ and so forth produces an expression for which a knowledge of $U_S(\alpha)$ is required for its solution. A simplifying (and limiting) case is the Ising model where the value of $U_S(\alpha)$ is zero at $0, \pi, 2\pi$, and so on, and infinite at all other angles. Under these circumstances the coefficients $\chi^{(2)}_{ZZZ}$ and $\chi^{(2)}_{ZXX}$ take on the following values at low field:

$$\chi^{(2)}_{ZZZ}(\text{Ising}) = 5\chi^{(2)}_{ZZZ}(\text{Isotropic}) \tag{4.39}$$

$$\chi^{(2)}_{ZXX}(\text{Ising}) = 0 \tag{4.40}$$

At high fields the alignment saturates in either case and

$$\chi^{(2)}_{ZZZ}(\text{Ising}) \rightarrow \chi^{(2)}_{ZZZ}(\text{Isotropic}) \tag{4.41}$$

In practice one would expect a less severe constraint on alignment than imposed by the Ising model. Nevertheless, substantial improvements in alignment in the Z direction might be anticipated in liquid-crystalline environments.

4.4 LANGMUIR–BLODGETT FILMS

Langmuir–Blodgett (LB) films consist of assemblies of molecules with polar head groups and long aliphatic tails deposited onto a substrate from the surface of water. A key feature of these films is that forces at the water surface and lateral surface pressure are used to condense a randomized set of such molecules from a gas-like phase to one that is highly organized and stabilized by van der Waals forces between molecules. These forces are sufficiently cohesive to allow the films to be transferred to a substrate as a coherent film. In this section we briefly review the preparation and properties of these layers and point out the key relationships between nonlinear optical properties of the constituent molecules and films.

The deposition behavior for a typical amphiphilic molecule used in LB film formation such as steric acid $C_{17}H_{35}COOH$ is shown in Figure 4.6. In the figure, the part of the molecule represented by • is the polar head group and the tail represents the aliphatic section. The first withdrawal of a polar substrate such as glass results in typical deposition behavior illustrated in the figure. Reimmersion results in a favorable interaction between aliphatic tails so that a tail-to-tail film is formed. The next withdrawal step results in head-to-head deposition. This process can be repeated numerous times and films with thicknesses in the range of $\sim 1\ \mu m$ are obtained. This type of deposition is commonly referred to as Y type. It should be apparent that deposition of this type leads to a film with overall centrosymmetric structure. Occasionally, a chemical structure will be encountered where deposition occurs only on immersion or withdrawal. These deposition types are referred to as X and Z, respectively, for historical reasons. They both lead to noncentrosymmetric structures but are often quite unstable, and rearrangement can occur, leading to centrosymmetric structures. A recent review of LB films and guide to the literature is given by Roberts (1988).

A number of chemical approaches have been devised for obtaining stable noncentrosymmetric LB film structures and these are reviewed in Chapter 7. For now we cite one of the first approaches for constructing multilayers for second-order nonlinear optics. Films consisting of alternating layers of a merocyanine dye and ω-tricosenoic acid were constructed by Girling et al. (1987). The arrangement of molecules is illustrated schematically in Figure 4.7. The general symmetry of other properties of this system serve to illustrate the relationship between microscopic and macroscopic nonlinearities in such films. From inspection it is apparent that the film illustrated in Figure 4.7 has a polar axis perpendicular to the plane of the film. In general, the molecular axes exhibit a tilt with respect to the Z axis defined in the figure. The distribution of tilt angles α may be sharply peaked about some average value α_0 or could exhibit

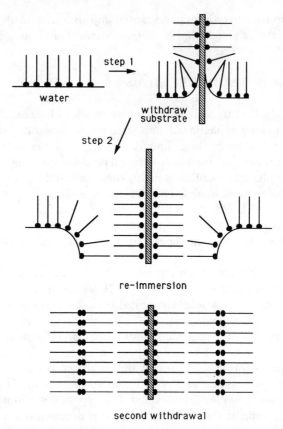

Figure 4.6 The Langmuir–Blodgett Y-type deposition process. In the first step a hydrophilic substrate is withdrawn from the surface of water and the hydrophilic heads of the compressed surface film adhere to the substrate. In the second step the substrate and film are reimmersed and tail-to-tail deposition occurs. The process is repeated to build thick films.

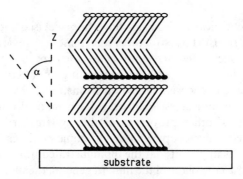

Figure 4.7 A Langmuir–Blodgett Y-type film made from two different amphophiles with an average tilt angel α relative to the surface normal.

a broad distribution, depending on the chemical nature of the system. A tilt angle α implies an azimuthal distribution ϕ as well. Experiments on LB films have generally supported a uniform distribution ϕ, at least on the scale of the physical dimensions of the beams employed in the experiment. Because of this, and for the sake of simplicity in interpreting experiments, we assume cylindrical symmetry for these systems. The analysis for LB films therefore follows closely that for poled polymer films.

In the case of an LB monolayer the nonlinear susceptibility for second harmonic generation is given by

$$\chi^{(2)}_{S,IJK} = N_S f_I(\omega) f_J(\omega) f_K(2\omega) b_{IJK} \tag{4.42}$$

where the subscript S refers to a surface layer. The surface concentration is given by N_S and b_{IJK} is the hyperpolarizability in the macroscopic frame as defined previously. If we continue to make the reasonable assumption that in rod-shape, highly polarizable chromophores β_{zzz} is the dominant component of the molecular hyperpolarizability tensor and that the distribution of angles α is reasonably sharp, then two values of b_{IJK} characterize the film (Heinz et al. 1983):

$$b_{ZZZ} = \langle \cos^3 \alpha \rangle \beta_{zzz} \tag{4.43}$$

and

$$b_{p(ZXX)} = \frac{\langle \cos \alpha \sin^2 \alpha \rangle \beta_{zzz}}{2} \tag{4.44}$$

where p indicates identical expressions for the permutations of the indices. The two independent components of $\chi^{(2)}_S$ are therefore given by

$$\chi^{(2)}_{S,ZZZ} = N_S f \langle \cos^3 \alpha \rangle \beta_{zzz} \tag{4.45}$$

and

$$\chi^{(2)}_{S,p(ZXX)} = \tfrac{1}{2} N_S f' \langle \cos \alpha \sin^2 \alpha \rangle \beta_{zzz} \tag{4.46}$$

where we have lumped the local field factors into f and f'. To determine the value of α we only need to measure the ratio of the two independent tensor components, assuming we know f and f' or are willing to make the reasonable assumption that they are close to unity. This can be seen by defining a ratio A as

$$A = \frac{2\chi^{(2)}_{S,p(ZXX)}}{\chi^{(2)}_{S,ZZZ} + 2\chi^{(2)}_{S,p(ZXX)}} \tag{4.47}$$

which is related to the molecular orientation according to equations 4.45 and

4.46 by

$$A = \frac{\langle \sin^2 \alpha \cos \alpha \rangle}{\langle \cos \alpha \rangle} \tag{4.48}$$

Under the assumption that the distribution of tilt angles is sharply peaked, then

$$\alpha \simeq \sin^{-1}(\sqrt{A}) \tag{4.49}$$

Therefore, second-harmonic generation is a powerful method for gaining insight into the structure and order in LB films. When multilayered films are fabricated we would expect relatively straightforward scaling of the harmonic signal intensity in an SHG experiment to apply. Consider the expression for second-harmonic intensity refleted from a surface (Bloembergen and Pershan 1962):

$$I(2\omega) = \frac{32\pi^2\omega^2}{c^2\varepsilon_\omega\varepsilon_{2\omega}^{1/2}\cos^2\theta}|e(2\omega)\cdot\chi_S^{(2)}:e(\omega)e(\omega)|^2 I^2(\omega) \tag{4.50}$$

where the quantities $e(2\omega)$ and $e(\omega)$ are the products of polarization vectors and Fresnel factors, θ is the angle of reflection of second-harmonic output, and ε are dielectric constants at frequencies ω and 2ω. When multiplied out, the quantity in brackets is proportional to an effective nonlinear coefficient $\chi^{(2)}$ for any combination of input and output polarizations (see Chapter 5 for a more complete explanation). It should be intuitively apparent that for incident light polarized parallel to the plane of the film only the coefficient $\chi_{p(ZXX)}^{(2)}$ will contribute to the term in brackets. For nonnormal incident light, polarized parallel to the plane of incidence, both $\chi_{p(ZXX)}^{(2)}$ and $\chi_{p(ZZZ)}^{(2)}$ will contribute to the effective coefficient and therefore the output signals. The use of equation 4.50 for determining structural information under various experimental conditions can be found in references by Heinz et al. (1983), Rasing et al. (1986), and Marowsky et al. (1987). Here we see that the nonlinear coefficients $\chi_{S,ZZZ}^{(2)}$ and $\chi_{S,p(ZXX)}^{(2)}$ should be proportional to the number of monolayers n (or bilayers if a spacer layer is used as in Y type deposition) so that

$$\chi_{ZZZ}^{(2)}(\text{film}) = n\chi_{S,ZZZ}^{(2)} \tag{4.51}$$

and we expect $I(2\omega)$ to vary as n^2 as long as the film thickness is substantially less than the coherence length. Since the coherence length is typically $10\text{--}20\,\mu m$ this condition is readily met in LB films. Deviations from a quadratic dependence on n can be taken as an indication of changing structural factors, either geometric or electronic, in multilayer films.

Information can also be obtained on the impact of various physical and chemical variations within a monolayer from SHG measurements. For instance, dilution of the nonlinear chromophore with an inert component within a mono-layer should exhibit a quadratic dependence on surface concentration, assuming geometrical structural factors remain the same. The ability to probe the orienta-tion and concentration dependence of the harmonic intensity makes SHG a useful structural probe.

The spectral dependence of the harmonic intensity also provides interesting and useful information. Resonance enhancement in the harmonic intensity when the fundamental beam is at one-half the frequency of an optical absorption band makes second-harmonic generation a very sensitive spectroscopic probe of the composition of a monolayer and the processes occurring at surfaces and inter-faces in general. Shen's group, for instance, has used this technique or minor variations of it to follow the polymerization of monolayers of vinyl stearate and octadecylmethacrylate (Berkovic and Shen 1988).

4.5   THIRD-ORDER SUSCEPTIBILITIES IN CRYSTALS, POLYMERS, AND LANGMUIR–BLODGETT FILMS

An analysis of the relation between the microscopic and macroscopic third-order optical nonlinearities is conceptually similar to that developed for the second-order processes discussed above. The difference is that the second hyperpolariz-ability (the microscopic nonlinear coefficient) γ_{ijkl} and the macroscopic third-order susceptibility $\chi^{(3)}_{IJKL}$ are fourth-rank tensors. They are related to each other as

$$\chi^{(3)}_{IJKL} = N f_I(\omega) f_J(\omega_1) f_K(\omega_2) f_L(\omega_3) c_{IJKL} \tag{4.52}$$

In the above equation

$$c_{IJKL} = \frac{1}{N_g} \sum_{ijkl} \left(\sum_{s=1}^{N_g} \cos\theta^{(s)}_{Ii} \cos\theta^{(s)}_{Jj} \cos\theta^{(s)}_{Kk} \cos\theta^{(s)}_{Ll} \right) \gamma_{ijkl} \tag{4.53}$$

describes transformation of the γ tensor from the molecular coordinate system $(i, i, k, l = 1, 2, 3)$ of sites to the external, for example, crystallographic, coordinate system $(I, J, K, L = 1, 2, 3)$. The meaning of the symbols here is the same as equation 4.3.

Let us consider a macroscopic system that has a center of symmetry. There would be a pair of sites (molecules labeled $s = 1$ and 2) related to each other by an inversion operation; the corresponding direction cosines are $\cos\theta^{(2)}_{Ii} = -\cos\theta^{(1)}_{Ii}$. An analogous expression holds for direction cosines with the sub-script indices Jj, Kk, and Ll. Substituting them into equation 4.53 gives

$$c_{IJKL} = \frac{1}{2} \sum_{ijkl} (\cos\theta^{(1)}_{Ii} \cos\theta^{(1)}_{Jj} \cos\theta^{(1)}_{Kk} \cos\theta^{(1)}_{Ll}$$

$$+ (-\cos\theta^{(1)}_{Ii})(-\cos\theta^{(1)}_{Jj})(-\cos\theta^{(1)}_{Kk})(-\cos\theta^{(1)}_{Ll})\gamma_{ijkl}$$

or

$$c_{IJKL} = \sum_{ijkl} \cos\theta_{Ii}^{(1)} \cos\theta_{Jj}^{(1)} \cos\theta_{Kk}^{(1)} \cos\theta_{Ll}^{(1)} \gamma_{ijkl} \qquad (4.54)$$

Contrary to the second-order processes where the corresponding tensor $b_{IJK} = 0$ for a centrosymmetric system, the tensor c_{IJKL} always has nonzero components.

In systems possessing other symmetry elements, the tensor components c_{IJKL} must reflect all of the symmetry operations involved. For crystals of all other point groups, appropriate expressions for fourth-rank tensor components are given in a book by Nye (1985). The number of independent coefficients of the c_{IJKL} tensor is 21 for the lowest symmetry in a triclinic system, and it is reduced to three in a cubic system. An isotropic medium, such as a liquid or an amorphous polymer, has only two independent components of the c (and hence $\chi^{(3)}$) tensor. Due to the complex nature of distortions involved during processing, a final product made of a polymeric material is frequently not isotropic. Furthermore, a polymer may be stretched to achieve a macroscopically ordered structure. This treatment leads to microscopic alignment of molecular chains which remain preserved in the material.

Using the methods and terminology of linear optics, one can examine and classify polymeric materials in one of the following groups: isotropic, uniaxial, or biaxial. Equations 4.52 and 4.53 are still useful, but the number of sites to be summed increases immensely due to their random distribution. Unfortunately, all the necessary details (direction cosines) for each site are not available to evaluate expressions 4.52 and 4.53. The use of macroscopically observed symmetry reduces the extent of the calculation, but generally the exact relationship between an average macroscopic tensor component $\langle c_{IJKL} \rangle$ and γ_{ijkl} is less evident.

The average $\langle c_{IJKL} \rangle$ can be related to the polymer chain orientational distribution function and the γ tensor defined within the polymer chain coordinate system. The subject of orientation distributions in polymer systems has been pursued for more than half a century (Hermans et al. 1946, 1948). Common experimental techniques like x-ray diffraction, birefringence, infrared dichroism, and Raman scattering, among others, have been used to obtain information on orientational distribution functions (Alexander 1969, Stein 1976, Bower 1976). The orientational distribution function is the probability density that the molecule is oriented in a direction specified by the angles with respect to the macroscopic (laboratory) frame. It can be simplified by imposing the symmetry of the macroscopic system. For example, in a uniaxially stretched polymer, the unique axis is visually determined by the machine draw direction and the polymer chains are predominantly aligned along this direction.

A frequently adopted assumption is that cylindrical symmetry exists along the long axis of the molecule. In such a case, the distribution function $F(\theta)$ will be a function of only the polar angle θ, measured between a given direction and the unique optical axis. $F(\theta)$ can be expressed in terms of Legendre polynomials:

$$F(\theta) = \sum_{\ell=0} \frac{(2\ell+1)}{2} \langle P_\ell \rangle P_\ell(\cos\theta) \qquad (4.55)$$

where $\langle P_\ell \rangle$ is the order parameter and $P_\ell(\cos\theta)$ is the ℓth Legendre polynomial, which is a known function of the polar angle θ for any ℓ. In systems with center of symmetry, as in the case with many polymers that have high $\chi^{(3)}$, the lowest nontrivial order parameters are $\langle P_2 \rangle$ and $\langle P_4 \rangle$. By measuring properties described by a second-rank tensor only $\langle P_2 \rangle$ can be estimated, whereas $\langle P_4 \rangle$ needs measurements that involve a fourth-rank tensor. In general, the relationship between the measured $\chi^{(3)}_{IJKL}$ and the hyperpolarizability γ_{ijkl} in terms of the order parameters is complicated. Also, further complication arises due to the local field effects, which would be anisotropic and also dependent on the order parameters. If one neglects the anisotropy of linear refractive index and the local field correction factor, and assumes that the polymer chains behave as one-dimensional systems for microscopic optical nonlinearity (γ), the third-order susceptibility component parallel to the director, $\chi^{(3)}_\parallel$, and the transverse component, $\chi^{(3)}_\perp$, are given by (Wong and Garito 1986)

$$\chi^{(3)}_\parallel = N f^4 (\tfrac{1}{5} + \tfrac{4}{7}\langle P_2 \rangle + \tfrac{8}{35}\langle P_4 \rangle)\gamma \qquad (4.56)$$

$$\chi^{(3)}_\perp = N f^4 (\tfrac{1}{5} - \tfrac{2}{7}\langle P_2 \rangle + \tfrac{3}{35}\langle P_4 \rangle)\gamma \qquad (4.57)$$

In the above equation, γ is the only active (nonzero) component of the polymer chain (γ_{zzzz} along the chain direction z) and f represents a collection of local field factors. For a disordered, isotropic material $\langle P_2 \rangle = \langle P_4 \rangle = 0$ and $\chi^{(3)}_\parallel = \chi^{(3)}_\ell = \tfrac{1}{5}\gamma N f^4$. In contrast, for a perfect alignment of polymer chains in the same direction (draw direction for example), $\langle P_2 \rangle = \langle P_4 \rangle = 1$, $\chi^{(3)}_\parallel = N\gamma f^4$, and $\chi^{(3)}_\perp = 0$.

In Langmuir–Blodgett films formed from polymeric materials, polymer chains may tend to preferentially align in the plane of the film. However, the orientation of the polymer chains in the plane of the film is, random. In this situation, the effective averaging is only in two dimensions. Consequently, one will have

$$\chi^{(3)} = \tfrac{3}{8} f^4 N \gamma \qquad (4.58)$$

Again the assumption is that the local field correction factor f is not anisotropic. The reduction of $\chi^{(3)}$ in this case is by a factor of $\tfrac{3}{8}$ compared to a perfectly unialigned system in which all chains align in the same direction.

4.6   RELATION OF $\chi^{(3)}$ BETWEEN FILM-BASED AND LABORATORY-BASED COORDINATE SYSTEMS

The third-order nonlinear optical response of a material is described by the projection of the electric field vector of the waves onto the $\chi^{(3)}$ tensor of the medium. The electric field vectors are defined in the laboratory framework, while the $\chi^{(3)}$ tensor is described by the coordinate systems (or symmetry axes) based in the material (film if one uses a polymeric film). To get information on the anisotropy of the film, one convenient way may be to rotate the film and measure the nonlinear response. This study would yield a polar plot of $\chi^{(3)}$ as a function of the rotation angle, the rotation angle being defined as an angle θ with respect to a laboratory principal axis, which may be chosen as the direction of polarization of the laser beam. If all beams are polarized in the same direction (for example, vertical), designated as 1, the nonlinear response will be described by the component $\chi_{1111}^{(3)}$ in the laboratory framework. To get the dependence of $\chi_{1111}^{(3)}$ on the rotation angle, one simply has to find how the laboratory coordinate-based $\chi_{1111}^{(3)}$ relates to the film-based fourth-rank tensor $\chi_{IJKL}^{(3),F}$. The transformation between the film-based fourth-rank tensor $\chi_{IJKL}^{(3)}$ and the laboratory-based fourth-rank tensor $\chi_{I'J'K'L'}^{(3)}$ is described as

$$\chi_{I'J'K'L'}^{(3)} = \sum_{IJKL} a_{I'I} a_{J'J} a_{K'K} a_{L'L} \chi_{IJKL}^{(3),F} \tag{4.59}$$

This expression for the case of an orthogonal symmetry of the assembly discussed here yields

$$\chi_{1111}^{(3)} = \cos^4\theta \chi_{1111}^{(3),F} + \sin^4\theta \chi_{2222}^{(3),F} + \sin^2\theta\cos^2\theta(2\chi_{1122}^{(3),F} + 4\chi_{1212}^{(3),F}) \tag{4.60}$$

In the above equation, the components $\chi_{1111}^{(3),F}$ and $\chi_{2222}^{(3),F}$ are the $\chi^{(3)}$ values along the film-based principal axes 1 and 2. For example, in the case of a uniaxially stretched polymer film, 1 and 2 may be the draw direction and the transverse direction. The terms $\chi_{1122}^{(3),F}$ and $\chi_{1212}^{(3),F}$ are the off-diagonal components. The expression (4.60) has successfully been used to describe the experimentally observed polar plots for $\chi^{(3)}$ in biaxially and uniaxially stretched polymers (Rao et al. 1986, Prasad 1988, Singh et al. 1988).

REFERENCES

Alexander, L. C. *X-Ray Diffraction Methods in Polymer Science*, Wiley, New York, 1969.

Berkovic, S., and Y. R. Shen, in P. N. Prasad and D. R. Ulrich (Eds.), *Nonlinear Optical and Electroactive Polymers*, Plenum, New York, 1988, pp. 157–168.

Bloembergen, N., and P. S. Pershan, *Phys. Rev.* **128**, 606 (1962).

Böttcher, C. J. F. *Theory of Electric Polarization*, Vol. 1, Elsevier, Amsterdam, 1973, p. 161.

Bower, D. J., *J. Phys. B* **9**, 3275 (1976).

DeGennes, P. G., *The Physics of Liquid Crystals*, Oxford University Press (Clarendon), London, 1974, p. 23.

Girling, I. R., N. A. Cade, R. J. J. Kolinsky, I. R. Peterson, M. M. Ahmad, D. B. Neal, M. C. Petty, G. G. Roberts, and W. J. Feast, *J. Opt. Soc. Am. B* **4**, 950 (1987).

Heinz, T. F., H. W. K. Tom, and Y. R. Shen, *Phys. Rev. A* **28**, 1883 (1983).

Hermans, J. J., P. H. Hermans, D. Vermass, and A. Werdinger, *Rec. Trav. Chem.* **65**, 427 (1946).

Hermans, P. H., J. J. Hermans, D. Vermass, and A. Weidinger, *J. Polym. Sci.* **3**, 1 (1948).

Marowsky, G., A. Gieralski, R. Steinhoff, D. Dorsch, R. Eidenschnik, and B. Rieger, *J. Opt. Soc. Am. B* **4**, 956 (1987).

Meredith, G. R., J. G. Van Dusen, and D. J. Williams, *Macromolecules* **15**, 1385 (1982).

Meredith, G. R., J. G. Van Dusen, and D. J. Williams, in D. J. Williams (Ed.), *Nonlinear Optical properties of Organic and Polymeric Materials*, ACS Symposium Series 233, ACS Publications, Washington, DC, 1984, pp. 109–133.

Nye, J. F., *Physical Properties of Crystals*, Oxford University Press, London, 1985.

Onsager, L., *J. Am. Chem. Soc.* **58**, 1456 (1936).

Oudar, J. L., and J. Zyss, *Phys. Rev. A* **26**, 2016 (1982).

Prasad, P. N., in P. N. Prasad and D. R. Ulrich (Eds.), *Nonlinear Optical and Electroactive Polymers*, Plenum, New York, 1988.

Rao, D. N., J. Swiatkiewicz, P. Chopra, S. K. Ghoshal, and P. N. Prasad, *Appl. Phys. Lett.* **48**, 1187 (1986).

Rasing, Th., G. Berkovic, Y. R. Shen, S. G. Grubb, and M. W. Kim, *Chem. Phys. Lett.* **130**, 2 (1986).

Roberts, G. G., in M. J. Bowden and S. R. Turner (Eds.), *Electronic and Photonic Applications of Polymers*, Advances in Chemistry Series, Vol. 218, ACS Publications, Washington, DC, 1988, pp. 225–270.

Singer, K. D., S. J. Lalama, J. E. Sohn, and S. D. Small, in D. S. Chemla and J. Zyss (Eds.), *Nonlinear Optical Properties of Organic Molecules and Crystals*, Vol. 1, Academic, New York, 1987, pp. 437–468.

Singh, B. P., P. N. Prasad, and F. E. Karaz, *Polymer* **29**, 1940 (1988).

Stein, R. S., *J. Appl. Phys.* **9**, 3275 (1976).

Williams, D. J., in D. S. Chemla and J. Zyss (Eds.), *Nonlinear Optical Properties of Organic Molecules and Crystals*, Vol. 1, Academic, New York, 1987, pp. 405–435.

Wong, K. Y., and A. F. Garito, *Phys. Rev. A* **34**, 5051 (1986).

Zyss, J., and D. S. Chemla, in D. S. Chemla and J. Zyss (Eds.), *Nonlinear Optical properties of Organic Molecules and Crystals*, Vol. 1, Academic, New York, 1987, pp. 97–99.

Zyss, J., and J. L. Oudar, *Phys. Rev. A* **26**, 2028 (1982).

5

SECOND-ORDER NONLINEAR OPTICAL PROCESSES

5.1 GENERAL CONSIDERATIONS

The origin of second-order nonlinear processes was discussed in some detail in the context of electromagnetic theory in Chapter 2. The relationship of the coefficients in the power series expansion of the induced polarization in powers of the field to molecular and bulk properties of materials was discussed in Chapters 3 and 4, respectively. In this chapter we focus on the various second-order nonlinear processes that occur for specific combinations of frequency, intensity, and phase relationships between fields and their manifestation in nonlinear media.

In general, there are two ways of visualizing second-order nonlinear processes. In the first, the field associated with the first beam is viewed as altering the refractive index of the medium and the propagation characteristics of the second beam are then modified appropriately. This view is quite appropriate for the linear electrooptic effect where a dc or static electric field takes the place of the first beam and alters the refractive index of the medium in proportion of the strength of the dc electric field. A variety of interesting and useful applications can result from this effect. The first-field component could also be at some frequency ω_1 and modulate the refractive index at that frequency. A second field passing through the medium at ω_2 would then be phase modulated and exhibit sidebands at the sum and difference frequencies. If the frequencies ω_1 and ω_2 are identical, then a harmonic overtone at 2ω is created. These processes are often termed parametric processes because they result from modulation of the parameters of the medium.

The second way of viewing second-order nonlinearities is the one we have used in previous chapters. Here nonlinear optical effects are seen as resulting from nonlinearities in the polarization response to incident fields at various frequencies. A power series expansion was used in equations 4.1 and 4.2 to account for the various orders of nonlinearity in the response of the molecule or medium. Oscillations occur at various frequencies as a result of the nonlinear response and these can be identified by relatively straightforward trigonometric arguments, as was shown in Chapter 2, or by the more sophisticated techniques of Fourier analysis. The oscillations occurring at other frequencies can be viewed as new sources of electromagnetic radiation and, hence, phenomena like second-harmonic generation at the molecular level can be readily visualized. The propagation requirements leading to buildup in intensity in the harmonic wave and termination of the nonlinear interaction at an interface with a linear medium are less intuitively obvious, but can be visualized and appreciated by relatively straightforward considerations. In this chapter we illustrate these processes at the conceptual level, with the emphasis on identifying and under-standing the relevant physical phenomena that are occurring.

The second-order nonlinear processes can be categorized according to the frequency, intensity, and phases of the field components. As introduced in Chapter 2, the second-order nonlinear tensor element can be written as

$$\chi_{IJK}^{(2)}(-\omega_3; \omega_2, \omega_1)$$

where $\omega_3 = \omega_2 \pm \omega_1$ and I, J, and K indicate the Cartesian components of the interacting fields and polarization waves. In general, each Cartesian component is associated with a frequency component in the argument of $\chi^{(2)}$. Under the assumption of Kleinman symmetry (also discussed in Chapter 2), which is applicable when the nonlinear process is purely electronic in origin and the frequency of light is far away from the absorption band of the molecule, the Cartesian indices and frequency components can be freely interchanged. For example, $\chi_{ZXX}^{(2)}(-2\omega; \omega, \omega)$ indicates a process where the second-harmonic frequency is generated in the Z direction as a result of field components in the X direction. The Kleinman symmetry condition says that this tensor element is equivalent to $\chi_{XXZ}^{(2)}(-2\omega; \omega, \omega)$ where the X and Z directions have been permuted. This is equivalent to assuming that there is no dispersion in the nonlinearity. For organic materials far away from resonances this assumption has been shown to be a good one (Morell and Albrecht 1979). With this assumption we can focus our attention on the frequency components.

In the case where $\omega_3 = 2\omega$ and $\omega_1 = \omega_2 = \omega$ we have one of the best known and highly utilized effects in nonlinear optics, second-harmonic generation. In this process, energy is redistributed between the fields as a result of interaction of the waves with the medium and no energy is lost to the medium. To have efficient redistribution of energy by second-order nonlinear processes, momentum must be conserved. This condition is referred to as phase matching. The implication here is that energy can only be transferred from the fundamental to the harmonic field, as the waves propagate through the medium, as long as

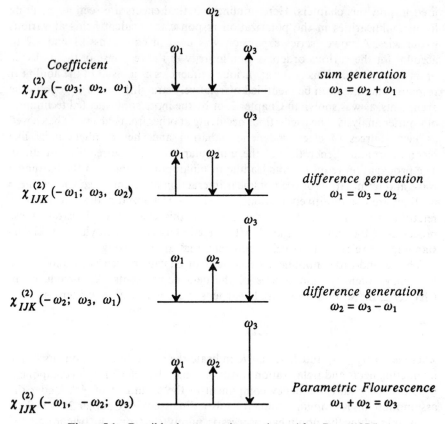

Figure 5.1   Possible three-wave interactions. After Byer (1974).

they maintain the required phase relationship. Second-harmonic generation is discussed in some detail in this chapter and serves to illustrate many of the key concepts of nonlinear optics.

When the frequencies ω_1 and ω_2 are not identical the processes of sum and difference frequency generation can occur. Rather than discussing each process in detail at this point, we diagram them schematically according to the scheme of Byer et al., (1974) (Fig. 5.1).

In this notation, as discussed in Chapter 2, the resultant frequency component is placed on the left of the semicolon in the argument. The significance of the longer arrow in each of these schemes is that even if one of the source fields is weak, as long as the other one (corresponding to the long arrow) is sufficiently strong, a response at the third frequency can occur. In the case of sum frequency generation, for instance, a weak signal in the infrared might be combined with a strong field at ω_2 in the visible to produce a new amplified signal at ω_3 in the visible which can be readily detected. This phenomenon is referred to as parametric amplification and has been demonstrated by Hulin et al. (1986) to

be useful for detecting subpicosecond pulses of weak near IR emission using
N-(4-nitrophenyl)-L-prolinol (NPP) as the nonlinear amplifying medium.

Definite rules govern the flow of power in nonlinear interactions and these
will be described in more detail later. for now we simply note that in the process
of sum frequency generation under phase-matched conditions the two lower-
frequency waves at ω_1 and ω_2 lose power to the sum frequency ω_3. In difference
frequency generation the source laser at ω_3 loses power not only to the
difference frequency ω_1 (in the case of the second scheme of Figure 5.1) but also
to the source at ω_2. Another way of looking at this process is that the source
photon at ω_3 is effectively split into the two lower-frequency photons. This
phenomenon has extremely important consequences and can lead to useful and
important nonlinear effects. If, for instance, ω_2 is a very weak source and is
passed through the nonlinear medium repeatedly it can build up intensity during
each pass. The weak signal at ω_2 need not be supplied by an outside source, but
can be built up out of the noise if partially reflecting mirrors at each end of
the cavity to reflect ω_2 are provided. If the gain per pass is higher than the loss,
the system can oscillate at the new frequency and produce a new coherent light
source. This process is referred to as parametric oscillation and is described in
more detail later in the chapter.

As was pointed out above, it is intuitively less appealing to use the frequency-
mixing approach to describe the linear electrooptic effect or a related process,
dc rectification. In the linear electrooptic effect a dc field is applied which alters
the refractive index and, hence, the propagation characteristics rather than the
frequency of light. This is written as $\chi_{IJK}^{(2)}(-\omega; 0, \omega)$. In optical rectification the
photon fields at frequency ω interact to produce a dc polarization in the
medium. Consistent with our notation, the coefficient would be represented as
$\chi_{IJK}^{(2)}(0; \omega, -\omega)$. From Kleinman symmetry considerations these coefficients
should be identical.

5.2 SECOND-HARMONIC GENERATION

A relatively simple trigonometric argument for second-harmonic generation
was presented in Chapter 2, Section 2.4, showing the reason for the polarization
response at 2ω. The polarization described by equation 2.29 contains a term
derived from $\chi^{(2)}$, which oscillates at twice the frequency of the fundamental
wave and propagates with twice the value of the wave vector. It is these new
polarization terms that act as a source of electromagnetic radiation in the
medium at the new frequency without loss of energy to the medium.

At this point some qualitative comments can be made on the difference in
propagation of light through a linear medium and light associated with harmonic
polarization through a nonlinear medium (Zernike and Midwinter 1973). In the
case of a linear dielectric medium, an incident electromagnetic wave causes the
electrons associated with the nucleii of the atoms to oscillate at the same
frequency as the incident wave. For the sake of simplicity we will assume that

the medium consists of an array of classical oscillators consisting of one electron and one nucleus. If the frequency of incident light is well below the resonance frequency of the oscillating dipole, the induced or forced oscillation will be driven in phase with the incident disturbance. At resonance the oscillation will be at 90° with respect to the incident wave and increases to 180° above resonance. The latter regime is the one we are most interested in. Consider the radiation pattern from an individual dipole. It radiates in all directions with a maximum intensity in the direction perpendicular to the radiating dipole. More precisely stated, the radiation pattern has a $\sin \theta$ dependence relative to the axis of the dipole, as in illustrated in Figure 5.2. For fixed-frequency incident radiation the phase of radiation of an individual dipole will be determined solely by the phase of the radiation incident upon it. It can be shown through a relatively complex analysis that phases of radiation from individual dipoles in anything but the forward direction of the incident wave will be randomized and intensity does not accumulate, while in the forward direction they will constructively interfere and determine the characteristics of propagated light through the medium. In this sense light propagating through a linear medium is compared to a phased array of antennas.

In the case of a second-order nonlinear medium the situation is distinctly different. Here the phase of a radiating dipole is determined not by the phase of a single wave at a single frequency incident upon it, but by the relative phases of two waves at differing frequencies. The oscillating dipole is now producing radiation at more than one frequency with a propagation velocity appropriate

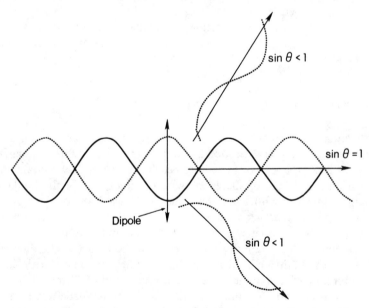

Figure 5.2 Radiation patterns in forward and other selected directions (dotted lines) from a dipole oscillating in a radiation field solid line.

for the refractive index at that frequency. Since appreciable dispersion exists in the refractive index in materials, particularly organic systems with delocalized π electrons, the opportunity for dephasing of the second-harmonic radiation in the forward as well as all other directions is appreciable. This situation arises because the harmonic polarization propagates with the velocity of the fundamental wave, while the second-harmonic radiation propagates at a velocity characteristic of that wavelength. The resulting situation is that the second-harmonic radiation is subject to destructive interference at the molecular level in randomly oriented media. In centrosymmetric media the radiation patterns generated by individual dipoles will be 180° out of phase with that from a counterpart in the forward direction related by inversion symmetry yielding a destructive interference. In noncentrosymmetric media dephasing occurs over much lager dimensions, $\sim 10\,\mu m$, resulting in periodic reversal of the direction of energy flow between fundamental and harmonic waves.

An understanding of the process of second-harmonic generation at a more fundamental level requires an analysis of Maxwell's equations and the constitutive relationships. These equations and relationships describe the spatial and temporal relationships between electric and magnetic fields, charges and currents, and the electric polarization and magnetism of a medium. These equations have been derived and explained in detail elsewhere (Bloembergen 1965, Flytzanis 1975) and are outside the scope of this book. Instead, we will highlight in a qualitative way the physical assumptions and insights this analysis produces.

An important starting point is the wave equation for propagation of an electric field and its resulting polarization in a nonmagnetic medium. This was defined by equation 2.42 and is restated here for the case of a plane wave propagating in the z direction:

$$\frac{\partial^2 E}{\partial z^2} = -\frac{\varepsilon}{c^2}\frac{\partial^2 E}{\partial t^2} - \frac{4\pi}{c^2}\frac{\partial^2 P}{\partial t^2} \tag{5.1}$$

In this equation P is assumed to be the first nonlinear polarization of the medium. The linear polarization is included in ε. In this problem we wish to examine the interaction of three waves, two at the fundamental frequency and one at the harmonic, in the nonlinear medium. At any point in the medium the increase or decrease in the amplitude of one of the components depends on the amplitude of the other waves. In the case of a linear medium, the amplitude of a wave is constant as it propagates through the medium and so E was used to define the amplitude of the field in equation 2.11 to within an arbitrary phase factor. Here we define a set of three interacting traveling waves:

$$E_1(z,t) = E_1(z)e^{-i(\omega_1 t - k_1 z)}$$
$$E_2(z,t) = E_2(z)e^{-i(\omega_2 t - k_2 z)} \tag{5.2}$$
$$E_3(z,t) = E_3(z)e^{-i(\omega_3 t - k_3 z)}$$

There are a similar set of equations for negative frequencies, associated with their complex conjugates. In (5.2) the subscripts refer to the frequencies of the waves. For second harmonic generation $\omega_1 = \omega_2 = \omega$, and $\omega_3 = 2\omega$. The fields $E_1(z)$ and so on are the complex fields defined by

$$E_1(z) = E_1^{(0)}(z)e^{i\phi_1(z)}$$

Here the arbitrary phase factor is included in the definition of the amplitude. The phase angle $\phi_i(z)$ ties the phase of the wave at any point in the medium to the phase it had when it entered the medium. The z dependence of the amplitude, resulting from the nonlinear interaction, is also implied. Similar definitions exist for $E_1^*(z)$, $E_2(z)$ and so on. A general set of expressions for the nonlinear polarization follow from the solution of Maxwell's equations and are simply stated here in terms of the field components defined above:

$$P_1(z, t) = 4dE_2^*(z)E_3(z)e^{-i[(\omega_3 - \omega_2)t - (k_3 - k_2)z]}$$

$$P_2(z, t) = 4dE_3(z)E_1^*(z)e^{-i[(\omega_3 - \omega_1)t - (k_3 - k_1)z]} \qquad (5.3)$$

$$P_3(z, t) = 4dE_1(z)E_2(z)e^{-i[(\omega_1 + \omega_2)t - (k_1 + k_2)z]}$$

Here the nonlinear coefficient is expressed as the d coefficient. Historically, experimentalists have used d as the coefficient for second-harmonic generalization and theoriticians have tended to use $\chi^{(2)}$. The coefficient are unambiguously related by $d = \frac{1}{2}\chi^{(2)}$.

Recognizing that the wave equation 5.1, satisfies each set of frequency components separately and substituting the appropriate derivatives obtained from (5.3) and (5.2), we obtain a set of coupled differential equation. Taking special care to define the polarization components properly for the case of second-harmonic generation leads to the expressions (Zernike and Midwinter 1973)

$$\frac{dE_\omega(z)}{dz} = -i\frac{8\pi\omega^2}{k_\omega c^2}dE_\omega^*(z)E_{2\omega}(z)e^{-i\Delta kz} \qquad (5.4)$$

and

$$\frac{dE_{2\omega}(z)}{dz} = -i\frac{16\pi\omega^2}{k_{2\omega}c^2}dE_\omega^2(z)e^{i\Delta kz} \qquad (5.5)$$

where $\Delta k = 2k_1 - k_2$ defines the phase mismatch between the incident fundamental and generated second-harmonic wave.

These are known as the coupled amplitude equations. Detailed analysis of these expressions leads to some profound consequences in nonlinear optics. One of these is the notion that a higher-frequency photon dissociates into two lower-frequency photons, which is manifested in the process of difference frequency generation. Further comments will be made on that point later. For now we examine the solution to these expressions in the small signal or low conversion limit. Note that the power per unit area (P^ω) for frequency ω in a material of index n is proportional to the magnitude of the field

$$P^{(\omega)} = \frac{cn}{8\pi} EE^* \qquad (5.6)$$

The solution to equations 5.4 and 5.5 thus leads to an expression for the power conversion efficiency η in the small conversion limit:

$$\eta = \frac{512\pi^5 d^2 l^2 P_\omega}{n_{2\omega} n_\omega^2 \lambda^2 c} \left(\frac{\sin x}{x}\right)^2 \qquad (5.7)$$

where $x = \Delta k l/2$, d is in MKS unis, and P_ω is in watts. The term $(\sin x/x)^2$ oscillates between one and zero depending on the degree of phase mismatch. If $\Delta k = 0$, then $(\sin x/x)^2 = 1$ and the conversion efficiency depends only on the square of the interaction length l^2, the square of the nonlinear coefficient d^2, and the incident power P_ω. As significant power is accumulated in the harmonic wave, equation 5.4 predicts some type of saturation behavior or steady-state condition since E_ω and therefore P_ω are gradually decreasing. If phase mismatch occurs, power is readily transferred from the harmonic to the fundamental wave.

The problem of achieving a phase-matched condition is best understood by considering the wavelength dependence of the refractive index. We start with a hypothetical example of a dye dissolved in an isotropic medium. Typically, the dye will exhibit absorption band in the visible or near-UV region of the spectrum. This medium will also exhibit dispersion in the refractive index associated with resonance enhancement of the imaginary part of the polarizability, as discussed for the harmonic oscillator in Chapter 2. The situation is illustrated schematically in Figure 5.3. The dispersion in the refractive index extends considerably beyond the absorption band. For wavelengths λ_1 and λ_2 we have refractive indices n_1 and n_2, respectively, with a difference Δn. Since we are discussing second-harmonic generation the phase shift for $2\lambda_2 = \lambda_1$ is given by

$$\Delta\phi = \Delta k l = \frac{2\pi l}{\lambda_2} \Delta n \qquad (5.8)$$

If there were no dispersion in the medium, Δn would equal zero regardless of the distance traveled in the medium. In general, this is not the case and it is not possible to retain the phase-matched conditon beyond a few micrometers in an isotropic medium. This is generally too short of a distance to obtain significant conversion of power from the fundamental to the harmonic frequency.

From a practical point of view, the truth of this last statement will depend on the magnitude of d, laser wavelength and power, and the absorption

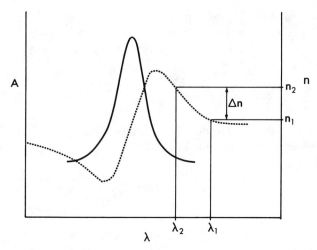

Figure 5.3 Optical absorption and dispersion and also refractive index mismatch Δn at wave wavelengths λ_1 and λ_2.

coefficient of the medium. Combinations of parameters may exist where useful levels of second harmonic might be generated for near-IR lasers with organic materials with d values in excess of $10^{-7}\,\mathrm{cm^2/esu}$.

In anisotropic media the situation is quite different. This can be seen by considering the dielectric tensor ε relating electric field components to the dielectric displacement components in an anisotropic medium, as discussed in Section 2.6. For a nonabsorbing medium, the dielectric tensor is symmetric, which, in the principal axes system (see Section 2.6), has only the diagonal elements as nozero. These principal dielectric constants $\varepsilon_x, \varepsilon_y$, and ε_z are related to the indices of refraction by

$$n_x = \sqrt{\varepsilon_x} \qquad n_y = \overline{\varepsilon_y} \qquad n_z = \sqrt{\varepsilon_z} \tag{5.9}$$

The total electric energy density W) in the medium is given by Zernike and Midwinter (1973) as

$$W = \frac{1}{8\pi}\left(\frac{D_x^2}{n_x^2} + \frac{D_y^2}{n_y^2} + \frac{D_z^2}{n_z^2}\right) \tag{5.10}$$

By substituting $x = D_x/\sqrt{8\pi W}$ and so on, we obtain the expression

$$\frac{x^2}{n_x^2} + \frac{y^2}{n_y^2} + \frac{z^2}{n_z^2} = 1 \tag{5.11}$$

which is the equation for an ellipse. The axes of the ellipse are parallel to the principal directions in the medium.

There are basically three variations of this ellipsoid that are encountered in crystalline materials and oriented poled polymer and LB films. In the first case $n_x^2 = n_y^2 = n_z^2$ and the ellipsoid is a spheroid. This is the situation for cubic crystals. When $n_x^2 = n_y^2 \neq n_z^2$ the ellipsoid is uniaxial. Here cross sections in the xy plane are circular and those in the xz and yz planes are ellipses. Materials with trigonal, tetragonal, and hexagonal symmetries are always uniaxial. For the third situation $n_x^2 \neq n_y^2 \neq n_z^2$ and the medium is biaxial. Orthorhombic, monoclinic, and triclinic systems are examples of biaxial symmetry.

In the following discussion we consider the problem of phase matching in a uniaxial medium. In this case the two independent values of the refractive indices are n_o and n_e and defined as $n_o = n_x = n_y$ and $n_e = n_z$ where n_o and n_e are termed the ordinary and extraordinary components of the refractive index, respectively. The significance of this terminology can be appreciated by referring to Figure 5.4.

The ellipsoid with a semimajor axis intersecting the z axis at $n_e(\theta)$ and semiminor axes intersecting the x and y axes at points n_0 contains the allowed propagation directions and values for the refractive index. For an incident beam S at an angle θ with respect to the z principal axis (optic axis or uniaxis) and in the xz plane, only two polarization directions are allowed for light propagating through the medium. One component is perpendicular to S and parallel to y. The refractive index associated with this direction is n_o. The value of n_o would

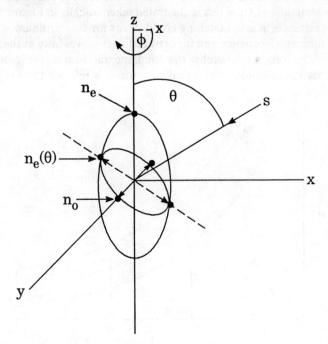

Figure 5.4 The refractive index ellipsoid showing the ordinary n_o the extraordinary $n_e(\theta)$ values of the refractive index for light incident along s.

also be appropriate if the incident beam was polarized along the y axis or any azimutthal angle (ϕ) in the xz plane. This is apparent since

$$x^2 + y^2 = n_o^2 \qquad (5.12)$$

The other allowed polarization component is orthogonal to S and n_0 and corresponds to $n_e(\theta)$. It is parallel to the dotted line in the figure. It is apparent that the value of $n_e(\theta)$ depends on the value of θ, whereas n_0 does not. If θ is equal to $\pi/2$ then $n_e(\theta) = n_e$; hence, the term extraordinary component of the refractive index. The situation illustrated in Figure 5.4 corresponds to a positive uniaxial crystal ($n_e > n_0$). If $n_e < n_0$, the crystal or medium is negative uniaxial.

Consider again the effect of dispersion on the refractive index. Both the ordinary and extraordinary components will experience dispersion. For the case of a positive uniaxial crystal, typical dispersion curves are illustrated in Figure 5.5. For n_0, one curve describes the dispersion, whereas for $n_e(\theta)$ a family of curves is needed to describe each value of θ. For this example the refractive index at n_e at frequency ω is equal to that for n_0 at 2ω at θ_1. We see, therefore, that for light incident at angle θ_1 with respect to z the extraordinary wave at 2ω and ordinary wave at ω are phase matched and the harmonic conversion efficiency will be governed by some effective value of the nonlinear coefficient which we term d_{eff}.

For a negative uniaxial medium the refractive index dispersion curves would look somewhat different and this is illustrated schematically in Figure 5.6. Under these circumstances phase matching could occur for two ordinary waves only at the fundamental frequency and for extraordinary waves only at the harmonic frequency. The process whereby the fundamental beams are colinear and polarized both as ordinary or extraordinary waves is referred to as type I phase

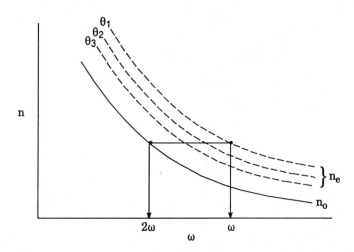

Figure 5.5   The refractive index dispersion for a positive uniaxial crystal.

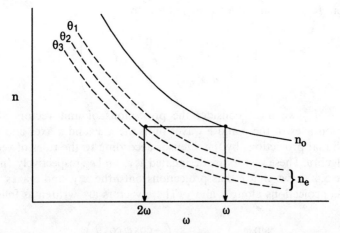

Figure 5.6   The refractive index dispersion for a negative uniaxial crystal.

matching. For two ordinary fundamental waves this process is represented as

$$n_e^{2\omega}(\theta) = n_o^{\omega} \qquad o + o \rightarrow e \tag{5.13}$$

A second type of phase matching can occur if the two fundamental beams are polarized orthogonal to one another and the harmonic beam is polarized at $45°$ with respect to the fundamental. The requirement for the refractive indices for the case where an o and e fundamental wave produce an e harmonic wave is

$$n_e^{2\omega} = \tfrac{1}{2}(n_e^{\omega}(\theta) + n_o^{\omega}) \qquad e + o \rightarrow e \tag{5.14}$$

This illustrates type II phase matching in a negative uniaxial crystal.

For the sake of illustration we discuss the determination of d_{eff} for type I phase matching in positive and negative uniaxial crystals. Furthermore, we specify that our material will have ∞mm symmetry so that we can specify the exact nature of the nonlinear tensor. In this case the tensor is given by

$$\begin{pmatrix} 0 & 0 & 0 & 0 & d_{15} & 0 \\ 0 & 0 & 0 & d_{15} & 0 & 0 \\ d_{31} & d_{31} & d_{33} & 0 & 0 & 0 \end{pmatrix}$$

where the contracted notation is being used. As a reminder $1 = x$, $2 = y$, $3 = z$ and the second two indices are contracted into one index according to the following scheme: $1 = 11$, $2 = 22$, $3 = 33$, $4 = 23$ or 32, $5 = 13$ or 31, $6 = 21$ or 12. In the case of the tensor shown $d_{15} = d_{xxz} = d_{xzx}$. If Kleinman symmetry is valid then $d_{31} = d_{15}$ and the tensor reduces to

$$
\begin{pmatrix}
0 & 0 & 0 & 0 & d_{15} & 0 \\
0 & 0 & 0 & d_{15} & 0 & 0 \\
d_{15} & d_{15} & d_{33} & 0 & 0 & 0
\end{pmatrix}
$$

To obtain d_{eff} we must consider the projections of unit vectors along the polarization axes of the o and e waves onto the $x, y,$ and z axes and multiply the individual projections by the d tensor according to the rules of vector and matrix algebra. These vectors are designated as e_o and e_e, respectively. Inspection of Figure 5.4 shows that e_e has projections onto the $x, y,$ and z axes, whereas the e_o has projections along x and y. These vectors are written as follows:

$$
\begin{pmatrix}
\sin \phi \\
-\cos \phi \\
0
\end{pmatrix} = e_o,
\qquad
\begin{pmatrix}
-\cos \phi \cos \theta \\
-\sin \phi \cos \theta \\
\sin \theta
\end{pmatrix} = e_e
\tag{5.15}
$$

For the situation where two ordinary waves are phase matched with an extraordinary wave, the value of d_{eff} for the d tensor written above is therefore

$$
d_{\text{eff}} = e_e \cdot d : e_o e_o = d_{eoo}
$$
$$
= (-\cos \phi \cos \theta \quad -\sin \phi \cos \theta \quad \sin \theta)
$$
$$
\times
\begin{pmatrix}
0 & 0 & 0 & 0 & d_{15} & 0 \\
0 & 0 & 0 & d_{15} & 0 & 0 \\
d_{15} & d_{15} & d_{33} & 0 & 0 & 0
\end{pmatrix}
\begin{pmatrix}
\sin^2 \phi \\
\cos^2 \phi \\
0 \\
0 \\
0 \\
-\sin \phi \cos \phi
\end{pmatrix}
$$
$$
= d_{15} \sin \theta
\tag{5.16}
$$

Going through a similar exercise for a positive uniaxial crystal shows that $d_{oee} = 0$. In other words phase-matched second-harmonic generation would not be possible for e polarized input waves.

An additional significant point is that d_{33} does not contribute to phase-matched second-harmonic generation. This is an important point since d_{33} is the largest nonlinear coefficient in poled polymers. This point is discussed further in Section 4.3.

This exercise could be carried out for all the noncentrosymmetric point groups to obtain general expressions for d_{eff}. Fortunately, Byer (1977) has done this for the case where Kleinman symmetry is valid. The results are reproduced in Table 5.1.

TABLE 5.1   Expressions for Effective Nonlinear Coefficient d_{eff} for Noncentrosymmetric Point Groups

Crystal Class	Two e Rays and One o ray	Two o Rays and One e Ray
6 and 4	0	$d_{15} \sin \theta$
622 and 422	0	0
6mm and 4mm	0	$d_{15} \sin \theta$
$\bar{6}$m2	$d_{22} \cos^2 \theta \cos 3\phi$	$-d_{22} \cos \theta \sin 3\phi$
3m	$d_{22} \cos^2 \theta \cos 3\phi$	$d_{15} \sin \theta - d_{22} \cos \theta \sin 3\phi$
$\bar{6}$	$\cos^2 \theta (d_{11} \sin 3\phi + d_{22} \cos 3\phi)$	$\cos \theta (d_{11} \cos 3\phi + d_{22} \sin 3\phi)$
3	$\cos^2 \theta (d_{11} \sin 3\phi + d_{22} \cos 3\phi)$	$d_{15} \sin \theta + \cos \theta (d_{11} \cos 3\phi + d_{22} \sin 3\phi)$
32	$d_{11} \cos^2 \theta \sin 3\phi$	$d_{11} \cos \theta \cos 3\phi$
$\bar{4}$	$\sin 2\theta (d_{14} \cos 2\phi - d_{15} \sin 2\phi)$	$-\sin \theta (d_{14} \sin 2\phi + d_{15} \cos 2\phi)$
$\bar{4}$	$d_{14}(\sin 2\theta \cos 2\phi)$	$-d_{14} \sin \theta \sin 2\phi$

Source: After Byer [1977].

An appropriate short-hand notation for d_{eff} that is often used is $d_{\text{eff}} = (e(\theta) \cdot d : oo)$ where $e(\theta)$ and o are unit vectors polarized in the $e(\theta)$ and o directions and sometimes referred to as polarization vectors.

Equation 5.7 would seem to imply that if we had a sufficiently long pathlength, a high degree of second-harmonic conversion could be achieved regardless of how small the d coefficient might be. Unfortunately, this is not the case, for both practical and fundamental reasons. If the incident fundamental beam is focused, the input angle θ will be not a discrete value, but an envelope centered at θ. Therefore, the phase-matching condition will be limited to some finite length in the crystal. For a phase-matching angle far from 90° the sensitivity of phase-matching angle to wavelength can be very high. This implies that input angular tolerance is low and spectral purity of the laser must be high. Unfortunately, many laser sources oscillate in a number of longitudinal modes, that is, frequencies varying by the speed of light divided by the round-trip length of the laser cavity. Thus, the various modes will exhibit slightly different degrees of phase matching, leading to more complex behavior. If a set of conditions can be found where the phase matching angle is 90°, the angular sensitivity is quite low so that large acceptance angles and a range of laser longitudinal modes could be phased matched simultaneously. This situation is sometimes referred to as noncritical phase matching.

A more fundamental limitation on the interaction length is a phenomenon called *walkoff*. The manifestation of walkoff is that, in general, o and $e(\theta)$ waves propagate in slightly different directions in birefrigent media. This is because the $e(\theta)$ wave polarizes the media differently along the $e(z)$ principal axis than along the o (x or y) principal axis. As a result the power flow is bent slightly from that associated with the o wave. For $\theta = 90°$ the $e(\theta)$ wave is parallel to the optic axis and there are no contributions from n_o. In this case the *walkoff angle* is zero. The formula for computing the walkoff angle is (Byer 1977)

$$\rho \simeq \tan\rho = \frac{n^o_\omega}{2}\left[\frac{1}{(n^e_{2\omega})^2} - \frac{1}{(n^o_{2\omega})^2}\right]\sin 2\theta \qquad (5.17)$$

For typical birefringent crystals at $\theta = 45°$, $\rho \sim 1$–$2°$. Distances on the order of 0.5 cm under these circumstances would lead to a substantial amount of walkoff.

Other schemes exist for achieving phase matching in nonlinear media. One of these is the use of dispersion in waveguides to achieve phase matching between guided modes. This is discussed fully in Chapter 11.

5.3   PARAMETRIC PROCESSES

Second-harmonic generation is a special case of the sum frequency generation process. In this section we introduce and discuss the general and special features of sum and difference frequency generation processes.

5.3.1   Sum-Frequency Generation

An analogous set of expressions to equations 5.4 and 5.5 can be written for the general three wave interactions. These are (Zernike and Midwinter 1973)

$$\frac{dE_1(z)}{dz} = \frac{-i8\pi\omega_1^2}{k_1 c^2}dE_2^* E_3 e^{-i\Delta kz} \qquad (5.18)$$

$$\frac{dE_2(z)}{dz} = \frac{-i8\pi\omega_2^2}{k_2 c^2}dE_1^* E_3 e^{-i\Delta kz} \qquad (5.19)$$

$$\frac{dE_3(z)}{dz} = \frac{-i8\pi\omega_3^2}{k_3 c^2}dE_2 E_1 e^{-i\Delta kz} \qquad (5.20)$$

where $\Delta k = k_i - k_j$ and k_i is related to the angular frequency ω_i, the refractive index at that frequency n_i, and the speed of light c by $k_i = n_i\omega_i/c$. By convention ω_3 is the highest frequency component. Multiplying both sides of equation 5.18 by E_1^* and doing similar to equations 5.19 and 5.20 and assuming further that $\Delta k = 0$ leads to the following set of equations:

$$\frac{n_1 c}{\omega_1}E_1^*\frac{dE_1}{dz} = -8\pi i dE_1^* E_2^* E_3 \qquad (5.21)$$

$$\frac{n_2 c}{\omega_3}E_2^*\frac{dE_2}{dz} = -8\pi i dE_2^* E_1^* E_3 \qquad (5.22)$$

$$\frac{n_3 c}{\omega_3} E_3^* \frac{dE_3}{dz} = - 8\pi i d E_3^* E_2 E_1 \tag{5.23}$$

It is evident from inspection that the right sides of equations 5.21 and 5.22 are equal and the right side of 5.23 is the complex conjugate of the other two. By straightforward inference we can now write

$$\frac{n_1 c}{\omega_1} \frac{d}{dz} (E_1 E_1^*) = \frac{n_2 c}{\omega_2} \frac{d}{dz} (E_2 E_2^*) = - \frac{n_3 c}{\omega_3} \frac{d}{dz} (E_3 E_3^*) \tag{5.24}$$

Since the power in a wave is proportional to EE^* (equation 5.6), we can interpret the consequences of equation 5.24 in the following manner (Zernike and Midwinter 1973):

$$\frac{\text{change in power at } \omega_1}{\omega_1} = \frac{\text{change in power at } \omega_2}{\omega_2}$$

$$= - \frac{\text{change in power at } \omega_3}{\omega_3} \tag{5.25}$$

This relationship was first recognized by Manley and Rowe (1959) and defines the direction of power flow in three wave mixing processes. For sum-frequency generation, power is lost by the pumping laser frequencies at ω_1 and ω_2 and gained by the sum frequency. In the case of the difference-frequency generation processes ($\omega_3 - \omega_2 = \omega_1$), power is lost not only to the generated frequency at ω_1 but also to the source at ω_2. In other words, a photon at frequency ω_3 is split into photons at ω_1 and ω_2. This is an interesting concept and has very significant practical consequences.

Sum-frequency generation could be exploited in a manner similar to second-harmonic generation. Here two beams similar in power would be directed into a nonlinear medium with the power generated at the sum frequency by

$$P_3 = \frac{512 \pi^5 l^2 d^2 P_1 P_2}{n_1 n_2 n_3 \lambda_3^2 c} \left(\frac{\sin x}{x} \right)^2 \tag{5.26}$$

As might be expected the generated power is proportional to input powers and the last term is associated with phase mismatch where $x = \Delta k l / 2$. The problem of phase matching is more complex than in the case of second-harmonic generation since now knowledge of the refractive index ellipsoid at three different frequencies is required. In general, however, it is often possible to find a direction in the medium where phase matching occurs. For the case of two o waves phase matched to an e wave, as was discussed for second-harmonic generation, the

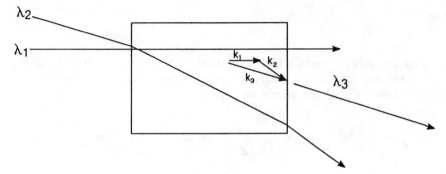

Figure 5.7   Noncolinear phase matching in a nonlinear medium.

phase-matching requirement is (Kurtz 1975)

$$\frac{n_1^o}{\lambda_1} + \frac{n_2^o}{\lambda_2} = \frac{n_3^e(\theta)}{\lambda_3} \tag{5.27}$$

Other alternatives to colinear phase matching exist. One way of achieving phase matching is to direct the incident beams into the medium at some angle with respect to each other. This is illustrated in Figure 5.7. This type of interaction is more important for other types of parametric processes, but it could, in principle, allow phase matching utilizing a contribution to d_{eff} that is not available for colinear phase matching with its particular polarization requirements.

5.3.2 Parametric Up-conversion

A special case of sum-frequency generation is the parametric up-conversion process. The distinguishing feature here is that one of the source frequencies, say ω_2, which we refer to as the pump, is much more intense than the second frequency ω_1. The power in the output frequency ω_3 is also zero at the start of the interaction. Since the field at ω_2 is so much more intense, very little energy is depleted from ω_2. This leads to the condition that $dE_2/dz = 0$ in equation 5.22 and relatively straightforward expressions for the electric field amplitudes (Zernike and Midwinter 1973):

$$E_3(z) = \sqrt{\frac{\omega_3^2 k_1}{\omega_1^2 k_3}} \; E_1(0) \sin\frac{z}{l_p} \tag{5.28}$$

where

$$l_p = \left(\frac{4\pi d}{c^2} \sqrt{\frac{\omega_1^2 \omega_3^2}{k_1 k_3}} E_2\right)^{-1} \tag{5.29}$$

and

$$E_1(z) = E_1(0)\cos\frac{z}{l_p}$$

where $E_1(0)$ is the incident amplitude of the weak field. Equation 5.28 states that the intensity of the weak photon can be shifted to a new frequency ω_3 where it might be detected more efficiently. This process can be extremely efficient if the interaction continues over a characteristic length l_c where $l_c = \pi l_p/2$. It is also appropriate to note that if the interaction continues to occur past l_c power is transferred back to the low-frequency signal. In fact, power will oscillate indefinitely between the two signals. This puts certain experimental constraints on the geometrical characteristics of the medium. Detectors in the visible region of the spectrum can be many orders of magnitude more efficient than infrared detectors making this process of potential considerable use for spectroscopic applications.

5.3.3 Difference-Frequency Generation

It was pointed out earlier in this section in conjunction with the discussion of the Manley–Rowe relations that the difference-frequency generation process has the unique characteristic that if a weak signal at a lower frequency is made to interact with the highest frequency signal, which we designated as ω_3, that a new frequency at ω_2 is created and both lower-frequency signals are amplified. The term for this process is *parametric amplification*. The signals ω_3, ω_2, and ω_1 have acquired designations as a result of their distinctive roles. The term pump is given to ω_3, and the generated signal, which could be either ω_2 or ω_1 is termed the idler. The lower-frequency input is referred to as the source. The designations ω_p, ω_i, and ω_s are often used in discussions of this topic.

It was hinted in our discussion of the Manley–Rowe relations that the difference frequency process could be viewed as splitting a photon at ω_3 into two photons at ω_1 and ω_2, respectively. In fact, a second frequency component is not required at all in order for this process to occur. Shining an intense beam at ω_3 into a second-order nonlinear material will spontaneously generate new frequencies at ω_1 and ω_2, the requirements being

$$\omega_1 + \omega_2 = \omega_3 \tag{5.30}$$

and

$$k_1 + k_2 \sim k_3 \tag{5.31}$$

Since there are many pairs of ω_1 and ω_2 that satisfy these conditions, one might expect a cone of radiation over some spectral region to be generated when ω_3 impinges on the material. This phenomenon is termed *parametric fluorescence* and was put on a sound quantum-mechanical footing by Yariv (1975). A striking color photograph obtained by Byer (1968) of parametric fluorescence generated in $LiNbO_3$ was published in a review article by Giordmaine (1969).

5.3.4   Parametric Oscillator

The process of parametric oscillation occurs when one or both of the frequencies generated in the parametric fluorescence process is fed back into the nonlinear medium. If provision for gain exists and it exceeds whatever loss processes that might occur, this device can oscillate. As was pointed out in the previous paragraph, the pair of photons that satisfies conditions (5.30) and (5.31) is not unique. Selection of a narrow band of frequencies in the parametric oscillation process through angle-tuned phase matching can lead to a tunable source of coherent radiation.

To discuss the process of parametric oscillation we again follow the treatment of Zernike and Midwinter (1973). In this case the pump frequency (ω_3) is expected to be the high-power one, so we expect that the loss of this field is small and we have from equation 5.20 that $\partial E_3^{(2)}/\partial z = 0$. Solution of (5.28)–(5.29) yields $E_2(l)$ and $E_1(l)$ where l is the interaction length in the crystal. The general solution under phase-matched conditions is

$$E_2(l) = E_2(0)\cosh\frac{l}{l_p} + \frac{\omega_1^2 k_2}{\omega_2^2 k_1} E_1(0)\sinh\frac{l}{l_p} \qquad (5.32)$$

where l_p is given by

$$l_p = \frac{\sqrt{n_1 n_2 \lambda_1 \lambda_2}}{8\pi d E_3} \qquad (5.33)$$

A similar solution exists for E_1.

We consider two different limiting situations. In the first case l and l_p are of the same order of magnitude and we have for the signal $E_2(l)$ and idler $E_1(l)$

$$E_2(l) = E_2(0)\cosh(\alpha l) \qquad (5.34)$$

and

$$E_1(l) = E_2(0)\sqrt{\frac{\omega_1^2 k_2}{\omega_2^2 k_1}}\sinh(\alpha l) \qquad (5.35)$$

where $\alpha = l/l_p$. This describes the situation for a parametric amplifier where the beams traverse the crystal only once. When l is much greater than l_p we have

$$E_2(l) = E_2(0)e^{\alpha l} \qquad (5.36)$$

$$E_1(l) = E_2(0)\sqrt{\frac{\omega_1^2 k_2}{\omega_2^2 k_1}}e^{\alpha l} \qquad (5.37)$$

This is the behavior exhibited by a parametric oscillator. The key feature to notice in these expressions is the appearance of exponential gain, whih can lead to significant power buildup if the wave traverses the crystal many times. This foregoing analysis was made for the phase-matched case. Considerable care must be taken in arriving at an analysis of the effects of phase mismatch on gain. Qualitatively we would expect the gain to decrease for some level of phase mismatch and this is indeed the case. The oscillator cavity, on the other hand, plays a role in selecting the phases of waves that build up out of the noise. Expression for power buildup for the small-gain case $\alpha l \ll l$, which illustrate the points, are

$$P_2(l) = P_1(0)(1 + 2\alpha l) \frac{\sin(\Delta kl/2)}{\Delta kl/2} \qquad (5.38)$$

and

$$P_1(l) = \frac{\omega_1}{\omega_2} P_2(l) \qquad (5.39)$$

There are basically two types of parametric oscillators: singly and doubly resonant. A schematic illustration of a doubly resonant parametric oscillator is shown in Figure 5.8. In this case a large amount of power is stored in the cavity formed by the nonlinear crystal and mirrors with the specified reflectivities at ω_1 and ω_2. Oscillation wil begin when the gain, which depends exponentially on E_3 is large enough to overcome the loss associated with the mirrors. For the doubly resonant device this threshold occurs at relatively low powers. For LiNbO$_3$ this is about 5.7×10^3 W/cm^2. Although we are not aware of attempts to make such a device from highly efficient organic crystals, we note that the threshold for oscillation is universally proportional to d^2 (Yariv 1975). Highly efficient organic crystals have d values 10–100 times larger than LiNbO$_3$ so that oscillation could occur in the W/cm^2 range. For the case of the doubly resonant oscillator the energy conservation and phase matching (momentum conservation conditions) become

$$\omega_1 + \omega_2 = \omega_3 \qquad (5.40)$$

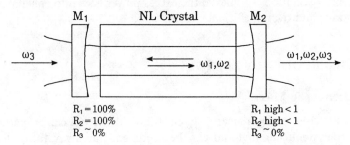

Figure 5.8 A doubly resonant parametric oscillator. Reprinted with permission from A. Yariv, *Quantum Electronics, 2nd ed.*, John Wiley & Sons, New York, 1975, 439.

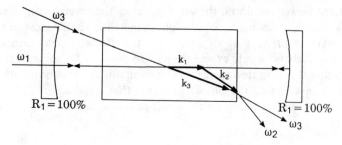

Figure 5.9   A singly resonant parametric oscillator.

and

$$\Delta k_1 l = \frac{\omega_1 n_1 l}{c} - m\pi + \frac{\phi_1}{2} \tag{5.41}$$

$$\Delta k_2 l = \frac{\omega_2 n_2 l}{c} - s\pi + \frac{\phi_2}{2} \tag{5.42}$$

which must be satisfied simultaneously. In equations 5.41 and 5.42 m and s are integers and ϕ_1 and ϕ_2 are any accumulated phase shift from the mirrors and inerfaces. These conditions place severe constraints on the stability of the resonator. Small vibrations or thermal fluctuations can change the length of the cavity slightly, which translates into large frequency excursions. If another set of frequencies ω_1' and ω_2' meets the phase requirements dictated by the fluctuations in the oscillator it can readily shift frequencies. Jumping between frequencies dictated by these fluctuations is the main manifestation of instability.

A second type of parametric oscillator is one having a singly resonant cavity. Here feedback is provided for at only one of the lower frequencies. This situation is illustrated schematically in Figure 5.9 (Falk and Murray 1969). The phase condition is the same as that for ω_1 in equation 5.41 but no restriction is put on the phase of ω_2. In this type of oscillator the problem of frequency hopping is eliminated but the threshold for oscillation is several orders of magnitude higher. Once oscillation is achieved the excess power goes into coherent output at ω_1 and ω_2 and overall efficiency is not sacrificed.

5.4   LOW-FREQUENCY EFFECTS

5.4.1   Electrooptic Effect

In the presence of a dc or nearly dc electric field the optical properties of a second-order nonlinear material can be modified. Far away from electronic resonances this effect results in a refractive index change that is linearly proportional to the electric field. This phenomenon, normally referred to as the electrooptic or Pockels effect, can be viewed as a three-wave mixing process where one of the fields is at dc. the nonlinear tensor element for this interaction is written with the appropriate frequency arguments as

$$\chi^{(2)}_{IJK}(\omega; 0, -\omega)$$

There are several ways to demonstrate the relationship between the three-wave mixing process description and electric field dependent refractive index approach to the linear electrooptic effect. The coupled amplitude equations (5.21)–(5.23) can be solved for the case where E_2 is the applied field and the relation between E_3 and E_1 is examined. When this is done, a phase shift appears in E_1 which is proportional to E_2. Using this approach it can be shown that

$$\frac{-\pi n^3 r E_2 l}{\lambda} = \frac{4\pi^2 l d E_2}{n\lambda} \tag{5.43}$$

and

$$r = \frac{-4\pi}{n^4} d \tag{5.44}$$

In this expression l is the length of the interaction and r is referred to as the electrooptic coefficient.

An alternative way of developing this relationship is to start with the power series of the polarization as a function of field, as we have discussed previously in Chapter 2, Section 2.4. From equation 2.33, retaining only linear terms in $E(0)$, we have

$$n^2 = 1 + 4\pi\chi^{(1)} + 8\pi\chi^{(2)}E(0) + 12\pi\chi^{(3)}E^2(0) \tag{5.45}$$

where $E(0)$ is the dc field component. The linear refractive index can be written as

$$n_0^2 = 1 + 4\pi\chi^{(1)} \tag{5.46}$$

The change in refractive index due to the linear electrooptic effect is commonly defined as (Yariv 1975)

$$\Delta\left(\frac{1}{n^2}\right) = rE(0) \tag{5.47}$$

The following algebra is presented to make a clear connection between r and $\chi^{(2)}$. The left side of equation 5.47 can be expanded as

$$\Delta\left(\frac{1}{n^2}\right) = \frac{1}{n_0^2} - \frac{1}{n^2} = \frac{n^2 - n_0^2}{n_0^2 n^2} \approx \frac{n^2 - n_0^2}{n_0^4} \tag{5.48}$$

Substituting for n^2 and n_0^2 from equations 5.45 and 5.46 leads to the relationship between r and $\chi^{(2)}$:

$$-r \simeq \frac{8\pi\chi^{(2)}}{n_0^4} \tag{5.49}$$

This effect has been studied extensively in organic crystals and polymer films and may prove to be one of the most technologically important applications of these materials. This application is described extensively in Chapters 7, 11, and 12.

5.4.2 Optical Rectification

One final effect that results from the interaction of a wave at frequency ω with itself, apart from generation of the second harmonic at 2ω, generates a dc polarization in the medium (Shen 1984). This is clearly predicted by equation 2.29. When we consider the predictions of Kleinman symmetry, which allows the indices in the nonlinear tensor element to be freely permuted, we have

$$\chi_{ijk}^{(2)}(\omega; 0, -\omega) = \chi_{ijk}^{(2)}(0; -\omega, \omega) \tag{5.50}$$

showing that the optical rectification and the dc electrooptic effects are clearly related.

The optical rectification effect was measured on a number of inorganic crystals by Bass et al. (1962) and Ward (1966). In their experiments they coated the surfaces of nonlinear crystals with Ag films and shone intense beams into the crystals. The voltages induced on the surfaces were measured. They compared their results with measurements of the electrooptic coefficient on the same crystals and showed the validity of equation 5.50.

While this effect can be expected to occur in organic systems, to the best of our knowledge measurements of optical rectifiation have not been reported at the time of this writing. This may be due to the perceived lack of technological importance for this effect and the fact that it gives little new information relative to other measurements.

REFERENCES

Bass, M., P. A. Franken, J. F. Ward, and G. Weinreich, *Phys. Rev. Lett.* **9**, 446 (1962).

Bloembergen, N., *Nonlinear Optics*, Benjamin, New York, 1965.

Byer, R. L. doctoral dissertation, Stanford University, Palo Alto, CA, 1968.

Byer, R. L., P. G. Harper, B. S. Wherret (Eds.), *Nonlinear Optics*, Academic, London, 1977, p. 82.

Conwell, E. M., *IEEE J. Quantum Electron.* **OE-9**, 867 (1973).

Falk, J., and J. E. Murray, *Appl. Phys. Lett.* **14**, 245 (1969).

Flytzanis, C., in H. Rabin, and C. L. Tang (Eds.), *Quantum Electronics: A Treatise*, Vol. 1, Academic, New York, 1975, p. 13ff.

Giordmaine, J. A., *Phys. Today* **28**, 38 (1969).

Hulin, D., A. Migus, R. Antonetti, I. Ledoux, J. Badon, and J. Zyss,*Ultrafast Phenomena 5*, Springer-Verlag, Berlin, 1986, p. 75.

Jain, K., and G. H. Hewig, *Opt. Commun.* **36**, 483 (1981).

Kurtz, S. A., in H. Rabin, and C. L. Tang (Eds.), *Quantum Electronics; A Treatise*, Vol. 1, Academic, New York, 1975, p. 239.

Manley, J. M., and H. E. Rowe, *Proc. IRE* **47**, 2115 (1959).

Morrell, J. A., and A. C. Albrecht, *Chem. Phys.* **64**, 46 (1979).

Shen, Y. R., *The Principles of Nonlinear Optics*, Wiley, New York, 1984, p. 57ff.

Ward, J. F., *Phys Rev.* **143**, 569 (1966).

Yariv, A., *Quantum Electronics*, Wiley, New York, 1975, p. 450ff.

Zernike, F., and J. E. Midwinter, *Applied Nonlinear Optics*, Wiley, New York, 1973.

6

MEASUREMENT TECHNIQUES FOR SECOND-ORDER NONLINEAR OPTICAL EFFECTS

6.1 ELECTRIC FIELD-INDUCED SECOND-HARMONIC GENERATION METHOD

The design of new materials for second-order nonlinear optical applications requires a knowledge of the contribution of the molecules constituting the medium to its nonlinear polarization as well as the way in which their geometrical arrangement in the medium determines the propagation characteristics of the resulting field components. The technique of electric field-induced second-harmonic generation (Levine and Bethea 1975), abbreviated as EFISH, was developed to determine the value of β for molecules from measurements on liquids or solutions. In this technique a strong dc electric field is applied to a liquid or solution causing a bias on the average orientation of the molecules due to the interaction of the field with the permanent dipoles of the molecules. The partial removal of the orientational average reduces the symmetry of the medium, allowing the second-harmonic field to propagate with the nonlinear polarization from the fundamental field over some distance in the medium. In this experiment the vector part β of the hyperpolarizability tensor β_{ijk} with components defined by

$$\beta_i = \sum_j \beta_{ijj} \tag{6.1}$$

is measured.

Before describing the relationship of β_i to the signals observed, the experiment will be described in some additional detail. A block diagram of the key elements

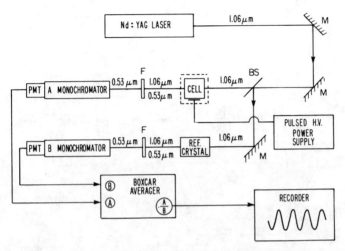

Figure 6.1 Schematic diagram of an experimental system for measurement of hyperpolarizability by the EFISH (electric field-induced second-harmonic generation) technique. M, mirror; BS, beam splitter; F, filters, PMT, photomultiplier tube. Lens and timing pulses are not shown.

of an experimental setup for EFISH is shown in Figure 6.1. More detailed descriptions of these systems and the experimental conditions and precautions that must be employed can be found in Bethea (1975), Oudar (1977), and Meredith et al. (1982).

In this example the 1.06-μm output of a Nd^{+3}:YAG laser is the light source. Dye lasers and stimulated Raman shifters employing H_2 or CH_4 gas (Sauteret et al. 1976, Wilke and Schmidt 1979) have also been used to generate alternative frequencies in the visible and near IR. Lens and mirrors and polarization control optics are used to direct the beam to the sample cell, which is shown in more detail in Figure 6.2. A pulsed high-voltage power supply capable of generating dc pulses in the 0 to 10-kV range and synchronized with timing pulses from the laser is essential for the experiment. A pulsed rather than dc source is used to minimize problems associated with electrolysis or polarization of the cell. Another key component is a translation stage that can be stepped or driven laterally with respect to the laser beam to observe the fringes associated with phase mismatch between the nonlinear polarization propagating with wave vector k_ω and second-harmonic field propagating with $k_{2\omega}$ through the medium. In designing the optics for focusing the beam into the sample cell great care should be taken to insure that the light intensity does not vary in the cell in the direction of propagation of the beams. In a practical sense this means that a long focal length lens must be used to focus the beam into the sample cell to avoid significant beam divergence in the cell and to avoid high powers at which effects such as self-focusing occur. A reference arm is shown in this setup to derive a second-harmonic signal from a sample of quartz or some other nonlinear

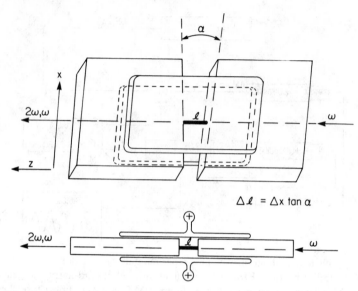

Figure 6.2 Top view (*above*) and edge view (*below*) of a cell used for EFISH measurements. The glass is about 3 mm thick and about 1 cm long. The gap in which the liquid is confined is 1–2 mm, and the electrodes extend about five times the gap spacing to avoid nonuniform electric fields at the glass–liquid interface. The cell is translated in the x direction with respect to the beam to produce the fringes described in Figure 6.3.

material. The signal from this arm is compared to the signal from the sample to avoid errors due to source beam intensity fluctuations. The ratio of the signals in the two channels should be independent of laser intensity for a given path length through the cell and serves as the basis for data analysis. Computer control over translation stage motion and synchronization with the laser timing can greatly simplify data collection and analysis.

The sample cell design is a key factor in the ease and accuracy of the EFISH experiment. The schematic shown in Figure 6.2 consists of two glass rectangles set at an angle α with respect to each other. Electrodes are placed above and below the glass surfaces. The electrodes extend well beyond the gap between the wedges to avoid nonuniformities due to fringe fields. For the same reason the electrode should be significantly wider than the distance to be covered by the translation. The cell should be designed in such a way that the liquid to be measured can be easily added or removed and conveniently cleaned. For reasons described below the total contribution to the nonlinear polarization at 2ω obtained in the EFISH experiment is given by

$$\gamma^0 = \langle \gamma \rangle + \frac{\mu \cdot \beta}{5kT} \tag{6.2}$$

where $\mu \cdot \beta$ is the scalar product of the dipole moment vector with the vector part of the hyperpolarizability tensor and $\langle \gamma \rangle$ is the scalar orientationally averaged part of the third-order hyperpolarizability tensor with components γ_{ijkl} and is given by

$$\langle \gamma \rangle = \tfrac{1}{5}(\gamma_{xxxx} + \gamma_{yyyy} + \gamma_{zzzz} + 2\gamma_{yyzz} + 2\gamma_{zzxx} + 2\gamma_{xxyy}) \qquad (6.3)$$

Because all materials have a nonzero value of $\langle \gamma \rangle$, an electric field-induced contribution to γ^0 will be provided by the solvents and solutes in the EFISH cell as well as from the glass in the region of the dc field. Where contributions from $\mu \cdot \beta$ to the signal at 2ω are weak these contributions must be carefully accounted for.

The EFISH experiment is conducted by directing the beam at fundamental frequency ω through the cell in the presence of the dc field and translating the cell with respect to the beam. Filters and/or monochromators are used to reject light at the fundamental frequency from the detector. The cell is then translated with respect to the beam and the harmonic intensity is recorded as a function of position. Translation of the cell in the x direction produces a path length variation Δl. The relationship between the translation distance Δx and Δl is given by

$$\Delta l = \Delta x \tan \alpha \qquad (6.4)$$

The need for the wedge geometry and measurement of signal intensity at 2ω as a function of path length is associated with phase mismatch between the fundamental and harmonic waves propagating through the medium. Second-harmonic generation is a phase-sensitive process so that the flow of energy between the fundamental and harmonic waves depends on the phase relationships between the waves. Refractive index dispersion at the fundamental and harmonic frequencies causes light to propagate at different velocities at the two frequencies. As the harmonic field propagates through the sample, overlapping the fundamental wave, energy will flow back and forth between these waves, depending on their phase relationships, until the nonlinear interaction is terminated (at a cell boundary for instance). On either side of the interface a free and reflected harmonic wave is created whose amplitude depends on the optical constants and other boundary conditions at the interface.

This process can be seen more clearly by considering the representation of the electric part of an electromagnetic wave:

$$E(z,t) = \tfrac{1}{2}(E^0 e^{(i\omega t - \phi(z))} + \text{c.c.}) \qquad (6.5)$$

E_0 is the amplitude of the field and the time and positional oscillatory behavior are described by the first and second terms in the exponent, respectively. The propogation of the nonlinear polarization $P_{2\omega}$, which is proportional to E_ω^2, is governed by the refractive index at ω, while that for the second-harmonic wave

$E_{2\omega}$ is governed by that at 2ω. This generates a phase angle difference between the two waves given by

$$\Delta\phi = \frac{2\pi l}{\lambda}\Delta n \qquad (6.6)$$

where λ is the wavelength corresponding to frequency ω.

The characteristic distance or coherence length l_c over which the waves will dephase is given by $l\Delta n$. For typical organic dyes this distance is on the order of 10–20 μm. The expected oscillatory behavior of the signal as a function of position is shown in Figure 6.3 for a simple organic liquid, $CHCl_3$.

The propagation of the various waves in a nonlinear medium in the EFISH experiment was analized in detail by Oudar (1977) and Levine and Bethea (1975). The intensity distribution between the fundamental, free, and reflected waves is determined solely by the phase relationships between the waves during the last coherence length where the interaction is terminated by an interface as determined by the boundary conditions at the interface. The term *free wave* is used to describe the electric field component propagating at the harmonic frequency with wave vector $k_{2\omega}$. In the absence of optical absorption the harmonic intensity $I(2\omega)$ is given by

$$I(2\omega) = 2I_m \sin^2\frac{\Delta\phi}{2} \qquad (6.7)$$

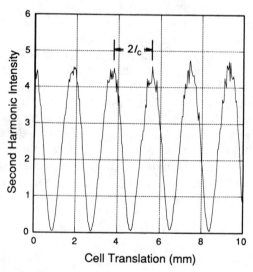

Figure 6.3 A Wedge fringe pattern obtained for chloroform using an apparatus and cell similar to that described in Figures 6.1 and 6.2. The spacing between fringes, as indicated, is twice the coherence length l_c.

The quantity I_m is related to the optical properties of the interface and nonlinear polarization in the media on either side of the interface. If the contribution from glass is small compared to the liquid of interest and $\gamma \ll \mu\beta/5kT$ (see equation 6.2) then $I_m \alpha (\mu \cdot \beta)^2$. Calibration of the output intensity with a sample of known nonlinear coefficient is generally used to obtain the magnitude of the coefficient.

To relate the experimental data obtained from the EFISH measurement to the vector part of β it is necessary to consider expressions for both macroscopic and microscopic nonlinearities. Since the experiment generates a field at 2ω from the interaction of three waves (two at the fundamental frequency and one at dc), the induced nonlinear polarization must arise from a third-order process. The polarization at 2ω is written accordingly as

$$P^{2\omega} = \Gamma_{IJKL} E_J^\omega E_K^\omega E_L^0 \tag{6.8}$$

where

$$\Gamma_{IJKL} = \chi_{IJKL}^{(3)}(-2\omega; \omega, \omega, 0) \tag{6.9}$$

We can define an effective second-order coefficient

$$d_{IJK} = \Gamma_{IJKL} E_L^0 \tag{6.10}$$

and write the expression for $P^{2\omega}$

$$P^{2\omega} = d_{IJK} E^\omega E^\omega \tag{6.11}$$

This expression now resembles one describing a second-order nonlinear effect. For a pure liquid we can define a microscopic hyperpolarizability γ^0 by

$$\Gamma = N\gamma^0 f_0 f_\omega^2 f_{2\omega} \tag{6.12}$$

where N is the density of molecules and f is a local field factor that provides a correction to the external fields to account for the influence of charges associated with neighboring molecules. Assuming the Kleinman symmetry relationships for an isotropic phase (see Chapter 2) yields

$$\Gamma = \Gamma_{ZZZZ} = 3\Gamma_{ZZYY} \tag{6.13}$$

If we designate the molecular axis parallel to the dipole moment as the z axis, (6.2) can be rewritten as

$$\gamma^0 = \gamma + \frac{\mu_z \beta_z}{5kT} \tag{6.14}$$

The meaning of the local field factor was discussed in Chapter 4. The value of those factors for the EFISH experiment for the sake of clarity is given by

$$f_0 f_\omega^2 f_{2\omega} = \varepsilon_0 \frac{\varepsilon_\omega + 2}{2\varepsilon_0 + \varepsilon_\omega} \left(\frac{\varepsilon_\omega + 2}{3} \right)^2 \frac{\varepsilon_{2\omega} + 2}{3} \tag{6.15}$$

For a mixture of nonassociating liquids, the total nonlinearity Γ_{mix} is expected to obey the relationship

$$\Gamma_{\text{mix}} = \sum_i (N\gamma^0 f_0 f_\omega^2 f_{2\omega})_i \tag{6.16}$$

For many molecules of interest the second term in (6.14) is larger than the first term so that

$$\sqrt{I(2\omega)} \propto N f_0 f_\omega^2 f_{2\omega} \frac{\mu_z \beta_z E_z^0}{5kT} \tag{6.17}$$

For a complete analysis of the EFISH method readers are referred to Oudar (1977) and Levine and Bethea (1975). Here we note the result expressed by Oudar for the intensity of the field at 2ω in terms of experimentally measurable quantities and the hyperpolarizability. Substituting the full expression for I_m into (6.7) gives

$$I(2\omega) = \frac{c}{2\pi} [t_{2\omega}^G (T_G E_b^G - T_L E_b^L)]^2 \sin^2 \frac{\Delta\phi}{2} \tag{6.18}$$

where superscripts and subscripts G and L refer to glass and liquid, E_b is the magnitude of the bound wave associated with the nonlinear polarization in the media, and c is the speed of light. The quantities $t_{2\omega}^G$, T_G, and T_L are essentially transmission coefficients defined by

$$t_{2\omega}^G = \frac{2n_{2\omega}^G}{1 + n_{2\omega}^G} \tag{6.19}$$

$$T_G = \frac{n_\omega^G + n_{2\omega}^L}{n_{2\omega}^G + n_{2\omega}^L} \tag{6.20}$$

$$T_L = \frac{n_\omega^L + n_{2\omega}^L}{n_{2\omega}^G + n_{2\omega}^L} \tag{6.21}$$

In cases where the harmonic field associated with the sample is large compared to glass and the contribution from γ^0, (6.18) can be expressed in terms of the incident beam intensity, dc field strength, and the other quantities we have discussed as

$$I(2\omega) = \frac{c}{2\pi}(t_{2\omega}^G T_L E_b^L)^2 \sin^2 \frac{\Delta\phi}{2} \tag{6.22}$$

where

$$E_b^L = -\frac{4\pi N \mu_z \beta_z E^0 (E^\omega)^2 f_0 f_\omega^2 f_{2\omega}}{5(n_{2\omega}^2 - n_\omega^2)kT} \tag{6.23}$$

As we discussed above, all components of a mixture of liquids contribute to the harmonic signal. Conditions were also discussed where these contributions to the signal are negligible.

While these conditions simplify the analysis of the data, it is important to be able to determine the sign of the hyperpolarizability. Levine and Bethea (1974) have discussed several techniques for absolute and relative determination of the sign of β. In one method the signal from the electroded region of the glass cell is allowed to interfere with the signal generated from the liquid. This can be accomplished by diluting a liquid of interest, nitrobenzene, for example, in a liquid such as benzene with a negligible value of β. The total signal emanating from the cell under these conditions is

$$I_{2\omega} \alpha \left[d_L \sin \frac{\Delta\phi_L}{2} + 2d_G \sin \frac{\Delta\phi_G}{2} \cos \frac{(\Delta\phi_G + \Delta\phi_L)}{2} \right]^2 \tag{6.24}$$

To analyze the sign of the liquid relative to glass the peak of the Maker fringe of the nitrobenzene, for instance, is identified. At the fringe $\Delta\phi_L$ is equal to $(\pi/2)m$, where m is an odd integer. Thus, we have

$$I_{2\omega} \alpha \left(d_L - 2d_G \sin^2 \frac{\Delta\phi_G}{2} \right)^2 \tag{6.25}$$

and the signal from the glass will subtract from the total intensity if their signs are the same and add if they are different. Levine and Bethea (1974) showed that glass and nitrobenzene have the same sign. Since the absolute sign of glass is known to be positive (Owyoung et al. 1972), the sign of β_z for nitrobenzene was established.

Analogously, the signs of other liquids can be determined by measuring the magnitude of signals from the individual liquids alone and comparing them to that obtained from a mixture. According to (6.16) the nonlinearity of a binary mixture is

$$\Gamma_{mix} = N_1 \gamma_1 (f_0 f_\omega^2 f_{2\omega})_1 + N_2 \gamma_2 (f_0 f_\omega^2 f_{2\omega})_2 \tag{6.26}$$

The sign for the unknown liquid is easily determined from this expression.

The development of new materials for second-order nonlinear optics requires experimental determination of β_z for molecules that do not exist as liquids near room temperature where measurements are most conveniently made. Singer and Garito (1981) developed an infinite dilution extrapolation method for determining β_z of a solute molecule. The method was shown to minimize the effects of solute–solute and solute–solvent interactions.

In applying the method the concentration dependence of Γ, the dielectric constant ε and the refractive index must be measured. The hyperpolarizability was shown, for Onsager local fields, to be related to these quantities by

$$\frac{N_A \mu_1 \beta_1}{5kT} = \frac{(2\varepsilon_0 + n_1^2)(2n_0^2 + n_1^2)^3 M_1}{(n_1^2 + 2)^4 n_0^6 \varepsilon_0} \left\{ v_0 \left. \frac{\partial \Gamma_L}{\partial w} \right|_0 + \Gamma_0 \left. \frac{\partial v}{\partial w} \right|_0 \right.$$
$$\left. + v_0 \Gamma_0 - v_0 \Gamma_0 \left[\frac{1}{n_0^2} \left. \frac{\partial n^2}{\partial w} \right|_0 + \left(\frac{1}{\varepsilon_0} - \frac{2}{2\varepsilon_0 + n_0^2} \right) \left. \frac{\partial E}{\partial w} \right|_0 \right] \right\} \tag{6.27}$$

where the subscript 0 refers to the solvent and the subscript 1 refers to the solute. Values for n_0, ε_0, and Γ_0 are obtained from the plots of the appropriate concentration dependences. The quantity w is the weight fraction, v is the specific volume of the solute, and the partial derivatives are determined from the slopes of the appropriate quantities at infinite dilution. This method is now widely used for obtaining the most accurate experimental values of β.

6.2 SOLVATOCHROMIC MEASUREMENTS

The two-level model was developed prior to more sophisticated computational methods to obtain qualitative aspects of the relationship of molecular structure of β in organic π systems having charge-transfer resonances (Oudar and Chemla 1977, Levine 1976). In such systems the nonlinear polarization can be visualized as arising from contributions from low-lying charge-transfer states which lead to asymmetry in the π-electron cloud and σ skeleton of the molecules as well as from field-induced mixing of excited-state polar character into the ground state. This method and the approximations involved are discussed in Chapter 3. Here we describe a useful approach for obtaining the parameters of the two-level model from relatively simple spectroscopic measurements and see that the results give intuitively correct values of β for certain classes of molecules as long as certain guidelines are followed for its applicability.

From equation 3.33, approximating the hyperpolarizability of a molecule as arising from the charge-transfer contributions from the two-level model, we have

$$\beta = \frac{3e^2}{2\hbar m} \frac{\omega_{eg} f \Delta\mu}{(\omega_{eg}^2 - \omega^2)(\omega_{eg}^2 - 4\omega^2)} \tag{6.28}$$

where f is the oscillator strength, $\Delta\mu$ is the difference between excited- and ground-state dipole moments, and ω_{eg} and ω are the frequencies of the charge-transfer resonances and excitation wavelength. The oscillator strength is related to the transition dipole moment r_{eg} by

$$r_{eg} = \sqrt{2.13 \times 10^{-30} v_m f} \quad (\text{esu·cm}) \qquad (6.29)$$

where v_m is the energy in cm^{-1} at the peak of the absorption band. The oscillator strength is determined from the absorption spectrum by the following relation:

$$f = 4.381 \times 10^{-9} \int \varepsilon_v dv \qquad (6.30)$$

where ε_v is molar excitation coefficient and the integration is carried out over the absorption band.

DeMartino et al. (1988) used solvatochromic measurements to evaluate the dipole moment difference $\Delta\mu$ between ground and excited states. In this method the optical absorption spectrum of the molecule of interest is measured in solvents of different polarity (as determined by their dielectric constants). The spectral shifts are related to $\Delta\mu$ and solvent dielectric constants ε by the following expression:

$$\Delta v = \frac{\mu_g}{l^3} \Delta\mu \frac{\varepsilon - 1}{\varepsilon - 2} \qquad (6.31)$$

where l is the molecular length and μ_g is the grround-state dipole moment. An

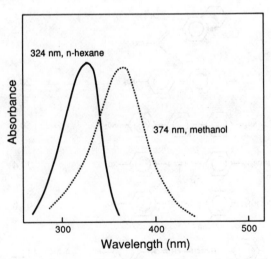

Figure 6.4 Salvotochromic shift of optical absorption band of 2-methyl-4-nitroaniline. After DeMartino et al. (1988).

example of the solvatochromic shift for 2-methyl-4-nitroaniline in two different solvents is shown in Figure 6.4. Thus, it is possible to use solvatochromic shifts (6.31), (6.29), and (6.28) to calculate values of β.

Due to the approximate nature of the two-level model as well as the complexities of specific solute–solvent interactions as pointed out by Levine and Bethea (1975), great care must be exercised in applying and interpreting results obtained from this method. DeMartino et al. (1988) pointed out that the most meaningful results are obtained for linear conjugated systems with uninterrupted conjugation. Using the method a variety of features of the relationship of molecular structure to β were illustrated and are shown in Table 6.1.

Relative to the reference compound p-nitroaniline, the effects of increasing acceptor strength *(ii)*, increasing planarity *(iii)*, and increasing conjugation length *(iv)* can be readily seen. While this method is by no means a substitute for the EFISH measurement in terms of its accuracy and reliability, it may

TABLE 6.1 Molecular Structural Control of β

	Structure	β at 1.9 μm ($\times 10^{-30}$ esu)	Comment
i		5.7	Standard
ii		21.4	Group
iii		41.8	Group
iv		20.1	Length
v		50.7	Length
vi		23.4	Planarity
vii		61.6	Planarity
viii		111.2	All

Source: DeMartino et al. (1988).

serve as a synthetic guideline and screening method to point the way to useful compounds.

6.3   METHODS FOR MEASURING $\chi^{(2)}$

A variety of methods, both absolute and relative, have been developed for measuring the tensor elements of $\chi^{(2)}$. They are reviewed thoroughly by Kurtz (1975) and are described only briefly here. The absolute methods include the phase-matched method, parametric fluorescence, and Raman scattering. The first two methods have accuracies of about $\pm 10\%$ for $\chi^{(2)}$. The Raman technique is probably as accurate, but can be used to determine the sign of the tensor elements in certain crystals.

In the phase-matched method the problem of small signals is avoided by choosing a fundamental and second-harmonic wavelength such that the wave vectors of the forced harmonic wave with wave vector $2k_\omega = 2(\omega/c)n_\omega$ and the free harmonic wave with wave vector $k_{2\omega} = (2\omega/c)n_{2\omega}$ have the same value. The effective interaction length of the waves increases from a few micrometers to distances exceeding several centimeters. Extremely careful consideration of a number of geometrical and optical factors is required to apply this method. Phase-matched second-harmonic generation is discussed thoroughly in Chapter 5.

The parametric fluorescence method (Byer and Harris 1968) uses the spontaneous photons emitted during a parametric down-conversion process to make an absolute measurement of the susceptibility

$$h v_p = h v_s + h v_i \tag{6.32}$$

In this process a higher-frequency pump wave at v_p gives up energy to photon fields at v_i and v_s. This will occur only for combinations of photon frequencies where the three wave vectors satisfy the conditions

$$k_p = k_s + k_i \tag{6.33}$$

Less stringent requirements on beam divergence and crystal quality and the linear relation between signal and pump powers make this method more useful than the phase-matched method. Zyss and Chemla (1987) report extensively on the use of this technique to characterize organic crystals and fabricate near infrared signal characterization devices.

The Raman scattering method, which was demonstrated by Faust and Henry (1966), relies on electric fields generated by certain lattice vibrations (longitudinal optical modes, ω_{LO}) in piezoelectric crystals to modulate electronic susceptibility. As a result of this process, incident light at frequency ω_L is Raman shifted to produce new frequencies $\omega_L \pm \omega_{LO}$. The effect vanishes in centrosymmetric crystals. The important point about this method is that it enables determination of the absolute sign of the nonlinear coefficient.

One of the most useful methods for determining the value of nonlinear coefficients is the Maker fringe method (Maker et al. 1962). It is a relative method in that the second-harmonic power generated in the sample can be used to obtain a value for a nonlinear coefficient $\chi_{IJK}^{(2)}$ only if it can be compared to that from a crystal with known $\chi_{IJK}^{(2)}$. In this method a plane parallel slab of material is rotated about an axis perpendicular to the laser beam, giving rise to a fringe pattern. The fringes are caused by the angular dependence of the phase mismatch Δk between the forced and harmonic waves.

$$\Delta k = 2k_\omega - k_{2\omega} = \frac{4\pi}{\lambda(n_\omega \cos \theta'_\omega - n_{2\omega} \cos \theta'_{2\omega})} \tag{6.34}$$

where θ'_ω and $\theta'_{2\omega}$ are the angles of refraction at the two frequencies. The relationship between these quantities can be understood by referring to Figure 6.5.

Fringes occur due primarily to the angular dependence of Δk, although it must be born in mind that l_c, the coherence length, is also angular dependent. This is in contrast to the closely related wedge method described for EFISH measurements, where the coherence length remained constant and fringes were associated with variations in the path length. The more straightforward analysis associated with the wedge method results in more accurate values for l_c. From a careful analysis of the fringe patterns, however, it is possible to obtain the coherence length at various angles. The harmonic power varies with angle as

$$P_{2\omega} = I_m(\theta) \sin^2 \psi \tag{6.35}$$

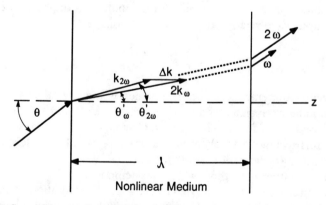

Figure 6.5 Second-harmonic generation for nonnormal incidence in an isotropic crystal illustrating that the wave vector mismatch is normal to the incident boundary. After Kurtz (1975).

where

$$\psi = \frac{\pi}{2} \frac{l}{l_c(\theta)} \qquad (6.36)$$

and l is the thickness of the slab. The angularly dependent coherence length is given by

$$l_c(\theta) = \frac{\lambda}{4(n_\omega \cos \theta_\omega - n_{2\omega} \cos \theta_{2\omega})} \qquad (6.37)$$

The quantity $I_m(\theta)$ is an envelope function and

$$I_m(\theta) \propto (\chi_{\text{eff}}^{(2)})^2 P_\omega^2 \qquad (6.38)$$

where $\chi_{\text{eff}}^{(2)}$ is the effective nonlinear coefficient and P_ω the incident power. The proportionality constant contains reflection and transmission coefficients and other optical factors for determining the magnitude of the optical fields inside the crystal. A thorough discussion of these factors is provided by Kurtz (1975) and is outside the scope of this chapter. Here we simply describe the physical origin of the fringes and note the similarity and distinctions with respect to the wedge method.

Because of the interest in organic thin films we briefly discuss the determination of the second-order nonlinear coefficients in uniaxial media, such as poled polymers. This problem was first treated for films poled parallel to the plane of the film by Meredith et al. (1982). Since poling perpendicular to the plane of the films has proved to be easier we will discuss this situation in some detail. The symmetry of the medium in this case is ∞mm and assuming Kleinman symmetry there are two independent nonlinear coefficients $d_\parallel$ and $d_\perp$ where $\parallel$ and $\perp$ designate parallel and perpendicular to the poling direction, respectively. A more detailed discussion of the second-order nonlinear tensor and its evaluation for ∞mm symmetry is presented later in this chapter on the determination of electrooptic coefficients.

The conversion efficiency to the second-harmonic frequency in a thin uniaxial film (thin relative to the coherence length) on a glass substrate is given by (Maker et al. 1962, Jerphagnon and Kurtz 1970, Singer et al. 1987b) as

$$\frac{I(2\omega)}{I(\omega)} = \frac{(128\pi^5) t_\omega^4 T_{2\omega} d^2 (t^0)^2 p^2 I(\omega) l^2}{1 - [4\sin^2 \theta / (n_\omega + n_{2\omega})]} \qquad (6.39)$$

where l is the film thickness, I the light intensity, θ the incidence angle, p an angular factor that projects the nonlinear susceptibility tensor onto a coordinate frame defined by the propagating electric field, t_ω and $T_{2\omega}$ are Fresnel-like transmission factors, and t^0 is the transmission factor for the second-harmonic through the glass substrate on which the film is cast. The factor $T_{2\omega}$ is given by

$$T_{2\omega} = 2n_{2\omega} \cos \theta_{2\omega} \frac{(n_\omega \cos \theta + \cos \theta_\omega)(n_{2\omega} \cos \theta_\omega + n_\omega \cos \theta_{2\omega})}{(n_{2\omega} \cos \theta_{2\omega} + \cos \theta)(n_{2\omega} \cos \theta_{2\omega} + n_0 \cos \theta_0)^2} \quad (6.40)$$

where θ_ω, $\theta_{2\omega}$, and θ are given by

$$n_\omega \sin \theta' = \sin \theta \qquad n_{2\omega} \sin \theta_{2\omega} = \sin \theta \qquad n_0 \sin \theta_0 = n_{2\omega} \sin \theta_{2\omega}$$

where the prime refers to angle of refraction of the fundamental beam in the film and n_0 is the refractive index of glass. The factor t_0 is given by

$$t_0 = \frac{2n_0 \cos \theta_0}{\cos \theta_0 + n_0 \cos \theta} \quad (6.41)$$

In a uniaxial medium the expressions for p-polarized second-harmonic power, $p_\parallel$, for s and p polarized incident light are given by

$$P_{\parallel(s \, incidence)} = d_\perp E_s^2 \quad (6.42)$$

and

$$P_{\parallel(p \, incidence)} = d_\parallel E_p^s \quad (6.43)$$

Measurements of p polarized harmonic intensity for the two input polarizations combined with (6.39) and the appropriate values of t_ω and p (projection factor) allow independent determination of $d_\parallel$ and $d_\perp$ (Singer et al. 1987b). For p polarized incident light and p polarized harmonic the factors t_ω and p

$$t_\omega = \frac{2 \cos \theta}{n_\omega \cos \theta + \cos \theta_\omega} \quad (6.44)$$

and

$$p = (\tfrac{1}{3} \cos^2 \theta_\omega + \sin^2 \theta_\omega) \sin \theta_{2\omega} + \tfrac{2}{3} \cos \theta_\omega \sin \theta_\omega \cos \theta_{2\omega} \quad (6.45)$$

For s polarized incident light and p polarized harmonic

$$t_\omega = \frac{2 \cos \theta}{n_\omega \cos \theta + \cos \theta} \quad (6.46)$$

and

$$p = 2 \sin \theta_\omega \cos \theta_\omega \cos \theta_{2\omega} \quad (6.47)$$

The expressions for p predict that the second-harmonic intensity should be zero for $\theta = 0$ and $90°$ and reach a maximum in the vicinity of $50°$. This behavior has been observed experimentally by Singer et al. (1987b) for films consisting of a dye, disperse red 1, dissolved in a polymethyl methacrylate host and has now been observed in a variety of uniaxial films. Using this method d values can be obtained in a straightforward fashion.

6.4   KURTZ POWDER TECHNIQUE

A widely used method for screening materials to determine second-harmonic generation activity is the powder technique developed by Kurtz and Perry (1968). This technique derives information concerning angular averages of second-order nonlinear tensor components, coherence lengths, and phase-matching behavior. It is of most value as a screening technique to identify materials with noncentrosymmetric crystal structures and the capability for phase matching. Taking other factors into account it might be used as a guideline for single-crystal growth.

In this method, crystalline materials are reduced to powders and the particles are classified into particles sizes. The particles are packed into a thin cell of thickness l. Addition of an index matching fluid can help reduce scattered light and put the experiment on a more quantitative footing. The sample cell is irradiated with a fundamental beam of diameter D and the harmonic intensity in the forward direction is measured with a photomultiplier tube. The results are compared with a signal from a quartz reference standard. The cell and beam geometries are chosen so that a large number of particles relative to the average particle size ($\hat{r}$) are traversed by the beam. This implies that $\hat{r} \ll l \ll D$.

Kurtz and Perry developed a semiquantitative theory of powders to explain the various behaviors. In the case of very fine powders where $\hat{r}$ is less than the average coherence length $\hat{l}_c$, the second-harmonic fields from the different particles are correlated with each other, provided that the particles are separated by a distance less than l_c. Under these circumstances the total intensity generated in the sample is given by

$$I \propto \frac{\hat{r}}{\hat{l}_c^2} \langle (d^{2\omega})^2 \rangle \tag{6.48}$$

From the equation it is clear that the harmonic intensity should increase as the particle size increases. Equation 6.48 was specifically derived for the non-phase-matched case. For the phase-matched case a detailed solution in this region is not available, but arguments can be made that $I_{ext}^{(2\omega)}$ should increase with increasing particle size. This behavior is illustrated in the left side of Figure 6.6. It is also clear that unless the particle size and coherence length are known, nothing can be concluded relative to the average nonlinear coefficient of the material.

In the case where $\hat{r}$ is much greater than the coherence length l_c, the situation becomes significantly different for the non-phase-matched and phase-matched cases. For the non-phase-matched case the total second-harmonic intensity is

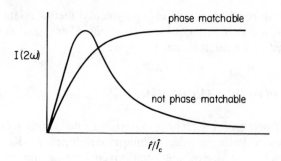

Figure 6.6   Dependence of the second-harmonic intensity on the dimensionless quantity $\hat{r}/\hat{l}_c$, where $\hat{r}$ is the average particle size and $\hat{l}_c$ is the average coherence length. After Kurtz and Perry (1968).

given by

$$I_{(2\omega)} \propto I_{(\omega)}(d^{2\omega})^2 l \frac{\hat{l}_c^2}{\hat{r}} \sin^2 \frac{\frac{1}{2}\pi\hat{r}}{l_c} \qquad (6.49)$$

Here the observed harmonic intensity is inversely dependent of particle size and goes to zero for very large particles. Since coherence lengths are typically 10–$20\,\mu m$, large means $>100\,\mu m$ and small means $<1\,\mu m$. Once again a knowledge of particle size is needed to determine whether a sample is described by this behavior.

If a material is phase matchable the physical situation is quite different. The major fraction of the angular dependence is now associated with coherence length rather than the average nonlinear coefficient. Here the second-harmonic intensity is given by

$$I_{(2\omega)} \propto I_{(\omega)}(d_{pm}^{2\omega})^2 l \qquad (6.50)$$

where $d_{pm}^{2\omega}$ is the value of the phase-matchable d coefficient.

The point to note here is that the second-harmonic intensity is particle size independent, as shown in Figure 6.6. For samples of this type a reasonable correlation exists between the phase-matchable coefficient and the observed harmonic coefficient. If all other experimental conditions are kept the same, the harmonic intensity observed under these conditions should provide a reasonable ranking among samples relative to d_{pm}.

6.5   ELECTROOPTIC MEASUREMENTS

A number of schemes have been discussed in the literature for measuring the electrooptic properties of second-order nonlinear materials. In this section, a general discussion of measurements of the electrooptic coefficient r_{ij} for poled polymer films, where special complexities arise, is given.

The linear electrooptic effect was discussed extensively by Yariv (1976) and more recently with special consideration to organic crystals by Singer et al. (1987a). Here we briefly review the essential concepts.

The propagation of a beam through a crystal or other anisotropic medium is governed by the requirement (or Maxwell's equations) that the electric field can be decomposed into two mutually orthogonal linearly polarized electric field components. These two components will generally be subjected to different refractive index values leading to birefringence. The relationship between the propagating components can be understood by considering the refractive index ellipsoid represented by equation 5.11 and illustrated in Figure 6.7.

The linear electrooptic effect in general produces a change in the refractive index of the medium resulting in a change of shape or rotation of the ellipse with new effective values of the refractive index and the introduction of off-diagonal components, which modifies equation 5.11 to

$$\left(\frac{1}{n^2}\right)_1 x^2 + \left(\frac{1}{n^2}\right)_2 y^2 + \left(\frac{1}{n^2}\right)_3 z^2 + 2\left(\frac{1}{n^2}\right)_4 yz + 2\left(\frac{1}{n^2}\right)_5 xz + 2\left(\frac{1}{n^2}\right)_6 xy = 1$$

$$(6.51)$$

The relationship between the new and old values of the indices is given by

$$\Delta\left(\frac{1}{n^2}\right)_i = \sum_{j=1}^{3} r_{ij}E_j \qquad \text{for } i = 1\text{--}6 \qquad (6.52)$$

The quantity r_{ij} is a contracted form of notation in the index i (by convention)

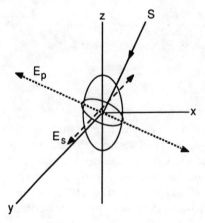

Figure 6.7 The refractive index ellipsoid for a unixal medium where s indicates the direction of an incident light beam of arbitrary polarization. The semimajor and semiminor axes of the ellipse formed by the intersection of the refractive index ellipsoid with a plane perpendicular to S define the allowed polarization directions for light propagating through the medium.

and has the following relationship to the uncontracted indices:

$$
\begin{vmatrix} r_{11j} \\ r_{22j} \\ r_{33j} \\ r_{23j} \\ r_{13j} \\ r_{12j} \end{vmatrix} = \begin{vmatrix} r_{1j} \\ r_{2j} \\ r_{3j} \\ r_{4j} \\ r_{5j} \\ r_{6j} \end{vmatrix} \tag{6.53}
$$

In the absence of a field, (6.51) reverts to (5.11). The matrix r_{ij} is subject to the same symmetry constraints as the second-order nonlinear coefficient and is directly related to it. The general expression relating the components of the electrooptic tensor to those of the $\chi^{(2)}$ tensor in noncontracted notation (cgs units) is

$$
r_{ij} = -\frac{8\pi\chi^{(2)}_{IJK}}{n_I^2 n_J^2} \tag{6.54}
$$

(See Section 5.4.1 for a derivation of this relationship.)

It is useful to write out the matrix version of the equation to appreciate the contracted notation.

$$
\begin{vmatrix} \Delta(1/n^2)_1 \\ \Delta(1/n^2)_2 \\ \Delta(1/n^2)_3 \\ \Delta(1/n^2)_4 \\ \Delta(1/n^2)_5 \\ \Delta(1/n^2)_6 \end{vmatrix} = \begin{vmatrix} r_{11} r_{12} r_{13} \\ r_{21} r_{22} r_{23} \\ r_{31} r_{32} r_{33} \\ r_{41} r_{42} r_{43} \\ r_{51} r_{52} r_{53} \\ r_{61} r_{62} r_{63} \end{vmatrix} \begin{vmatrix} E_1 \\ E_2 \\ E_3 \end{vmatrix} \tag{6.55}
$$

From the rules of matrix multiplication we see, for instance, that

$$
\Delta\left(\frac{1}{n^2}\right)_4 = r_{41}E_1 + r_{42}E_2 + r_{43}E_3 \tag{6.56}
$$

Assuming that one of the matrix elements is nonzero, the term $(1/n^2)_4$ would contribute to the rotation of the ellipsoid in (6.51). A testing of the nonzero and independent electrooptic coefficients for the various polar point groups is given by Yariv (1976). A discussion of the electrooptic tensor for several specific crystals can be found in Yariv (1976). Lipscomb et al. (1981) discuss measurements of the various tensor components of the polar monoclinic crystal

2-methyl-4-nitroaniline. Because of the current importance and interest in the properties of poled polymer films we discuss the electrooptic coefficient and its components in some detail and methods for evaluating the components.

When an amorphous polymer film containing dipolar species is subjected to a strong electric field, the average orientation of dipolar species is perturbed and a polar axis (parallel to the poling direction) is induced. This is an infinite rotational axis with an infinite number of mirror planes. We designate this as $\infty mm\,(C_{\infty v})$. The nonzero and equivalent electrooptic tensor elements are similar to those for $4mm$, $6mm$. The electrooptic tensor has the following form.

$$
\begin{vmatrix}
0 & 0 & r_{13} \\
0 & 0 & r_{23} \\
0 & 0 & r_{33} \\
0 & r_{42} & 0 \\
r_{51} & 0 & 0 \\
0 & 0 & 0
\end{vmatrix}
\tag{6.57}
$$

Because of the symmetry of the medium, two pairs of indices are equivalent: $r_{13} = r_{23}$ and $r_{42} = r_{51}$. In fact, if we assume a dispersionless electrooptic medium, which has been shown to be a reasonable one for organic media (Lipscomb et al. 1981), then Kleinman symmetry holds and $r_{13} = r_{23} = r_{42} = r_{51}$. In other words, we now have two independent components of the electrooptic $r_{\parallel} = r_{33}$ and $r_{\perp} = r_{13}$, which are parallel and perpendicular to the polar axis, respectively.

Now consider the application of an electric field along the z axis. Multiplying the tensor (6.55) by the electric field in the $3(z)$ direction gives

$$
\Delta\left(\frac{1}{n^2}\right)_1 = \Delta\left(\frac{1}{n_o^2}\right) = r_{13}E_3
$$

$$
\Delta\left(\frac{1}{n^2}\right)_3 = \Delta\left(\frac{1}{n_e^2}\right) = r_{33}E_3
$$

$$\tag{5.58}$$

The notation n_o and n_e are adopted according to convention and refer to the ordinary and extraordinary components of the refractive index ellipsoid described by (6.51). For a uniaxial medium n_e is parallel to the uniaxis and n_o is perpendicular to it. This terminology arises because of a point made at the beginning of this section. An incident light beam with arbitrary direction propagates through a medium with two orthogonal polarizations components E_p and E_s. The propagation of E_s is governed by n_o and E_p is governed by $n_e(\theta)$, which is, in turn, a function of both n_o and n_e. For the reasons stated in equation (5.11) and the discussion that follows, equation (6.51) in the absense of a field is usually written as

$$\frac{x^2 + y^2}{n_o^2} + \frac{z^2}{n_e^2} = 1 \tag{6.59}$$

With application of a field along the z direction the changes in n_o and n_e are given by

$$\Delta\left(\frac{1}{n_o^2}\right) = r_{13}E_3 = r_\perp E_\parallel \tag{6.60}$$

$$\Delta\left(\frac{1}{n_e^2}\right) = r_{33}E_3 = r_\parallel E_\parallel \tag{6.61}$$

The ellipsoid now becomes

$$\left(\frac{1}{n_o^2} + r_\perp E_\parallel\right)x^2 + \left(\frac{1}{n_o^2} + r_\perp E_\parallel\right)y^2 + \left(\frac{1}{n_e^2} + r_\parallel E_\parallel\right)z^2 = 1 \tag{6.62}$$

The interpretation is unusually simple for this combination of field direction and symmetry. The ellipsoid changes ellipticity as indicated by the length changes of the semimajor and semiminor axes. Since no off-diagonal components appear, the ellipse was not rotated.

Applying the field in the direction perpendicular to the uniaxis creates a somewhat different situation. Applying the field in the y direction gives

$$\Delta\left(\frac{1}{n^2}\right)_4 = r_{42}E_y = r_\perp E_\perp = r\left(\frac{1}{n^2}\right)_5 \tag{6.63}$$

The equation for the ellipsoid now becomes

$$\frac{x^2 + y^2}{n_o^2} + \frac{z^2}{n_e^2} + r_\perp E_\perp xz = 1 \tag{6.64}$$

From inspection it is clear that rotation about the y axis to a new coordinate system with

$$y = y'$$
$$x = x' \cos 45 - z' \sin 45 \tag{6.65}$$
$$z = x' \sin 45 + z' \cos 45$$

will diagonalize the equation for the ellipsoid. Making the substitution we have

$$\left(\frac{1}{n_o^2} + r_\perp E_\perp\right)x'^2 + \frac{1}{n_o^2}y'^2 + \left(\frac{1}{n_e^2} - r_\perp E_\perp\right)z'^2 = 1 \tag{6.66}$$

The ellipsoid in this case is rotated and no longer uniaxial in character. Persuing the analysis further, we can obtain the refractive indices n_x' and n_y' in the new coordinate system. The starting point here is the expression

$$\left(\frac{1}{n_{x'}^2}\right) = \frac{1}{n_o^2} + r_\perp E_\perp \tag{6.67}$$

Making the assumption that $r_\perp E_\perp \ll (1/n_o^2)$ we can write

$$\Delta\left(\frac{1}{n^2}\right) = \frac{1}{n_{x'}^2} - \frac{1}{n_o^2} = \frac{n_o^2 - n_{x'}^2}{n_{x'}^2 n_o^2} \sim \frac{n_o^2 - n_{x'}^2}{n_o^4} \sim \frac{2(n_o - n_{x'})}{n_o^3} \tag{6.68}$$

Combining this result with (6.66) gives

$$n_{x'} = n_o - \frac{n_o^3 r_\perp E_\perp}{2} \tag{6.69}$$

and

$$n_{y'} = n_y \tag{6.70}$$

For the previous case where the dc field was applied along the polar axis the values of the refractive indices are

$$n_{x'} = n_o + \frac{n_o^3 r_\perp E_\parallel}{2} \tag{6.71}$$

$$n_{y'} = n_o + \frac{n_o^3 r_\perp E_\parallel}{2} \tag{6.72}$$

$$n_{z'} = n_e + \frac{n_e^2 r_\parallel E_\parallel}{2} \tag{6.73}$$

Measurement of the electrooptic coefficients in polymer films can be performed by a number of relatively straightforward techniques, some of which require light guiding and others that do not. The general principle is to allow light to propagate over some well-defined distance and direction relative to the polar axis and compare the electric field-induced phase shift for the polarization components relative to light in some path external to the field. These methods

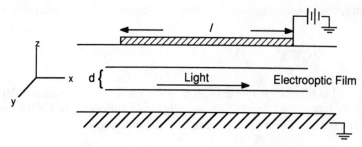

Figure 6.8 A waveguide consisting of a poled polymer with upper and lower cladding layers of low refractive index on a conductive substrate with an upper surface electrode.

require the use of a coherent light source with control over the polarization and a method for analyzing the amount of phase shift induced by the applied field.

Waveguide methods, discussed in detail in Chapter 11, are particularly useful because they provide a wealth of useful information regarding film quality, coupling efficiency, and absorption losses. Also, since light can be propagated over long distances, relatively low voltages produce large easily measured phase shifts. In a typical geometry (Figure 6.8) light coupled into a poled polymer waveguide of thickness d propagates with its wave vector in the x direction beneath an electrode of length l. Various modes or electromagnetic field distributions are allowed by Maxwell's equation to propagate in the waveguide. These features are discussed in Chapter 11.

The designation TE and TM refer to transverse electric and transverse magnetic modes (see Chapter 11 for details). Consideration of the lowest order TE and TM modes shows that $r_\perp$ is relatively easily determined by this experiment. For TE modes, the electric vector is polarized in the plane of the film (along the y axis). Application of a field along the z direction produces a refractive index change, which from (6.72) is

$$\Delta n_\perp = n_{y'} - n_o = \frac{n_o^3 r_\perp}{2} E_z \qquad (6.74)$$

The phase shift associated with this refractive index change over the length of the electrode is given by

$$\Delta \phi_\perp = \frac{\omega l}{c} \Delta n_\perp \qquad (6.75)$$

$$\Delta \phi_\perp = \frac{\omega l}{c} n_o^3 r_\perp E_z \qquad (6.76)$$

In the TM mode the electric field is polarized in the xz plane so that an effective electrooptic coefficient with contributions from $r_\perp$ and $r_\parallel$ governs the phase retardation. Since $r_\perp$ and $r_\parallel$ are not independent quantities, as is discussed in Section 5.3.2, a knowledge of $r_\perp$ is sufficient to infer $r_\parallel$. In the case of modest degrees of poling $r_\perp = \frac{1}{3}r_\parallel$.

One convenient method for measuring the phase shift is to recombine the outcoupled beam from the waveguide and allow it to interfere with a reference beam to produce a fringe pattern. The voltage required to bring a fringe from maximum brightness to maximum darkness corresponds to $\Delta\phi = \pi$, and with knowledge of n_o the coefficient can be computed.

A simpler method for measuring electrooptic coefficients in poled films involves the use of an ellipsometer. In this technique a multilayer structure is formed with the poled polymer on a metallic reflector, and a thin semitransparent conductor is used as the second electrode. In practice, the polymer film is cast on an indium tin oxide-coated glass slide and a metal electrode is evaporated onto the surface of the film. In the experiment the electric field dependence of the ratio of s to p polarized light is measured and from that $r_\perp$ and $r_\parallel$ can be extracted. This method is particularly convenient since simple structures without the need for buffer layers can be used and the measurement is suitable for screening and process development studies (Schildkraudt in press). The ellipsometric constant ψ and Δ are related to the reflectance R_p of p and R_s of s polarized light by

$$\frac{R_p}{R_s} = \tan\psi e^{i\Delta} \tag{6.77}$$

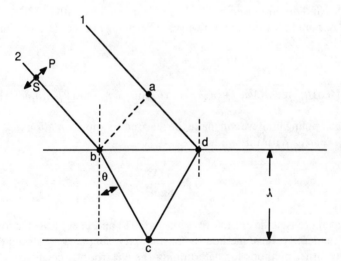

Figure 6.9 Schematic illustration of two parallel incident light waves into a nonlinear optical medium. The phase difference between the two rays at point d is the basis for determination of the electrooptic coefficient as described in the text.

where ψ is the relative attenuation in the film and Δ the phase shift (McCrakin et al. 1963). In (6.77) only Δ is sensitive to the electrooptic coefficient of the film. Designating R_p^0, R_s^0, and Δ^0 as the values of the parameters at zero field and assuming changes in ψ due to the electric field are small we have

$$\frac{R_p/R_p^0}{R_s/R_s^0} = e^{i(\Delta - \Delta^0)} = e^{i\delta\Delta} \tag{6.78}$$

The problem of calculating the phase shift of light in a uniaxial medium was addressed by Dignam et al. (1971). The problem is illustrated in Figure 6.9 where two parallel rays of light (1 and 2) impinge on another surface of a uniaxial film of thickness l. At points a and b the rays had identical phases. Ray 2 accumulates a phase shift due to interaction with the uniaxial dielectric medium relative to ray 1 by the time they interfere at point d. It is clear that the s polarized wave is influenced only by $r_\perp$, whereas the p polarized wave interacts with both $r_\perp$ and $r_\parallel$. Schildkraudt (in press) has incorporated these phase shifts into the expressions for R_p, R_p^0, R_s, and R_s^0. The result for a field in the uniaxial direction $E_\parallel$ is

$$\frac{R_p}{R_p^0} = \exp\frac{4\pi}{\lambda l[\cos\theta_p \delta n_\perp + n_\perp/n_\parallel \tan\theta_p \sin\theta_p \delta n_\parallel]} \tag{6.79}$$

$$\frac{R_s}{R_s^0} = \exp\frac{4\pi}{\lambda l[\cos\theta_s \delta n_\perp + \tan\theta_s \sin\theta_s \delta n_\perp]} \tag{6.80}$$

where $\delta n_\perp$ and $\delta n_\parallel$ are given by (6.71)–(6.73) in analogy with (6.72) and θ is the angle of refraction for the s and p waves. The change in the ratio of phase shifts is given by

$$\delta\Delta = \frac{4\pi}{\lambda l[(\cos\theta_p - \cos\theta_s)\delta n_\perp + (n_\perp/n_\parallel \tan\theta_p \sin\theta_p \delta n_\parallel - \tan\theta_s \sin\theta_s \delta n_\perp)]} \tag{6.81}$$

In the weak poling limit we can make the reasonable approximations $n_\perp \sim n_\parallel \sim n$ and $\delta n_\parallel \sim 3\delta n_\perp$. With this we have

$$r_\parallel = \frac{3\delta\Delta\lambda}{8\pi l n^3} \tag{6.82}$$

Note that there is sufficient information in (6.79) and (6.80) to solve for $r_\parallel$ and $r_\perp$ independently. This method is particularly simple and straightforward relative to other methods for determining the electrooptic coefficient. Caution must be used, however, in neglecting interferences arising from partial reflections at surfaces and interfaces, which can lead to systematic errors in r value. Corrections to the simple predicted behavior can be made and a high level of accuracy achieved. For a more detailed discussion of this method the reader is referred to Schildkraudt (in press).

REFERENCES

Bethea, C. G., *Appl. Opt.* **14**, 1447 (1975).

Byer, R. L., and S. E. Harris, *Phys. Rev.* **168**, 1064 (1968).

DeMartino, R. N., E. W. Choe, G. Khanarian, D. Haas, T. Leslie, G. Nelson, J. Stamatoff, D. Stutz, C. C. Teng, and H. Yoon, in P. N. Prasad and D. R. Ulrich (Eds.), *Nonlinear Optical and Electroactive Polymers*, Plenum, New York, 1988, pp. 169–187.

Dignam, M. J., M. Moskovits, and R. W. Stobie, *Trans. Faraday Soc.* **67**, 3306 (1971).

Faust, W. L., and C. H. Henry, *Phys. Rev. Lett.* **17**, 1265 (1966).

Jerphagnon, J., and S. K. Kurtz, *J. Appl. Phys.* **41**, 1667 (1970).

Kurtz, S. W., in H. Rabin and C. L. Tang (Eds.), *Quantum Electronics*, Academic, New York, 1975, p. 238.

Kurtz, S. K., and T. T. Perry, *J. Appl. Phys.* **39**, 3798 (1968).

Levine, B. F., *Chem. Phys. Lett.* **37**, 516 (1976).

Levine, B. F., and C. G. Bethea, *J. Chem. Phys.* **60**, 3856 (1974).

Levine, B. F., and C. G. Bethea, *J. Chem. Phys.* **63**, 2666 (1975).

Lipscomb, G. F., R. S. Narang, and A. F. Garito, *J. Chem. Phys.* **75**, 1509 (1981).

Maker, P. D., R. W. Terhune, M. Nisenoff, and C. M. Savage, *Phys. Rev. Lett.* **8**, 21 (1962).

McCrakin, F. L., E. Passaglia, R. R. Stromberg, and H. L. Steinberg, *J. Res. Natl. Bur. Stand. Sect.* Article 7, 363 (1963).

Meredith, G. R., J. G. Van Dusen, and D. J. Williams, *Macromolecules* **15**, 1385 (1982).

Oudar, J. L. *J. Chem. Phys.* **67**, 446 (1977).

Oudar, J. L., and D. S. Chemla, *J. Chem. Phys.* **66**, 2664 (1977).

Owyoung, A., R. W. Hellwarth, and N. George, *Phys. Rev. B* **5**, 628 (1972).

Sauteret, C., J. P. Hermann, R. Frey, F. Pradere, J. Ducing, R. Baughman, and R. R. Chance *Phys. Rev. Lett.* **36**, 956 (1976).

Schildkraudt, J. S., *Appl. Opt.*, in press.

Singer, K. D., A. F. Garito, *J. Chem. Phys.* **75**, 3572 (1981).

Singer, K. D., S. J. Lalama, J. E. Sohn, and S. D. Small, in D. S. Chemla and J. Zyss (Eds.), *Nonlinear Optical Properties of Organic Molecules and Crystals*, Vol. 1, Academic, Orlando (1987a), pp. 437–468.

Singer, K. D., M. G. Kuzyk, and J. E. Sohn, in P. N. Prasad and D. R. Ulrich (Eds.), *Nonlinear Optical and Electroactive Polymers*, Plenum, New York (1987b), pp. 189–204.

Wilke, V., and W. Schmidt, *Appl. Phys.* **18**, 177 (1979).

Yariv, A., *Introduction to Optical Electronics*, Chapter 9. Holt, Rinehardt, Winston, New York, 1976.

Zyss, J., and D. S. Chemla, in D. S. Chemla and J. Zyss (Eds.), *Nonlinear Optical Properties of Organic Molecules and Crystals*, Vol. 1, Academic, Orlando, FL, 1987, p. 150ff.

7

A SURVEY OF SECOND-ORDER NONLINEAR OPTICAL MATERIALS

7.1 PERSPECTIVE

In previous chapters the semiclassical description of nonlinear optics that accounts for all optical phenomena in condensed media is discussed. The source of the nonlinear polarization in an organic molecule was shown to be accounted for by calculating the response of a molecule to an electric field by several quantum mechanical approximations. The propagation characteristics of the linear and nonlinear polarization and resultant fields were shown to be intimately related to the symmetry of the medium and the relationship of the microscopic to macroscopic framework.

A considerable amount of research has taken place to synthesize and characterize new molecules for second-order nonlinear optical applications. Structure property relationships have been explored and a reasonably well-understood physical organic framework now exists for accounting for and predicting trends in β as a result of substituent patterns, chromophore proprieties, and spacer lengths between donor and acceptor units in the chromophore. There are still considerable gaps in understanding of the effects of local fields, weak association with diluents and inert species, and molecular association on the value of β. These influences can cause variations in β of greater than $2 \times$ depending on the circumstances of the measurement.

Much effort has gone into the design and preparation of noncentrosymmetric condensed phases where the macroscopic nonlinear properties are expressed. New materials engineering methods are being explored and developed to optimize the value of desired nonlinear tensor components, while simultaneously controlling the wide range of ancillary materials properties required for specific device applications. These properties include transparency, birefringence, refractive index, dielectric constant, thermal, photochemical, and chemical stability, among others.

The main approaches to the development of materials for second-order nonlinear optics have been crystal growth, including inclusion complexes; and growth in confined structures, such as waveguides and fibers; poled polymer systems, including polymer crystal composites; and monolayer assemblies, typified by the Langmuir–Blodgett (LB) approach. Each of the methods has advantages and disadvantages and these will be discussed briefly before proceeding with a more detailed survey of these approaches.

The advantages of organic crystals lie in the highly specific arrangements of molecules and density of packing in the crystalline state. The first advantage can result in the highest levels of bulk nonlinearity, provided that optimized geometrical arrangements and crystal symmetries can be achieved. The high packing density also contributes to the magnitude since $\chi^2 \propto N$, where N is the molecular density. Crystals have the potential for being cut and polished and shaped into useful devices for various applications.

On the minus side, there is very little predictability to organic crystal structures. Minor changes in molecular structure almost always lead to unpredictable results, making the process of systematic structural optimization nearly impossible. It should be pointed out that nitroaromatic amines have been discovered which exhibit nearly optimized structures for that class of molecular species. On the other hand, organic crystals tend to be fragile, difficult to grow in large sizes, and susceptible to degradation from light and ambient environments.

A fruitful approach to the development of materials for second-order nonlinear optical application is the use of poled polymers discussed in Section 4.3. The advantages of the poling approach are numerous. First of all, great flexibility exists in the choice of molecular and polymeric constituents, allowing for systematic design and optimization of materials. Solvent-coated polymers lend themselves well to planar fabrication approaches so that waveguide and integrated optical devices can be fabricated. Variations of this approach involve the formation of molecular aggregates of nonlinear dyes through solvent treatment or doping with noncentrosymmetric crystallites followed by electric field poling.

Disadvantages of this approach include the need for the relatively complex and difficult to control electric field poling process, the dilution of the nonlinear chromophore in the matrix, the broad orientational distribution function, and concerns about long-term stability of the metastable poled state.

An approach that may potentially offer the orientational advantages of the crystalline state, with an enhanced degree of materials design flexibility, utilizes the Langmuir–Blodgett as well as other approaches to monolayer assembly. In the LB method a lattice is constructed layer by layer from either monomeric or polymeric materials of appropriate chemical structure. Rational chemical design principles have resulted in considerable progress toward increasing the quality and stability of films fabricated by these approaches. In principle, these approaches can lead to noncentrosymmetric films with uniaxial symmetry and well-defined thickness. Polymeric LB materials have contributed considerably to film stability and quality. On the other hand, much progress remains to be made in molecular and polymer design and synthesis as well as in deposition processes before this approach constitutes a practical alternative for second-order nonlinear materials.

In this chapter we survey the state of the art in materials development for second-order nonlinear optics. The focus of this survey is to identify trends and structure property relationships rather than to provide exhaustive compilations of molecules and solids where nonlinear optical measurements have been made.

7.2 STRUCTURAL REQUIREMENTS FOR SECOND-ORDER OPTICAL NONLINEARITY

Molecules containing delocalized π electronic systems and asymmetry in their response to an applied electric field tend to exhibit large values of β. The largest values are obtained when the molecules have low-lying charge-transfer resonance states which can be mixed with the ground state by the action of an external field. This process combined with substituent-induced charge asymmetry in the π electronic system is the main requirement for large values of β. A variety of theoretical calculations of β as well as experimental determinations of β or $\mu\beta$ have been made for organic liquids and molecules dissolved in solution. Based on these calculations and measurements, a reasonable qualitative framework has been developed and many molecular π electronic systems have been identified for incorporation into crystals, polymers, or LB films. In this section we summarize the influences of structural trends on β.

Some of the earliest measurements of β were performed on monosubstituted benzenes (Levine and Bethea 1975, Oudar 1977). The equivalent internal field model (EIF) (see Chapter 3.3) was developed to explain the observed trends (Oudar and Chemla 1975). In this model the substituent is postulated to induce an "electromeric" dipole $\Delta\mu$ in the π electronic system and the substituent's influence is expressed in terms of an equivalent field.

The result is a relationship between β, γ, α, and the induced dipole

$$\beta = \frac{3\gamma\Delta\mu}{\alpha} \tag{7.1}$$

derived in Chapter 3. This correlation has been verified by Oudar (1977) for both monosubstituted benzenes and stilbenes and is shown in Figure 7.1 for monosubstituted benzenes. A reasonable correlation appears to have been established between $\Delta\mu$, using the mesomeric moments of Everard and Sutton (1951), and measured values of β. A more sophisticated approach was developed by Morley et al. (1987a), who performed CNDO/2-CI all-valence electron

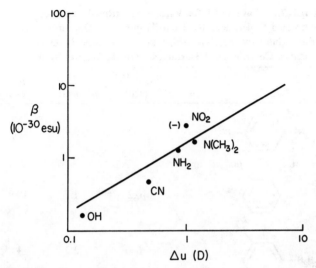

Figure 7.1 Log–log plot of β versus the mesomeric dipole moment for some monosubstituted benzene derivatives (Oudar (1977).

calculations on monosubstituted benzenes and compared values of β_z (the vector component of β) to the most relevant experimental quantity β determined by EFISH. In this study it was found that $N(CH_3)_2$ was by far the strongest electron donating substituent, followed by NH_2 and OCH_3. Among the electron acceptors considered in this series NO is the strongest acceptor, followed by NO_2 and CHO, although experimental numbers were not available to verify this trend.

Of more importance from a practical point of view are the structure property relationships for multiply substituted π-electron structures, where donor and acceptor substituents provide a charge-transfer resonance interaction. An example of such a system is p-nitroaniline:

$$H_2N-\langle\bigcirc\rangle-NO_2$$

where the $-NH_2$ group is the electron donor and $-NO_2$ is the acceptor. The importance of this type of structure is illustrated by the data in Table 7.1. Experimental values for the various isomers of nitroaniline and monosubstituted compounds are compared with numbers obtained from a simple additive model. In the study it was found that β could be separated into two parts:

$$\beta = \beta_{add} + \beta_{CT} \tag{7.2}$$

where β_{add} is the vector sum of the values for the monosubstituted species and

TABLE 7.1 Values of β for Various Substituted Benzenes Determined Experimentally, Contributions to β Determined from Additive Substituent Relationships, and the Charge Transfer Resonance Contribution Determined from the 2-Level Model

	$\beta_{exp} \times 10^{30}$ (esu)	β_{add} (esu)	β_{CT} (esu)
H_2N—⟨benzene⟩—NO_2 (para)	34.5	3.4	19.6
H_2N, NO_2 (ortho) benzene	10.2	1.7	10.9
NO_2 / H_2N (meta) benzene	6	3.3	4
O_2N—⟨benzene⟩	2.2	—	—
⟨benzene⟩—NH_2	1.1	—	—

TABLE 7.2 The Longest Wavelength Optical Transition and Values of β for Various Distributed Benzenes

$$A-\langle benzene \rangle-D$$

λ_{CT} (nm)
[β(esu $\times 10^{30}$)]

A	D			
	CH_3	OCH_3	NH_2	$N(CH_3)_2$
CN	232 [2.92]	247 [4.83]	269 [13.34]	297 [14.24]
CHO		[5.67]		352 [19.5]
$COCH_3$			310 [2.40]	
NO_2	280 [9.12]	314 [17.35]	378 [47.67]	418 [52.75]

β_{CT} is the time-dependent perturbation theory value of β for a simple two-level model for a nonpolar ground state and polar excited state. The expression for β_{CT}, which is explained more fully in Chapter 3, is (Oudar and Chemla 1977)

$$\beta_{CT} = \frac{3\hbar e^2}{2m} \frac{\omega_{eg} f \Delta\mu}{(\omega_{eg}^2 - \omega^2) - (\omega_{eg}^2 - 4\omega^2)} \qquad (7.3)$$

where ω_{eg} is the frequency of the optical transition, f the oscillator strength, and $\Delta\mu$ the difference between ground- and excited-state dipole moment. The trends exhibited by the table are consistent with physical organic notions of the relative strengths of ortho and para resonance interactions and with the small value obtained for the meta derivative.

The effect of variation in the relative strengths of donor and acceptor substituents on β for p-disubstituted benzene derivatives was investigated by Dulcic and Sauteret (1978). Table 7.2 shows measured values of β (EFISH) in brackets and values for the wavelength of the longest wavelength electronic transition. The trends are consistent with donor and acceptor strengths for the various substituents and indicate that within a series where the bridging group is constant, in this case the phenyl ring, an increase in β is accompanied by a red shift in the optical absorption spectra.

The effect of the length of the conjugated linkage between donor and acceptor substituents has been the subject of considerable experimental and theoretical interest. The effect of increasing the conjugation length from that of p-nitroaniline to the corresponding stilbene derivative was shown to have a drastic increase on β (Oudar and LePerson 1975, Oudar 1977), increasing it by more than an order of magnitude. Further experimental work by Dulcic et al. (1981) on the series I showed a quadratic dependence of β on the length of the chain for the range of n examined.

$$(CH_3)_2 - N - \langle\text{benzene}\rangle - (CH=CH)_n - \langle\text{benzene}\rangle - CN$$

I

Not only does the length of the conjugation chain between the donor and acceptor groups critically influence β, but, as might be expected, the subtleties of the electronic structure of the chain does as well. A case in point is illustrated by extending the conjugation length of nitroaniline in several ways. Oudar (1977) compared experimental results and predictions of the two-level model for **II** and **III**. Although **II** has 6 carbon atoms between the donor phenyl ring and the nitrogroup and **III** has only 4, β is approximately 50% larger for **III**. It would therefore appear that the ethylenyl unit is superior to the phenyl ring in transmitting charge from D to A. One can speculate that the formation of a quinonic state in **III** is more facile than in **II** because not as much energy is required to break the aromaticity of the phenyl ring. The impact of increasing conjugation length on β was examined for a dimethylaminonitropolyene series

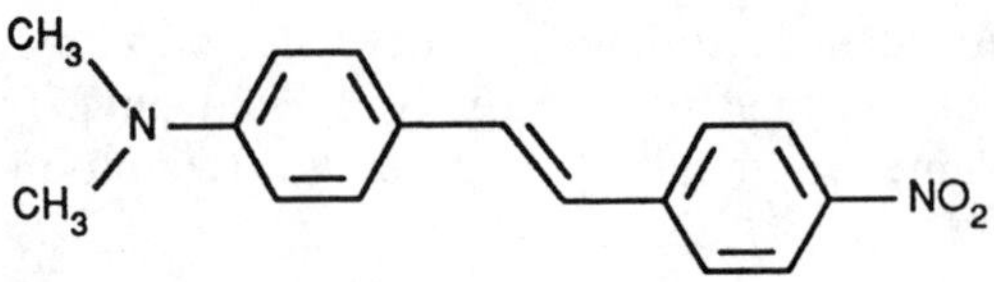

II

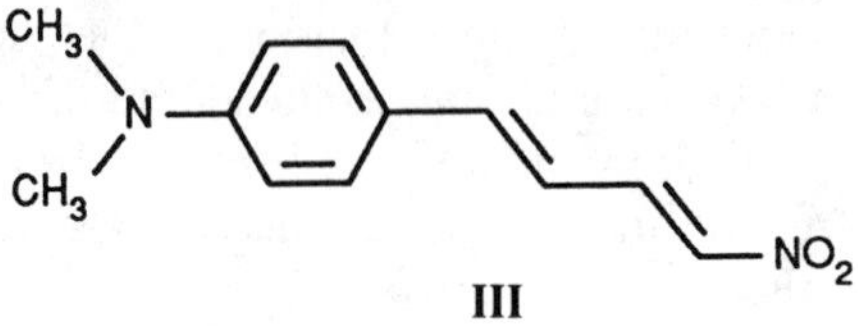

III

$(CH_3)_2N \text{---} (CH = CH)_n \text{ } NO_2$

IV $n = 2 - 22$

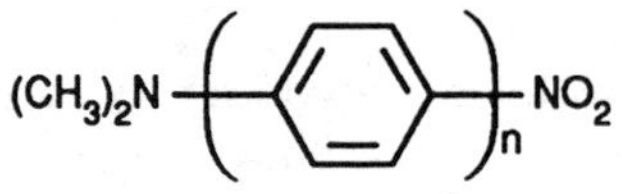

$(CH_3)_2N \text{---} \left(\text{phenyl} \right)_n \text{ } NO_2$

V $n = 1 - 8$

$(CH_3)_2N \text{---} (CH = CH)_n \text{ } CHO$

VI $n = 1 - 6$

IV, a dimethylamino polyenal series **VI**, and a dimethylaminopolyphenyl series **V**, by Morley et al. (1987). In this paper the hyperpolarizability density, ρ_h, and β were computed for series **IV** and **V** and both calculated and measured for **VI**. Several important features were noted. For the polyphenyl series **V** β increased monotonically with increasing value of n but ρ_h reached a maximum between $n = 2$ and 3 and then decreased rapidly with increasing n. In this series molecular volume increases faster than β. This effect may be attributed to ring twisting in the polyphenyl system, which reduces the efficiency of π-electron conjugation from one phenyl ring to another (Berkovic et al. 1987). For series **IV** it was found that β increases faster than molecular volume with increasing n so that ρ_h increased monotonically and saturated in the neighborhood of $n = 20$. The increase in ρ_h is accompanied by a significant red shift in λ_{max} so that for $n = 14$, $\lambda_{max} \sim 830$ nm. Theoretical results were compared with experimental values for series **VI** and a trend similar to that for series **IV** was observed (Morley et al. 1987b). Additional experimental verification of this trend was provided by Barzoukas et al. (1989), who compared a series derived from **VII** and **VIII**.

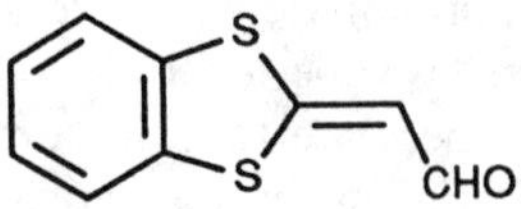

VII

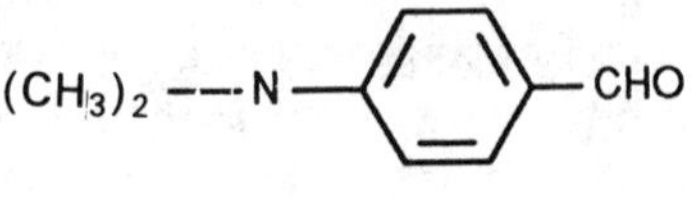

VIII

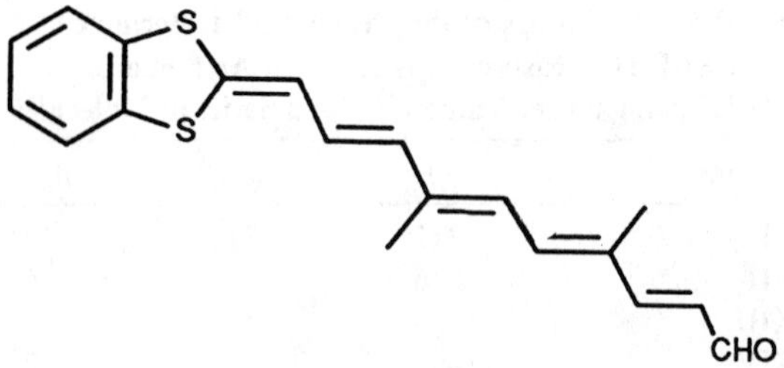

IX

For up to an additional 8 double bonds the susceptibility $\mu\beta$ increased as $\sim n^{2.4}$, with the major part of the increase attributed to increasing β. These works also examined the effect of insertion of a triple bond in the conjugated chain as in **X**. The measured $\mu\beta$ was less than that expected, based on extrapolation for a 6 double-bond chain. This led to the conclusion that the triple bond can be viewed as a potential barrier contributing to some additional diffusion in the electronic wave function. The effect of a triple bond was also noted by Buckley et al. (1986), who compared experimentally determined values of β at $1.9\ \mu$m for **II** and **X**.

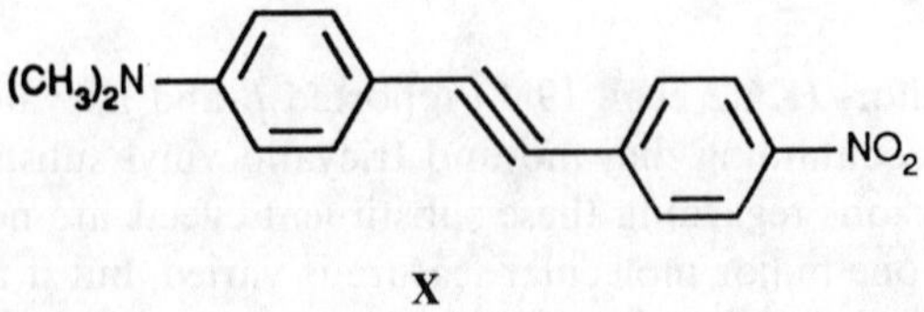

X

Efforts have gone into finding new donor and acceptor substituents that could replace $(CH_3)_2$–N, and $-NO_2$ in p, p'-phenyl and p, p'-stilbene derivatives. Katz et al. (1987) reported measurements of β_0 (the dispersion free value) on phenyl and stilbene derivatives with novel donors and acceptors. Substitution of dicyanovinyl **XI** and tricyanovinyl **XII** for the nitro group in structure **V-1** resulted in an increase in β with an excellent correlation with the Hammett sigma constant σ_R^- for the acceptor substituent (Table 7.3). Replacement of the dimethyl amino group with a dithiolylidenemethyl group **XIII** results in an increase in β far greater than one would except for σ_R^- constant associated with that substituent. The correlation found for the acceptor substituents confirms the validity of the two-level model for that series, whereas the effectiveness of the dithiolylidinemethyl group may have to do primarily with excited- state properties.

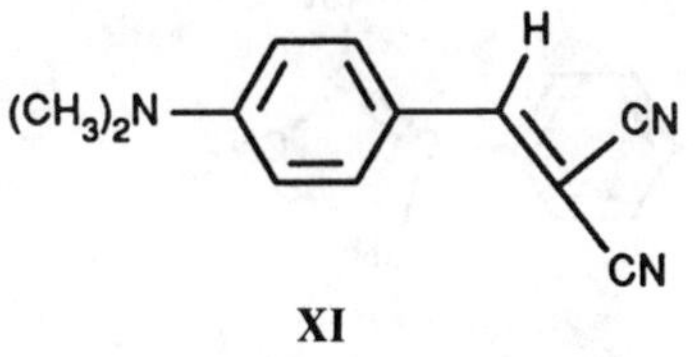

XI

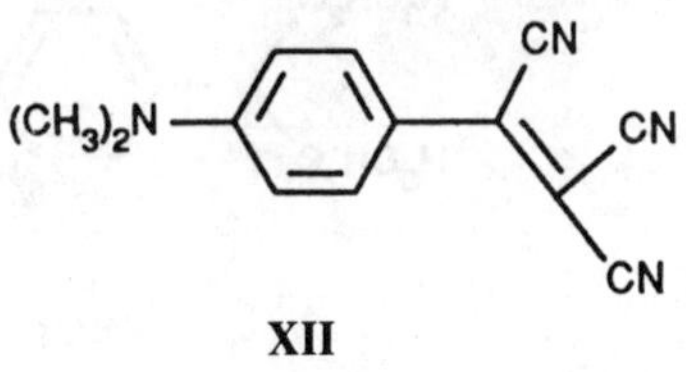

XII

TABLE 7.3 Values of the Product $\mu\beta$ Determined from EFISH Measurements As Well As β and β_0 (the Dispersion Free Value of β) for a Series of Molecules

Compound	$\mu\beta$	β	β_0
XI	271	31	16
XII	846	78	26
XIII	358	52	25
XIV	1090	125	47
XV	2650	323	133
XVI*	4110	390	154

Note. β and β_0 in 10^{-30} esu, μ in debyes, $\lambda = 1.356\,\mu m$, *$\lambda = 1.58\,\mu m$ after Katz et al. (1988).

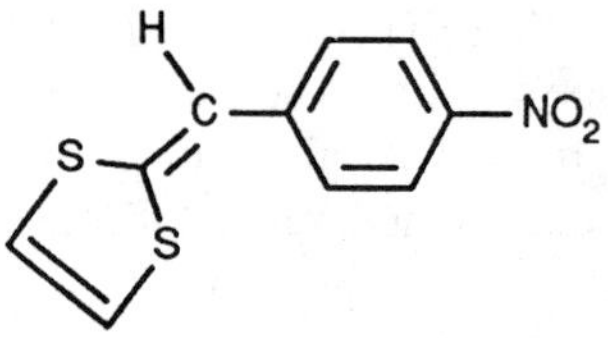

XIII

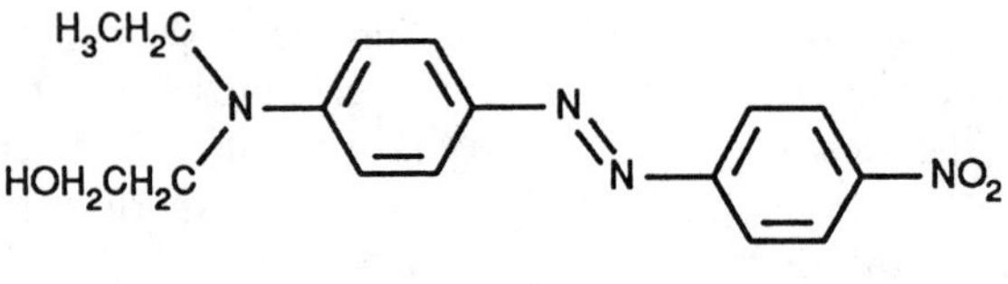

XIV

The same authors (Katz et al. 1988) reported β and $\mu\beta$ values for stilbenes and azostilbenes containing dicyano and tricyano vinyl substituents, **XV** and **XVI**. The conclusions regarding these substituent effects are not unambiguous since more than one major molecular feature is varied, but it appears that the diazo linkage has little effect on β and may increase $\mu\beta$ slightly. Comparing **XV** and **XVI** with **XI** and **XII** one would also conclude that the effect of increasing the acceptor strength is diminished as the conjugation path increases.

Another major category of compounds are the dicyanoquinodial structures typified by **XVII** that were first investigated both theoretically and experimentally

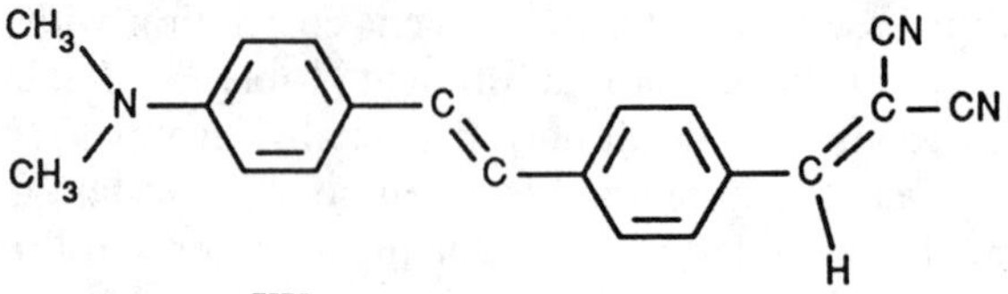

XV

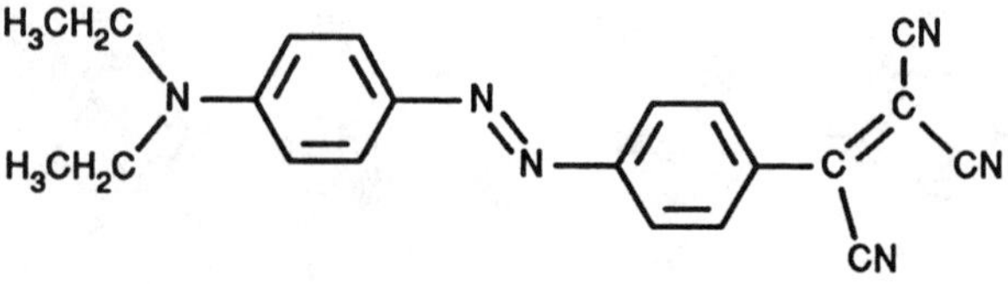

XVI

XVIIa   XVIIb

by Lalama et al. (1981). Calculations indicate a substantial transfer of charge from the –CN to the –NH region of the molecule upon excitation to the first excited state. In the ground state substantial accumulation of positive charge exists in the amino region and corresponding negative charge exists on the cyano portion. The valence bond structure **XVIIb** would appear to make a substantially higher contribution to the ground state than to the excited state. Since excitation decreases the dipole moment one would expect a negative value of β. In fact $\beta_0 = -240 \times 10^{-30}$ esu for **XVII**, which is comparable to the stilbene derivatives having a substantially larger conjugation length.

XVIIIa   XVIIIb

Related to this are the merocyanine and hemicyanine structures. The merocyanines dyes represented by **XVIIIa, b** are very solvatochromic and the basicity of the oxygen makes interpretation of experimental results complex since it is easily protonated. Large negative values of β ($\sim 10^{-27}$ esu), for the same reason discussed above were reported by Levine et al. (1978) and Dulcic and Flytzanis (1978). In the hemicyanine structure **XIX** the basic phenoxy $-O^-$ is replaced by the neutral but basic dimethylamino unit resulting in a highly colored salt with **XIXb** the likely valence bond structure for the ground state (Meredith 1983). The fact that this is a charged species made EFISH measurements problematic, but powder measurements on noncentrosymmetric crystals containing the chromophore indicate a very large β, at least as large as the merocyanine dyes. Also related to the merocyanines are the spiropyans **XX** which, in their ring opened form **XXI** photolytically produced, exhibit a merocyanine structure. While species **XXI** is not sufficiently stable with respect to thermal conversion back to **XX** so that it can be isolated and characterized by EFISH, a cocrystalline complex with **XX** can be prepared by photochemical means which exhibits a highly efficient SHG signal (Meredith et al. 1983b).

A final example that is worthy of mentioning in the context of molecular design for second-order nonlinear optics is 3-methyl-4-nitropyridine-1-oxide (POM) **XXII**. The pyridine oxide system was chosen owing to the push–pull nature of the N-oxide bond (Zyss et al. 1981). This property of the bond leads

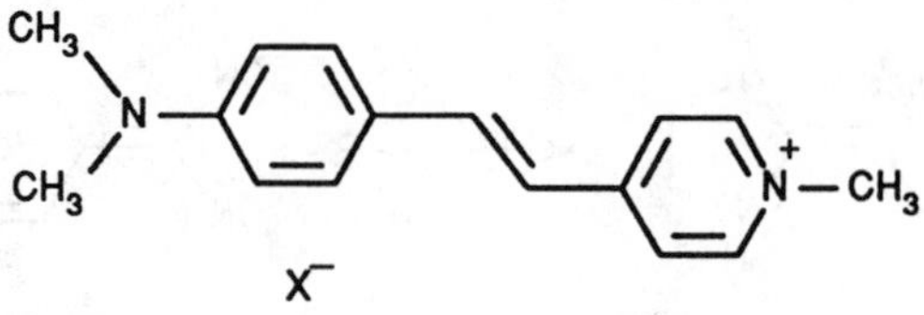

XIXa

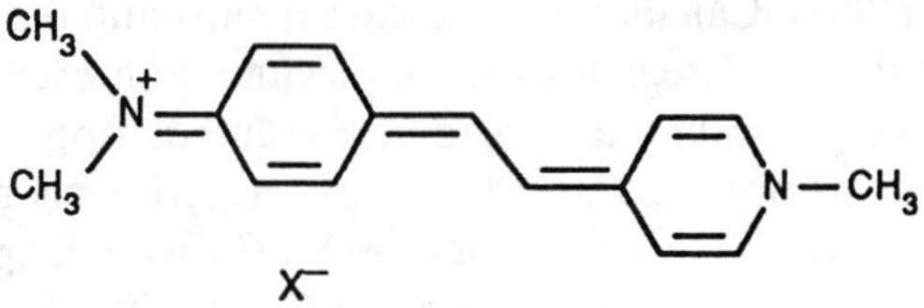

XIXb

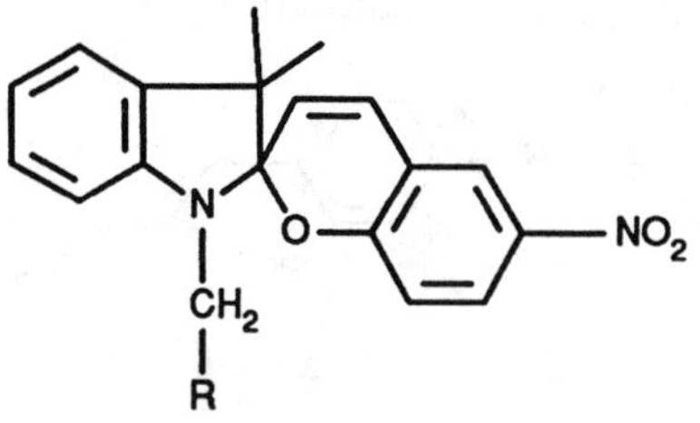

XX

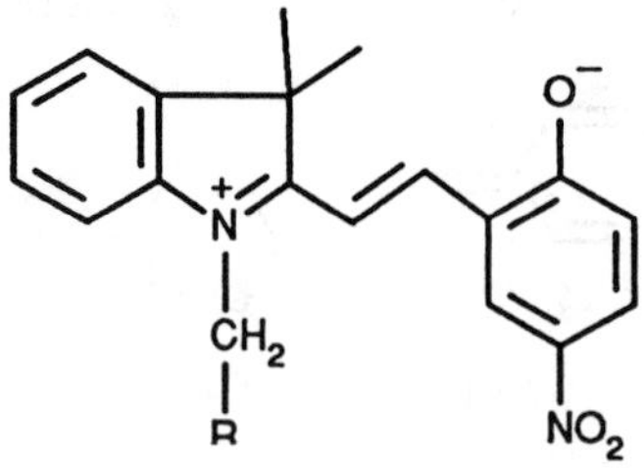

XXI

to a large degree of cancellation of the molecular dipole moment in the ground state for either donor or acceptor substituents in the 4 position. In the example **XXII** the total dipole moment in the ground state is approximately the vector sum of that for nitrobenzene (4.01 D) and pyridine-1-oxide (4.24 D), making the total dipole moment approximately zero for the ground state. A strong charge-transfer mechanism exists which leads to a high value of β.

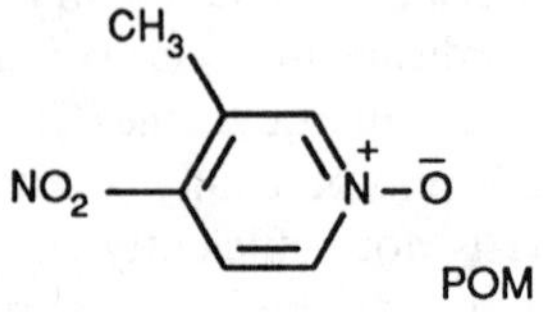

POM

XXII

Part of the rational for exploring this type of molecule was that crystal structure might be readily controlled by varying the topology of the molecule through substituents on the 2 or 3 positions and that the tendency for dipolar cancellation to influence crystal packing should be minimized. A variety of structural

XXIIa   XXIIb

variations of **XXII** were investigated, but only **XXII**, which has a β comparable to p-nitroaniline, was found to exhibit efficient powder SHG.

In closing this section of the materials survey we note that an extensive compilation of hyperpolarizability and spectral data was published by Nicoud and Twieg (1987a). Using this compilation combined with the structure property relationships in this section should make it possible to select or further optimize chromophores for specific applications. While it is impossible to give a complete list of molecular structural requirements for specific applications, since this is an ongoing topic of basic and exploratory research, several general features can be noted. For almost any conceivable application the largest value of β obtainable is a desired characteristic. Spectral transparency is a necessity at the wavelengths of interest for the application. Electrooptic devices operating at 825 nm, 1.3 μm, and 1.55 μm can take advantage of the extremely large molecular nonlinearity available in these chromophores, assuming a suitable solid-state structure can be obtained. For second-harmonic generation of diode lasers to the 412-nm range, transparency is a problem. Molecules with $\beta \gtrsim 20 \times 10^{-30}$ esu tend to exhibit tails in the visible, the electronic absorption bands giving them a slightly yellowish color at best. Higher β values tend to be accompanied by further shifts to the red. This would inevitably lead to device failure through heating. Systematic work will be required in the future if electronic structures are to be manipulated to optimize the λ_{max}, β trade-off for particular applications. Other critical requirements are oxidative and photochemical stability of a chromophore. Once again these requirements can be specified only in the context of a specific device since devices may be used under widely varying conditions. Suffice it to say that these considerations should not represent major limitations for the utilization of a particular material for the intended application.

7.3 CRYSTALS AND INCLUSION COMPLEXES

The design and fabrication of crystals for second-order nonlinear optical applications and studies is a complex and ongoing task. Early efforts in this area centered around two distinctive approaches. The first involved studies of large numbers of powdered samples by the Kurtz powder technique described in the previous chapter. Powder samples exhibiting enhanced second-harmonic signals were identified and considered as candidates for crystal-growth studies. Single crystals were grown as a result of this approach and their physical and nonlinear optical properties were characterized. The second approach, which is more rational, involves altering various topological and electronic features of desirable chromophores and using powder assessments to determine whether noncentrosymmetric structures were formed. Several valuable candidates for crystal growth were identified by this procedure and large high-quality crystals prepared. It is fair to say, however, that organic crystal growth is in general a long and arduous task requiring considerable skill. Much progress needs to be made in this area for organic crystals to emerge as candidates for widespread application.

Considering the requirements for second-order NLO applications, a list of desirable characteristics can be developed. The first set relate to mechanical and chemical considerations. The crystal should have long-term chemical stability, ruggedness to withstand processing operations, such as cutting, polishing, and electrode deposition, and resistance to ambient and laser radiation. In cases where crystal surfaces degrade under ambient conditions, immersion in inert fluids or other potting and packaging arrangements may be required.

A second set of requirements is related to the exact device function to be performed. Some of the properties that must be considered in this regard are transparency, dielectric constant, figure of merit for the nonlinear function to be performed, damage threshold, and phase-matching considerations. Each of these properties will in some way influence device performance and some properties can be traded off against others. For instance, if a material with a low damage threshold has a high figure of merit and exhibits noncritical phase matching, the power incident on the crystal for the required performance may be significantly lower than for an alternative material with a higher threshold. With few exceptions it is not possible to discuss these requirements in general.

The requirement of transparency is critical for one of the applications envisaged for organic NLO materials: second-harmonic generation of diode lasers. It has proved to be a difficult task to identify π-electronic systems with large hyperpolarizability that exhibit adequate transparency at 412 nm. At this time a number of crystals have been identified with good transparency at 532 nm for doubling of the Nd^{+3}:YAG 1.06-μm line.

In this section we briefly review some of the crystals that have been extensively studied and whenever appropriate comment on their distinguishing features. We also comment on efforts to grow single crystals in various waveguide formats. Finally, some novel approaches to obtaining noncentrosymmetric structures are discussed.

One of the first organic crystals to be studied as a single crystal was 2-methyl-4-nitroaniline (MNA) **XXIII** (Levine et al. 1979). The crystal is monoclinic (m) which is one of the most useful point groups, according to Table 4.1 and Zyss and Oudar (1982). The chromophore and its counterpart related by a twofold

XXIII

rotation are tipped away from the optic axis by about 21°. MNA exhibits two nonlinear coefficients. The first, d_{31}, is 5.8 times larger than d_{31} LiNbO$_3$ (or $\sim 9 \times 10^{-8}$ esu) and has a phase-matchable figure of merit $45 \times$ LiNbO$_3$. The second coefficient, d_{11}, is 40 times larger than that for LiNbO$_3$ and has a figure of merit 2000 times larger. Unfortunately, d_{11} is not phase matchable. This early success led other workers to investigate MNA as both a slab wave-guide material and a fiber core waveguide material. Umegaki et al. (1987) grew single crystals of MNA as a fiber-optic core and demonstrated weak frequency doubling of 1.06 μm radiation. The reason for the low conversion efficiency is that the x axis of the crystal, which has the largest nonlinear coefficient d_{11}, was parallel to the fiber axis. Unfortunately, the electric fields components of guided light tend to be perpendicular to that direction, making the interaction inefficient.

The development of MNA somewhat typifies the development of other organic crystalline materials. That is, an interesting material is identified, single-crystal growth follows, and attempts are then made to grow them in wave-guided formates.

One of the next systems to be studied was 3-methyl-4-nitropyridine-1-oxide (POM) **XXII** (Zyss et al. 1981). As was discussed in the previous section, the POM chromophore exhibits a very low dipole moment due to the push–pull nature of the N-oxide and it was felt that topological effects of substituents might lead to noncentrosymmetric crystal structures. Powder SHG studies on a variety of derivatives of the parent chromophore showed substantial SHG intensity only for POM. Crystal-growth studies produced a single crystal with an orthorhombic structure (point group 222), which according to Table 4.1 should have a substantially lower theoretically maximum d value than MNA. It is important to point out that within a given point group the geometrical arrangement of chromophores can drastically influence the d values. When Kleinman symmetry is applied this point group produces only one nonvanishing d coefficient. The value of that coefficient was determined by SHG measurements at 1.06 μm to be 2.3×10^{-8} esu. Since the β values for the POM molecule and p-nitroaniline are similar, the large differences in d must be attributable to crystal structural factors. One aspect of POM that makes it desirable for application in second-harmonic generation and parametric amplification is its near noncritical phase-matching behavior over a range of wavelengths between 1 and 2 μm. This means that, although the d value is only a modest $\sim 2 \times$ that for LiNbO$_3$, the walkoff problem is minimal for second-harmonic generation. Because of its yellowish color, POM is not of value for frequency doubling of 825- to 875-nm diodes laser.

A strategy for insuring noncentrosymmetric crystal structures is the incorporation of chiral substituents or constituents into the molecular structure. Here we describe three applications of this approach with various degree of success.

The first attempt to this approach involved protonation of a merocyanine dye with (+)-camphor-10-sulfuric acid (Ziolo et al. 1982) to form **XXIV**. Although the crystal structure was noncentrosymmetric (monoclinic P_{21}) there was a plane of pseudo symmetry within the unit cell corresponding to a head-to-tail dimer of dye units. Powder SHG studies showed very little discernible signal in that case.

Considerably more successful was the study of methyl-2-(2,4-dinitroanilino)-propionate (MAP) **XXV** (Oudar and Zyss 1982).

XXIV   XXV

The crystal structure is monoclinic (point group 2) with two molecules per unit cell related by a twofold symmetry axis. The orientation of the molecular z axis passing through the amino and nitro substituents is approximately 37° from the unique axis of the crystal structure. Oudar and Zyss performed an extensive analysis of the relationship between the microscopic and macroscopic tensors for this system and concluded that for optimum orientation of the molecule in the unit cell, a phase-matchable coefficient of up to 6 times larger might be obtained for the same chromophore. For the simplified case described above this would correspond to a tilt of the molecular z axis from the crystal Z axis of 54.74°.

Probably the most successful example of the use of chirality in crystal design and preparation is that of N-(4-nitrophenyl)-L-prolinol (NPP) **XXVI** (Zyss et al. 1984). This system is once again monoclinic with a $P2_1$ crystal structure. The combination of the chiral substituent and hydrogen-bonding character of the prolinol group leads to nearly optimal orientation of the molecule within the unit cell for phase-matched second harmonic generation. A simplified representa-

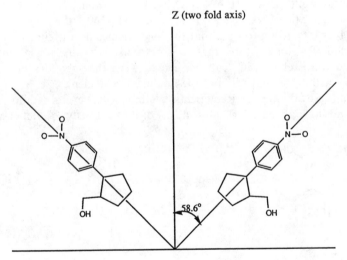

tion of point group 2 which contains $P2_1$ is shown in Figure 7.2. Although the value of β for the NPP chromophore is nearly identical to that for MAP, the coefficient d_{21} is 200×10^{-9} esu, which is approximately five times larger than for MAP. This difference is mainly attributable to the near-optimum chromophore orientation in the unit cell: 58.6° versus the optimum value of 54.74°. High-quality single crystals and thin-film crystals of NPP were produced and found to exhibit noncritical phase matching at 1.33 μm (Ledoux et al. 1985). These authors also showed the utility of these crystals to do second-harmonic and parametric amplification in the femtosecond time regime (Hulin et al. 1986).

The suitability of materials such as NPP and MAP for growth of crystalline waveguide fiber-optic cores was discussed in detail by Vidakovic et al. (1987). It was found that when an evenfold axis grows parallel to the fiber axis this orientation is very inefficient for SHG, while the optimum case is where the

Figure 7.2   A simplified representation of the unit cell structure for *N*-(4-nitrophenyl)-L-prolinol (NPP), which belongs to point group 2.

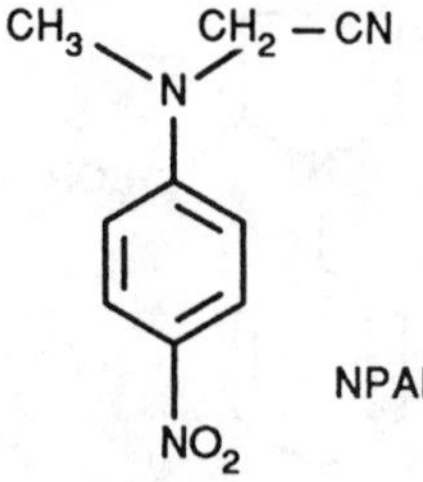

$$CH_3 \diagdown \quad CH_2 - CN$$
$$N$$

NPAN

$$NO_2$$

XXVII

evenfold axis is perpendicular to the fiber direction. Most organic materials exhibiting orthorhombic (*mm*2 point group) symmetry possess unfavorable orientation for waveguide SHG. NPP and MAP having lower monoclinic symmetry were felt to be good candidates since the lower symmetry might be expected to promote growth with the twofold axis in an intermediate direction. These materials apparently show a tendency toward cleavage when grown in confined geometries leading to waveguide quality problems. Vidakovic et al. reported a new nonlinear crystal N-(4-nitrophenyl)-N-methylaminoacetonitrile (NPAN) **XXVII**. This material has orthorhombic (*mm*2) symmetry but apparently does not exhibit cleavage in confined geometries. Excellent high-quality single crystals were grown and an angle of 58.6° (similar to NPP, Figure 7.2) with respect to the Z axis was exhibited.

An interesting comparison was made by Nicoud and Twieg (1987a) between three nitrobenzene derivatives all exhibiting the same space group ($P2_1$) and with similar electronic and linear optical properties. The three compounds were NPP (**XVI**), 4-nitrodimethylaniline (NDMA) **XXVIII**, and 4-N, N-dimethylamino-2-acetamido-4-nitroaniline (DANA) **XXIX**. Their powder SHG intensities differ greatly with a ratio of NDMA/DAN/NPP as 1/115/150 × the value for the urea crystal. It is interesting to note that the angle depicted in Figure 7.2 is 89.89°, 70.84°, and 58.6° for the three materials, respectively. Thus, a very simple and straightforward relationship can be seen between the subtleties of molecular geometry within the unit cell and the nonlinear coefficient.

Crystal growth and characterization studies on several additional new compounds have appeared recently and are discussed briefly. Günter et al. (1987)

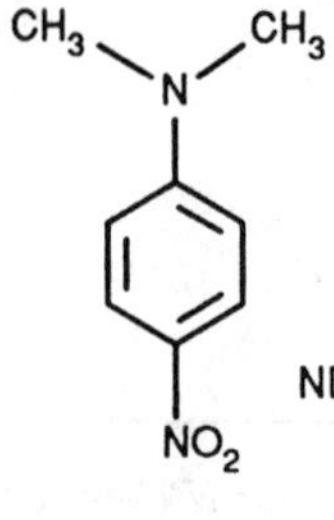

$$CH_3 \diagdown \quad \diagup CH_3$$
$$N$$

NDMA

$$NO_2$$

XXVIII

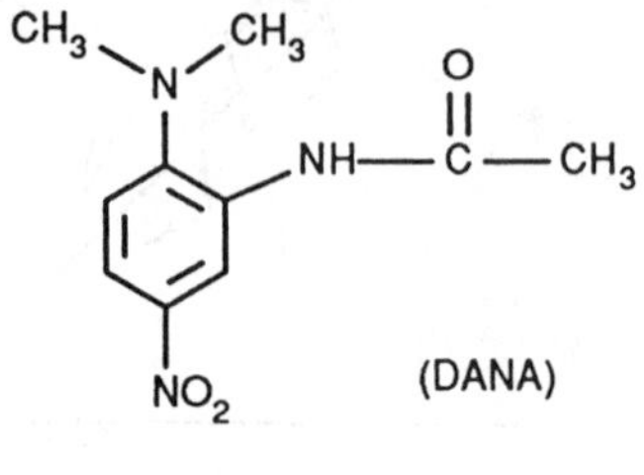

$$CH_3 \diagdown \quad \diagup CH_3$$
$$N$$
$$\qquad \overset{O}{\underset{\parallel}{C}}$$
$$NH - C - CH_3$$

(DANA)

$$NO_2$$

XXIX

CONAP

XXX

PNP

XXXI

reported on 2-cycloacetylamino-5-nitropyridine (CONAP) **XXX**. The crystal has the familiar orthorhombic $PCa2_1$ ($mm2$) symmetry with $d_{33} = 2.5 \times 10^{-8}$ esu and $d_{31} = 3.8 \times 10^{-8}$ esu. The lower coefficients (relative to NPP and NPAN) would seem to indicate a less than optimal orientation within the unit cell. The pyridine ring was undoubtedly selected to push the transparency toward the ultraviolet to open a window for frequency doubling of diode lasers. Twieg and Jain (1983) have noted an approximately 50-nm blue shift in the optical absorption spectrum in pyridine heterocyclic structures and an additional 50 nm in pyrimidines. Continuing to explore this avenue, Twieg and Dirk (1986) reported measurements on 2-(N-prolinol)-5-nitropyridine (PNP) **XXXI**. This compound is isostructural with NPP and appears to have a slightly increased second-harmonic efficiency.

$$\overset{\displaystyle S}{\underset{\displaystyle \text{CdCl}_2 \cdot \text{H}_2\text{O}}{\underset{\displaystyle \|}{\text{NH}_2\text{C}-\text{NH}-\text{NH}_2}}}$$

XXXII

The search for organic crystals with high nonlinearity and excellent transparency in the ultraviolet was given impetus by the work of Halbout et al. (1979) on crystals of urea. These crystals showed transparency and were phase matchable down to 229 nm, making them useful for frequency doubling the argon ion laser. It was possible to grow large single crystals with a d_{14} of $\sim 3.5 \times 10^{-9}$ esu. A recent exciting result was found for crystals of thiosemicarbozide cadmium chloride monohydrate **XXXII** (Xutang et al. 1988). In a preliminary publication it was reported that, for a nonoptimum phase-matching direction, this crystal gave a SHG conversion efficiency 14 × KDP.

Based on single-crystal studies, cation characterization, and the oriented gas model, Meredith (1983) concluded that *trans*-4′-dimethylamino-N-methyl-4-stilbazolium methyl sulfate (DMSM) **XXXIII** possessed an enormous non linearity, with a $d \sim 5 \times 10^{-6}$ esu. Recently, electrooptic measurements were performed on single crystals of this material (Yoshimura 1987). The electrooptic

XXXIII

coefficient is 4.3×10^{-10} m/V and the figure of merit is $5 \times$ that of LiNbO$_3$ and $6 \times$ that for MNA. This is the largest value ever reported in an organic crystal.

An enormous number of organic compounds exist and could be potential candidates for powder measurement assessments and crystal-growth studies. We make no attempt to analyze or even summarize the data that have been gathered and examined in this regard. An extensive listing of powder SHG data obtained by the Kurtz technique with references to crystal data was compiled by Nicoud and Twieg (1987c) and serves as an excellent guide to the literature. These authors developed a classification scheme for organizing the data into the following categories: (a) simple alcohols, amines, and so on, (b) amino acids and derivatives, (c) ureas and thioureas, (d) aromatic compounds, (e) polarized olefin, imine, and azo compounds, (f) miscellaneous compounds, and (g) salts. This is a useful guide to organic crystalline materials that one may wish to consider for future crystal-growth studies.

Several approaches outside of conventional melt or solution crystal growth of pure crystalline compounds have emerged for preparing noncentrosymmetric crystal structures. In one approach a spiropyran **XXXIV** is photolyzed in nonpolar solvents in an electric field to its ring opened merocyanine form **XXXV**. In the process globules about one-tenth of the wavelength of light in diameter with noncentrosymmetric crystalline cores are formed and were termed quasicrystals. When formed in the presence of an electric field the globules align themselves into threadlike structures along the field lines. These threads exhibit efficient

XXXIV   **XXXV**

SHG of 1.06 μm fundamental radiation. The globules exhibited complex electric field-dependent SHG behavior. For instance, the SHG signal of quasicrystals could be enhanced or decreased by an externally applied field after the initial formation, depending on the direction of the field. Due to complex behavior, the rigorous experimental conditions required to produce these species, and thermal stability issues these materials were not pursued further.

In another approach, inclusion complexes between host materials known to form such structures and chromophores with high values of β were shown to form noncentrosymmetric crystal structures. This was first demonstrated by Tomaru et al. (1984) for complexes of β-cyclodextrin (β-CD) **XXXVI** and various nitroaniline derivatives such as N-methyl-4-nitroaniline (HMNA) **XXXVII**. A

XXXVI

XXXVII

powder of the guest molecule alone shows no detectable signal and β-CD showed <0.1 relative to urea. The 1:1 complex showed 5 × urea. The host compound β-CD is a polysaccharide exhibiting spatial voids or channels in its crystal structure. The channels are large enough to accept a guest molecule without destroying the crystal structure of the host. These results suggest a preferential interaction of the guest molecule within the void. Additional examples of inclusion complexes with β-CD were generated by Wang and Eaton (1985) along with physical and chemical characterization of the complexes. It was concluded in this study that of the possible types of inclusion hosts, the ones with channels rather than cages were likely to produce the preferential interactions leading to noncentrosymmetric structures.

Several additional inclusion hosts were investigated with a variety of organic and organometallic guest structures (Tam et al. 1989). Besides β-CD, complexes of thiourea as host with benzene Cr(CO)$_3$ as guest (3:1) gave the most favorable SHG powder results. The number of guest structures that were converted into noncentrosymmetric host–guest complexes is about 63%, comparing quite favorably with 24% found in the nature. To the best of our knowledge, large single-crystalline materials based on this approach have not yet been reported.

A final example of unconventional methods for generating noncentrosymmetric structure was described by Weissbuch et al. (1989). Here acentric guest molecules are occluded into centrosymmetric host structures. This process is best described by considering one of the guest–host pairs that was studied. In this case the

XXXVIII   XXXIX

host is p-(N-dimethylamino)benzylidene-p'-nitroaniline **XXXVIII** and the guest is p, p'-dinitrobenzylideneaniline **XXXIX**. The host molecule from its chemical structure should have a large value of β. The guest molecule, on the other hand, is nearly centrosymmetric. The crystal structure of pure **XXXVIII** is centro-symmetric and the crystal exhibits no SHG. The 010 face of the growing crystal of **XXXVIII** contains both nitro and amino groups in equal proportion. As a guest molecule **XXXIX** approaches that crystal face, a nitro–amino interaction is favored and the guest is incorporated into the crystal structure. When a nitro group of the guest approaches the nitro group of the host, the guest is rejected. In the region of the crystal where the guest is incorporated, the host now finds itself in a noncentrosymmetric environment. Several examples of this approach were cited. The SHG signals that were obtained were extremely small, but the principle of reducing the symmetry of the medium was demonstrated and thus a new approach to molecular engineering of crystals.

7.4 POLED POLYMERS

One of the most promising approaches to the development of new materials for second-order nonlinear optical applications is that of poled polymers, discussed in Section 4.3. In this approach a film of a polymer is fabricated and subjected to an external electric field, usually at an elevated temperature where segmental mobility in the polymer is significant, and then cooled to room temperature before the field is removed. Dipolar species in the film can reorient during this process provided viscous forces can be overcome, and a net alignment is then retained in the film.

The main approach to developing materials for this purpose has been to either attach a nonlinear chromophore as a side chain in a polymeric structure, or to dissolve a dopant chromophore in a glassy host polymer. This approach has many favorable attributes that could lead to early utilization of polymers in devices and lends itself well to the optimization of device dependent parameters. This approach takes advantage of the broad base of polymer science for tailoring optical, mechanical, rheological and other properties required for the fabrication of precision thin-film structures. Films made in this way also hold the promise of being relatively inexpensive compared with conventional substrates derived from materials such as $LiNbO_3$. Very significant is the potential compatibility of a solvent-based fabrication system with semiconductor (Si, GaAs) planar processing as well as photolithography and other patterning methods for defining fine structural features.

As in the case of crystals, it is difficult to make general statements about optimum material properties since these properties are device dependent. One can say, in general, however, that low light scattering and absorption losses at the wavelength of operation are essential to most devices. To avoid heating which could result in significant loss of light due to scattering and changes in waveguide parameters, the attenuation should be $\gtrsim 1$ db/cm. The electrooptic coefficient of $LiNbO_3$ (r_{33}) is 32×10^{-12} m/V. A useful polymer film should have a similar, if not larger, coefficient. There are several device-related figures of merit that point out advantages as well as disadvantages for organic materials. The figure of merit for electrooptic phase retardation is $n^3 r_{ij}$ (see equation 6.76). The refractive index n for $LiNbO_3$ is 2.3 and around 1.55 for many organic films, giving $LiNbO_3$ a roughly threefold advantage for this particular figure of merit. For high-speed electrooptic phase retardation in lumped element devices (devices where electrical and optical signals propagate together) the figure of merit is $n^3 r_{ij}/(\varepsilon - 1)$. The appropriate ε (the low-frequency dielectric constant) for $LiNbO_3$ is 33 versus 3.5 for typical side-chain acrylics. This factor results in a considerable advantage for organic systems for this type of application. For many other types of devices where capacitive loading is an important consideration the low ε is a considerable advantage. The fact that most polymers are poled perpendicular to the plane of the film to achieve sufficiently large poling field tends to limit their use to waveguide devices. Bulk electrooptic devices require that the useful value of r_{ij} be parallel to the surface and poled polymers may not be useful for these applications.

For second-harmonic generation the figure of merit is $\sim d^2/n^3$, giving nearly a fivefold advantage to an organic material with a d value similar to that for $LiNbO_3$. Since d values on the order of d_{13} for $LiNbO_3$ might be expected in suitably designed and poled films, a tremendous potential performance advantage exists for polymer films. Freedom in tailoring the linear optical properties of polymer films and in how they are processed may enable unique and efficient phase-matching schemes in polymer thin films.

The discovery that polymer films could be poled to induce large second-order nonlinear coefficients was reported by Meredith et al. (1982b). The films

XL

consisted of p-(dimethylamino)-p'-nitrostilbene **II** dissolved up to 2% by weight in an acrylic copolymer host. **XL**. A number of important findings were reported in this initial study. First it was shown that polar $C_{\infty V}$ symmetry was induced along the poling direction which has been verified in many subsequent studies and which gives rise to two distinct tensor components $\chi^{(2)}_{ZZZ}$ and $\chi^{(2)}_{ZXX}$. A statistical thermodynamic analysis of the orientational distribution function accurately predicted the magnitude of $\chi^{(2)}_{ZZZ}$ and $\chi^{(2)}_{ZXX}$ as a function of the dopant susceptibility $\mu\beta$ and the poling field E. It was also shown in that study that homeotropic alignment of the nematic host could be achieved, free from domain-induced light scattering, and that an enhancement of up to a factor of 5 could be achieved in $\chi^{(2)}_{ZZZ}$ as a result of the anisotropy in the dopant local environment due to the nematic host. Second-harmonic generation was used to measure the $\chi^{(2)}$ tensor elements. A comparison made with **II** doped at 2% in PMMA showed a large enhancement in the liquid-crystalline system. The value of $\chi^{(2)}_{ZZZ}$ was determined by SHG to be 3×10^{-9} esu ($d_{33} = 6 \times 10^{-9}$ esu) for a poling field of ~ 1 V/μm.

The polar alignment that was induced in this system was shown by monitoring SHG intensity to be stable for long periods of time. However, several complexities in the poling process were uncovered in this study. It was found that the induced SHG dropped precipitously as T_g was approached from below. This was initially speculated to be due to some equilibrium process involving dimers or aggregates, but could also be due to polarization of the electroded structure due to mobile ionic impurities in the system.

In a subsequent study Singer et al. (1986) carried out a careful analysis of the system Disperse Red **XIV** dissolved at up to about 15% by weight in poly-methyl methacrylate. Using a sandwiched electroded structure with poling fields of about 0.65 MV/cm they were able to obtain a $\chi^{(2)}_{ZZZ}$ value of 1.2×10^{-8} esu ($d_{33} = 6 \times 10^{-9}$ esu). The observed d values were somewhat low compared with the predictions of the thermodynamic model, which illustrated again the complexity of the poling process. They cited complexities due to the non-negligible contribution of γ_{ijkl} to the second-harmonic signal of the dye molecule in EFISH causing a possible over estimation of β as a possibility. They also speculated that trapped charges at the interfaces (presumably due to mobile ionic charge) may lower the effective poling field.

Studies of the time dependence of the SHG signal induced by poling in a series of similar systems were reported by Hampsch et al. (1988). In this study DANS II, Disperse Red XIV, and MNA XXIII were doped into the host polymers polystyrene and PMMA and the SHG signal was followed over many hours after the poling field was removed. Typically the films exhibited decay of the signal with two distinct time regimes associated with the process. An initial rapid decay on the order of several hours was observed in the case of DANS in PMMA, followed by a much slower decay over a period of weeks.

Several additional interesting effects were noted in this study. First, it was found that the SHG signal decayed more rapidly in PMMA relative to poly-styrene, which was somewhat surprising since their glass transition temperatures are similar. Although this is true, PMMA exhibits significantly more sub-T_g motion than polystyrene, implicating the β relaxation as playing a role in the stability of induced alignment. Second, it was found that the stability was related to the physical size of the dopant molecule with MNA having the lowest stability. The combination of observations point out the role of excess free volume at the molecule level as well as local thermal motion on the stability of alignment. It was also shown recently (Hampsh et al. 1989) that the aging process could be controlled by retaining the alignment field on the film for several days following the initial heating and cooling of the film. It is suggested that this process allows for the relaxation of excess free volume in the system. One would therefore conclude that the temporal and thermal characteristics of the poling process must be well understood and controlled to maximize poling-induced nonlinearity and stability.

XLI

Further progress was made on the stability issue by Ye et al. (1988) who explored poling-induced SHG on polymeric systems capable of forming hydrogen bonds. Copolymers **XLI** were prepared by functionalizing a preformed para-chlorosubstituted polymer with n ranging from 0.15 to 0.48. In this study a considerable reduction of loss of SHG signal was observed. Holding the poling field on at elevated temperatures for long periods of time to anneal the system was found to be very helpful as well. This process would accelerate free volume relaxation and bake out casting solvents to enhance the stability. Hydrogen bonding between –OH groups and the electron-rich amine unit are also felt to contribute to the enhanced stability in this system.

Results on a number of side-chain-substituted and molecularly doped poled polymer systems have appeared in the literature. Copolymers with methyl methacrylate or side-chain-substituted methacrylates containing dyes such as **XV** and **XVI** were reported by Sohn et al. (1989). One such copolymer **XLII** exhibited a $\chi^{(2)}_{ZZZ}$ value of 1.5×10^{-7} esu ($d_{33} = 7.4 \times 10^{-8}$ esu), which is one of the largest reported values to date. The high nonlinearity is undoubtedly due to the high value of β for **XV** and the high molar ratio of the chromophore in the copolymer. DeMartino et al. (1988) have reported preparation of acrylic polymers with 4,4'-oxynitrobiphenyl as the active chromophore **XLIII**. Although $\mu\beta$ for the chromophore was not reported it is not expected to be particularly

$$m = .66$$
$$n = .33$$

XLII

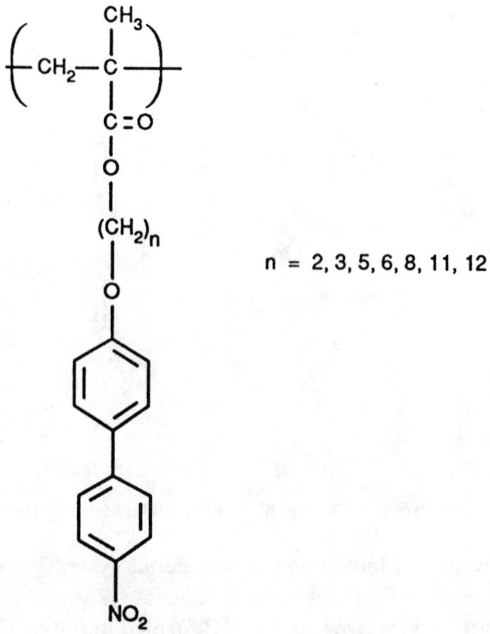

$$n = 2, 3, 5, 6, 8, 11, 12$$

XLIII

large. Other reports, particularly in the patent literature indicate interest in chromophores derived from diaminodicyanoquinodimethanes (**XLIV–XLVII**). Reports of large electrooptic coefficients from polymers of undisclosed structure may be attributable to the incorporation of chromophores with these or related structures into polymeric systems.

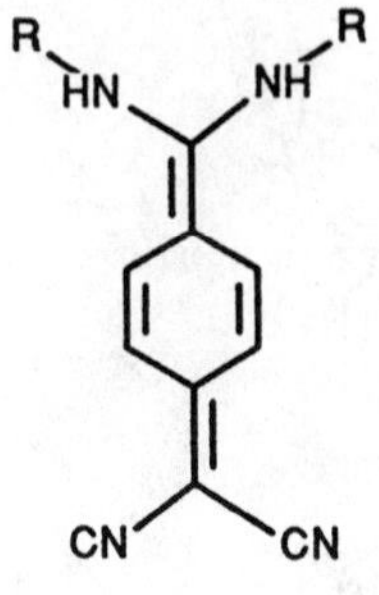

XLIV   **XLV**

Choe and Khanarian (1988)   Choe et al. (1987)

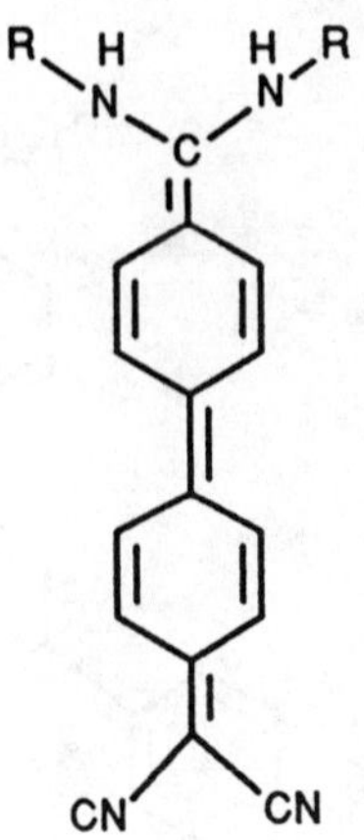

| XLVIIa | XLVI | XLVIIb |
| Choe et al. (1987) | Choe et al. (1988) | DeMartino (1988b) |

Several studies were conducted on heterogeneous systems where an aggregate or crystallite was isolated in a polymer matrix. Wang (1986) showed that aggregates of a thiapyrylium dye salt **XLVIII** formed in polycarbonate exhibited second-harmonic generation in a manner similar to other efficient organic particles. Due to the random orientation of the aggregates, these films were not useful for applications. Calvert and Moyle (1988) showed that the spatial distribution of second-harmonic light generated from crystallites of MNA in PMMA was strongly dependent on the orientation of the needles in the host matrix. Processing conditions leading to needle-like morphologies led a high fraction of light to be scattered in the forward direction. Daigo et al. (1988) showed that composites of MNA in PMMA could be aligned in an electric

XLVIII

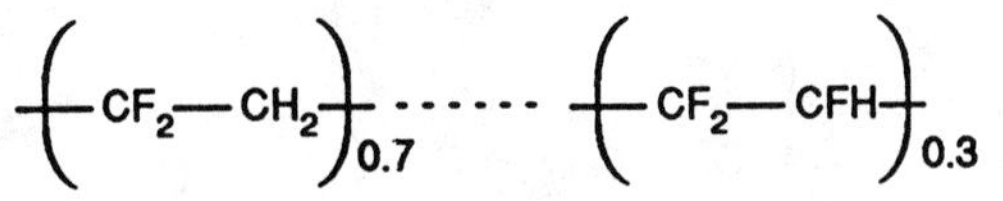

$$\left(CF_2-CH_2\right)_{0.7} \cdots\cdots \left(CF_2-CFH\right)_{0.3}$$

XLVIX

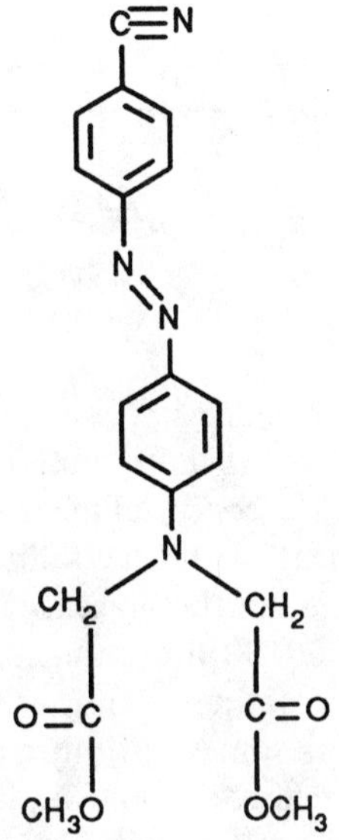

L

field at elevated temperatures. Powder-type signals $18 \times$ urea were observed but scattering of light due to inhomogeneities and discontinuities in the refractive index makes this approach questionable for device applications. One advantage of this approach, however, is the virtually indefinite stability of the polar alignment because of the tremendous stability of the particles in the polymer host. Several recent examples have been reported on the influence of the host-aligning medium on the degree and stability of host alignment. Pantelis et al. (1988) reported experiments on dyes dissolved in ferroelectric polymers. In this example, films of the host polymer **XLVIX** and guest **L** are poled at room temperature with fields in excess of 1×10^5 V/cm. The C–F bonds of the polymer in the crystalline regions (as well as the amorphous regions) align in the direction of the poling field, creating an enormous internal field that provides for stable orientation of dye molecules dissolved in the amorphous regions. Limitations on the total fraction of amorphous content and the solubility of the dye in this region of the polymer make utilization of the advantages of this approach doubtful at this point.

Two final examples of host–guest systems where the host–polymer interaction plays an important role in the nonlinearity induced in the films are *p*-nitroaniline dissolved in polyethyleneoxide (PEO) **LI** (Watanabe et al. 1988) and *p*-nitroaniline dissolved in poly(ε-caprolactone) (PεC) **LII** (Miyazaki et al. 1988).

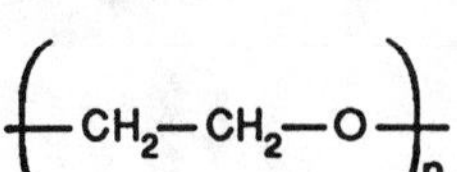

$$\left(CH_2-CH_2-O\right)_n \qquad \left(CH_2-CH_2-CH_2-CH_2-CH_2-\overset{\overset{\textstyle O}{\|}}{C}-O\right)_n$$

LI 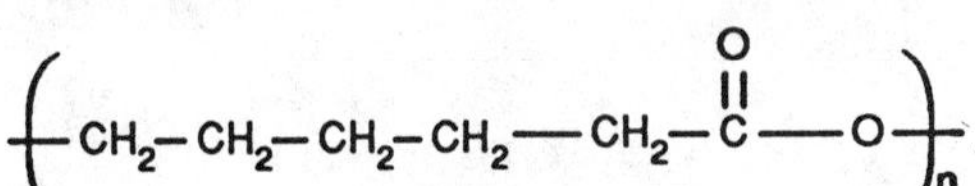   **LII**

LIII

In the former case, p-nitroaniline, a material that normally exhibits a centro-symmetric crystal structure, crystallizes in the PEO host, forming a noncentro-symmetric cocrystalline complex with the host polymer when held at 80 °C. Alignment of the crystallites could be achieved when the growth process was conducted in the presence of a weak electric field, ~ 1 V/M. Powder-type SHG signals 20–30 times that of urea were observed but the signal deteriorated in time, indicating decomposition of the complex into its components. The second system is somewhat more complex. Here the p-nitroaniline concentration was varied relative to PɛC and the system was allowed to crystallize from benzene at 80 °C without application of fields of any type. The resulting films exhibited SHG powder signals 115 times that of urea or 3.3 times that of MNA. Once again a cocrystalline complex is implicated, but the system was found to be stable indefinitely. Although not stated, it would appear that this system is composed of random orientations of crystallites and one would expect light scattering and inhomogenities to pose a problem.

Another concept in its early stages of exploration for obtaining large values of $\chi^{(2)}$ through the poling process as well as long-term stability is polar main-chain systems. Willand and Williams (1987) reported studies of the polar main-chain copolymers system **LIII**. This copolymer, which may have substantial blocks of the comonomers in the main-chain structure, has a chromophore with a substantial nonlinear coefficient incorporated directly into the main chain. Moreover, cooperativity between neighboring units in the main chain can result in correlation distances substantially larger than the individual monomer unit, creating an effective main chain dipole over that length. It was anticipated that this would lead to an enhanced value of poling-induced $\chi^{(2)}$ relative to an un-correlated side chain or doped host–guest system. In solution EFISH measurements the expected result was obtained, indicating a correlation length of about 20 units. Extending these measurements to melts of the polymer did not produce the expected enhancements and this was attributed to lack of understanding of the proper poling conditions in this system. If alignment of major main-chain segments could be achieved, the entropy of activation for randomizing the system would be extremely large and the system would be expected to be quite stable.

7.5 LANGMUIR–BLODGETT FILMS

Given the progress in crystal growth and poled polymers for second-order non-linear optical applications, films formed by molecular assembly techniques must be regarded as future generation materials. The appeal of this approach is the design flexibility associated with being able to construct a film layer by layer. In principle, films with thicknesses on the order of the wavelength of light can be prepared with tolerances on the order of the dimensions of a monolayer. Since the wave vector of a propagating mode in a waveguide is thickness dependent, films prepared by these techniques might be capable of preserving the phase relationships between modes over considerable distances. Since nonlinear optical interactions are phase dependent, the ability to control phases over long distances is an obvious potential advantage for these films.

There are two distinctive approaches to preparing molecular assemblies. The first of these is the Langmuir–Blodgett (LB) technique. Here the nonlinear chromophore is derivatized with a long hydrocarbon chain, often terminated with a carboxylic acid group on the tail. The intent is to make one part of the molecule hydrophobic and the other part hydrophylic. Molecules of this type are referred to as amphiphiles. When molecules of this type are spread onto the surface of water and compressed laterally, they collapse into an oriented film with polar head groups at the water surface and tails perpendicular to the surface. These layers can then be transferred onto a substrate with an orientation determined by the wetting characteristics of the substrate. The balance of a variety of forces determines the orientation and packing of the first and subsequently deposited layers (Swalen 1986). Considerable effort has gone into the design and synthesis of amphiphiles, including low molecular weight molecules and polymers, construction and characterization of mono- and multilayered films, and measurement of their second-order nonlinear optical properties.

A related but distinctly diffferent approach is that of self-assembled monolayers (SAM). This approach bears some relationship to the LB approach, but has potential advantages. The utility of this approach for nonlinear optics was demonstrated by Ulman et al. (1988). SAMs generally consist of a head group, a long alkyl chain, and a reactive tail. A typical head group is the trichlorosilyl group, which reacts with hydroxylated surfaces such as Si to form a covalently bonded polymerized monolayer. The long alkyl group provides physical stability for the layer via van der Waals interactions. A typical tail group is an ester, which can be reduced with $LiAlH_4$ to form a new hydroxylated surface, which can be subsequently deposited with a new layer. Studies of the formation of self-assembled monolayers were conducted by Moaz and Sagiv (1984). Extensive studies of orientation, packing, and multilayer formation showing that this method could give high-quality multilayer films were conducted by Tallman et al. (1988).

The first report of SHG measurements from LB monolayers was made by Aktsipetrov et al. (1983). They prepared monolayers from the aminoazostilbene amphiphile **LIV** and measured SHG in reflection from the surface. In this type of experiment the reflected harmonic intensity $I_{ij}^{2\omega}$, where i specifies the polarization of the incident beam and j that of the reflected beam, are measured. Light beams polarized in the plane of incidence and perpendicular to it are referred to as p and s, respectively. For a film with $C_{\infty v}$ symmetry, that is, one with a polar axis perpendicular to the film but isotropic in the plane of the film, one would expect $I_{pp}^{2\omega}$ to be large and $I_{ps}^{2\omega}$ to be somewhat smaller, depending on the average angle θ the z molecular axis makes with the polar axis. For this case $I_{ss}^{2\omega}$ should be ~ 0. Under the assumptions discussed by Heinz et al. (1983) and Rasing et al. (1986) (see Chapter 4 for discussion and references), it is possible to obtain the average orientation of the chromophore in the films from the ratio of $I_{pp}^{2\omega}/I_{ps}^{2\omega}$. In the present study a significant amount of $I_{ss}^{2\omega}$ was observed, leading the authors to conclude that the monolayer consisted of microcrystalline domains tilted in the direction that the substrate was drawn through the water interface. A $\chi^{(2)}$ in the surface layer of 2.8×10^{-8} esu was reported. This is a somewhat small number, considering the magnitude of β for the chromophore and the proximity of the harmonic frequency to the absorption band of the dye.

In a subsequent study, Aktsipetrov et al. (1985) showed that films of both Y and Z type (see Section 4.4 for definitions of Y and Z) could be obtained from **LIV**. Films formed by withdrawing a fused-SiO_2 substrate through a water surface containing the dye gave Y deposition and centrosymmetric packing. Films formed by inserting a hydrophobic substrate through the water surface gave Z-type noncentrosymmetric multilayer films. The second-harmonic intensity scaled quadratically with the number of layers deposited.

Ledoux et al. (1987) and Loulerque et al. (1988) conducted extensive studies on the related amphiphile **LV**. The conditions of deposition were discussed in detail here and significantly different results were obtained, depending on whether Cd^{2+} was present in the aqueous phase. These workers obtained Z-type deposition in either case, but without Cd^{2+} present the dependence of $I^{2\omega}$ on layer number n was sublinear. With Cd^{2+} present, a subquadratic dependence on n due to the cumulative effect of defects on the layers was obtained. These

CH$_2$
|
CH
|
(CH$_2$)$_{20}$
|
COOH

LVI   **LVII**

workers also reported a value for $\chi^{(2)}$ of 6.7×10^{-6} esu, the largest ever reported for any material. The magnitude of the coefficient was attributed to resonance enhancement at $2\omega = 532$ nm.

Due to its very high value of β the merocyanine chromophore **XVIII** has been incorporated into amphophilic structures and LB films. Girling et al. (1985) reported fabrication of noncentrosymmetric multilayers of **LVI** and ω-tricosenoic acid **LVII**. In the alternating multilayer, the oxygen of the merocyanine moiety is in the proximity of the carboxylic acid unit, causing the dye to be in the considerably less polarizable protonated form. Exposure of the films to ammonia vapore deprotonated the dye, resulting in much higher relative SHG intensity. Modeling the films as uniform axially isotropic dielectric media (as described in Chapter 4) resulted in a value of β_Z for the dye in the monolayer of 1000×10^{-5} C^{3}m^{3}J^{-2} $(2.42 \times 10^{-27}$ esu), an extraordinarily large value, perhaps reflecting resonance enhancement at the harmonic wavelength 532 nm.

Multilayers containing up to three layers each of the dye and arachidic acid were prepared. The ratio of harmonic intensity was 1:3.2:6.2, which appeared to be systematically less than the expected value of 1:4:9. This was attributed to the cumulative effect of imperfections in the layers.

Stroeve et al. (1987) conducted studies on mixed monolayers of **LVI** with polymethyl methacrylate. They observed that layers with high dye content exhibited Y-type deposition, whereas those with high PMMA content exhibited Z-type deposition with considerably higher SHG than the Y-type systems. Once again ammonia vapors were required to keep the films deprotonated, raising questions regarding the practical value of this chromophore.

The chromophore that has received the most attention for fabrication of LB mono- and multilayers is that of the hemicyanine **XIX**. Salts containing this chromophore were shown to exhibit extremely large powder SHG signals (Meredith 1983), leading to the conclusion that β was similar to that of the merocyanine   hemicyanine

merocyanine chromophore. Comparison of the two chromophores exemplifies their similarities as well as differences. Both chromophores contain the electron-deficient pyridinium ring as the acceptor. The negatively charged phenoxy ring and electron-rich dimethylamino group are both strong donors. On the other hand, the phenoxy group is a much stronger base than the dimethylamino group, hence its ease of protonation. Marowsky et al. (1988a) studied the optical absorption spectra of hemicyanine monolayers and found that films prepared under neutral pH conditions were entirely deprotonated. Films formed in the presence of substantial concentrations of H_2SO_4 did exhibit protonation as evidenced by a blue shift in their spectrum and a substantial decrease in SHG activity. This behavior parallels that for the merocyanine systems, but at a much higher pH.

Girling et al. (1985) were the first to report on the hemicyanine amphophile **LVIII** in monolayers and as alternating multilayers with ω-tricosenoic acid **LVII**. Under the assumptions described earlier in this chapter and in Chapter 4 they were able to extract values for β and the average molecular tilt away from the polar axis. A value for β_z of $95 \times 10^{-50}\,C^3m^3/J^2$ (2.29×10^{-28} esu) and average tilt angle of $37°$ was obtained from the isotropic distribution assumption and subsequent analysis. A supralinear but subquadrate dependence on bilayer number was obtained (1:2.9:5.9). It was noted that $I_{ps}^{2\omega}$, which was substantial in the first layer, decreased in the double and triple bilayer, indicating that the film becomes more isotropic with increasing layer number.

Undoubtedly, the scaling of harmonic intensity with layer number is related to the ordering within a layer and the quality with which subsequent layers can be deposited. The quality of packing within a layer should be related to weak nonbonded interactions and stronger interactions such as hydrogen bonding. Neal et al. (1986) combined the concept of using hydrogen bonding

LVIII   LIX

within a layer and making bilayer structures with active chromophores in each layer to make noncentrosymmetric films via Y deposition. The bilayers consisted of alternating layers of the hemicyanine **LVIII** with **LIX** (Ahmad et al. 1986). In a bilayer, the aliphatic tails point toward each other so that the donor acceptor sense of each chromophore now points in the same direction. Monolayers of **LVIII** and **LIX** gave β values and average tilt angles of $116 \times 10^{-50}\,C^3m^3/J^2$ (2.8×10^{-28} esu), $21.5 \times 10^{-50}\,C^3m^3/J^2$ (5.2×10^{-29} esu), $24°$, and $30°$, respectively. On the other hand a bilayer of the two gives $316 \times 10^{-5}\,C^3m^3/J^2$ (7.64×10^{-28} esu) and $23°$. It was concluded that hydrogen bonding in the stilbene layer improved the ordering within the hemicyanine layer.

Subtleties in the degree and nature of interactions within a monolayer can have a major impact on the second-order nonlinear optical properties within a layer. This is illustrated very graphically when SHG measurements were performed on mixed monolayers of **LVIII** with arachidic acid **LX**. Studies showed that SHG intensity of $I_{pp}^{2\omega}$ was maximum when the ratio of the hemicyanine to inert spacer was 1:1. For 1.06 μm fundamental light the intensity was diminished

LX

by approximately 50% in a film of the pure material. Considering the dilution factor the SHG per molecule is reduced even further. It was speculated that collapsed regions or aggregation effects might account for these observations. Schildkraut et al. (1988) showed that the optical absorption spectrum of the film was very concentration dependent. Films of the pure dye exhibit substantial absorption at 350 nm, with very little absorption at 460 nm where the diluted dye absorbed. Based on this and SHG measurements at 1.06 and 1.217 μm they concluded that an H aggregate was formed in the pure dye layer. H aggregates exhibit a spectral blue shift relative to the free dye molecule. Using a simple two-level electronic model they estimated that the nonresonant value of β was three times larger in the dye than in the aggregate. Marowsky and Steinhoff (1988) confirmed these observations and the conclusion regarding aggregate formation. They also performed SHG measurements corresponding to $2\omega =$ 420 nm and found considerable enhancement over measurements obtained at $2\omega = 530$ nm and conclude that the aggregate is more effecient than the monomer. Unfortunately, there is considerable resonant contribution from the dye and the aggregate to the electronic absorption spectrum in this region pointing to the need for caution in interpreting the results.

Continuing to pursue noncentrosymmetric bilayers as an approach for constructing stable Y-type films with high nonlinearity in each layer Cross et al. (1988) synthesized **LXI** and compared results for monolayers of **LXI** with **LVIII**. Here the long aliphatic chain is attached to the amino function. It was found that $I_{pp}^{2\omega}$ was nearly an order of magnitude larger for **LXI** than **LVIII**. Their average tilt angles were similar: 23.7° versus 22°. The authors were at a loss to explain this discrepancy between similar chromophores and similar films. Marowsky et al. (1988a) reported experiments on a similar system **LXII**. They

LXI   LXII

noted a reduction in extinction coefficient at 450 nm in the hemicyanine **LVIII** relative to **LXII**. They attribute the enhancement in SHG and changes in optical absorption spectrum to inductive effect differences associated with the alkyl chain substituent patterns. That is, the longer alkyl substituents on the nitrogen in **LXI** and **LXII** make them stronger donors relative to **LVIII** with similar effects on the pyridinium nitrogen. On the other hand, they do not note significant shifts in λ_{max} for **LVIII** relative to **LXII**, which one would expect to accompany the large apparent reduction in the extinction of **LVIII** if the inductive effects were as large as they imply. These explanations do not take into account the observations of aggregate formation noted in the preceding paragraphs. We favor the explanation that aggregate formation in **LXI** and **LXII** would explain the differences between these monolayers and the ones formed from **LVIII**.

The long alkyl tails on **LXII** and related materials were shown to significantly improve deposition characteristics and other physical properties of films. Cross et al. (1987) reasoned that the twin aliphatic tails would help to match the cross-sectional areas of the hydrophobic and hydrophilic parts of the molecule. This should improve mobility and assist the deposition. The C_{18} compound **LXII** was shown to exhibit much better vertical orientation than counterparts with shorter chains and a tenfold reduction in light scattering relative to previously reported hemicyanines.

A systematic study of factors influencing second-harmonic efficiency showed relations among orientation, chemical structure, and nonlinear polarizability (Marowsky et al. 1987). In this study SHG was measured for a series of molecules containing the pyridene ring and comparison was made with compounds containing the nitrophenyl unit. A key conclusion from this study was that the pyridine ring is strongly bound to a polar surface, leading to excellent orientation within a monolayer.

Hayden and Kowel (1987) examined conditions for deposition of films of PMMA containing **LVIII**. Using mixtures containing approximately 58% dye, these workers were able to obtain Z-type deposition. It is not clear at this point why Z deposition is observed, but the trend is persistent up to 22 layers, where the experiment was terminated. A slightly supralinear dependence of SHG intensity versus layer number was observed. The reason for the less than quadratic dependence was not speculated upon in this work. It is significant, however, that noncentric ordering could be obtained by a nonalternate deposition method.

Selfridge et al. (1988) showed that **LVIII** could be deposited onto optical fibers and exhibit SHG of fundamental light guided in the wire. A single layer of the dye coated on 5 cm of a 600-μm-diameter wire showed detectable signal. An experiment was undertaken to coat multilayers of the dye onto the fibers. The signal was shown to increase with the number of dipping cycles in a complex way. For the first 20 dipping cycles the increase in signal was modest but small. Deposition ratio studies showed that for small numbers of layers, material was deposited on both the up and the down dipping cycle. After the film thickness increased, less material was deposited on the down stroke, resulting in a more

LXIII

Z-like deposition process in the later number of cycles. While this work shows promise for devices based on fiber optic geometries, it is clear that considerably more fundamental work must be done to learn to control the deposition and symmetry of the resulting layers.

In the hope of finding approaches to more stable higher-quality LB films for second-order nonlinear optical applications, several groups have explored the use of preformed polymers incorporating various structural features to affect some aspect of the process. Carr and Goodwin (1987) were the first to demonstrate the use of polymers for this purpose by synthesizing the functionalized polysiloxane **LXIII**. The alcohol function is known to be one of the best polar head groups from the point of view of monolayer stability. Since the system is polymeric a long hydrocarbon tail was not deemed necessary for monolayer formation. SHG measurements at $1.06\,\mu m$ fundamental frequency led to a value of $\beta_Z = 3.5 \times 10^{-49}\,C^3 m^3/J^2$ (8.4×10^{-29} esu) for the repeat unit.

Taking a somewhat different approach, Hall et al. (1988) synthesized polymers with hydrophilic polyether backbones and hemicyanine chromophores **LXIV** and **LXV**. The polymers were about 50% functionalized with chromophore. Unlike in the previous approach, the hydrophilic backbone is at the water surface and the dye extends toward the hydrophobic region. The polymers are deposited alternatively in Y fashion so that the β contributions of the dyes have an additive rather than a canceling effect. For up to four hemicyanine bilayers, the films showed excellent quadratic behavior of SHG intensity with layer number.

LXIV   LXV

Popovitz-Biro et al. (1988) demonstrated in a fundamental way the require-ments for Z-type deposition of multilayers of LB films. The requirement, as they state, for Z deposition is that the advancing water contact angle with the film be near 90° and the receding one be close to 0. This behavior indicates that a hydrophobic surface can be converted to a hydrophilic one during the deposition process. In general, such a situation is not achievable with amphiphiles exposing hydrophilic ends. To demonstrate this, a series of amphiphiles having a polar head group, such as α-amino or carboxylic acid, and two amide groups along the chain were investigated and found invariably to give Z deposition. An example is **LXVI**. This and related molecules showed remarkably low receding contact angles in spite of their hydrophobic tail. Excellent quadratic dependence of the SHG signal with layer number was demonstrated up to 10 layers. This behavior was attributed to the ability of the amide groups in the interior to bind water, rendering the surface more hydrophilic.

In all of the molecular systems described thus far in this section, the polar axis of the molecule is parallel to the long molecular axis and this in turn tends to be oriented perpendicular to the film plane. An exception to this type of system was reported recently by Marowsky et al. (1988b). The system they described consisted of arachidic acid (**LXI**) and quinquethienyl (QQ) **LXVII**.

In this chromophore the polar C_2 axis is perpendicular to the long axis of the molecule. In fact, $\beta_{zzz} = 0$ in this molecular system in the absence of other

LXVI   **LXVII**

factors. Studies of the polarization dependence of the SHG signal intensities led these authors to speculate that symmetry-breaking influences in the monolayer environment might result in a nonzero β_{ZZZ}. These influences were assumed to be due to the fact that one end of the molecule extends into the air, whereas the other end is at the glass surface. This effect was simulated by CNDO studies of QQ at various distances from the $Si_2O_6H_4$ moiety as a model for the glass surface.

REFERENCES

Ahmad, M. M., W. J. Feast, D. B. Neal, M. C. Petty, and G. G. Roberts, *J. Mol. Electron.* **2**, 129 (1986).

Aktsipetrov, O. A., E. D. Akhmediev, E. D. Mishina, and V. R. Novak, *JETP Lett.* **37**, 207 (1983).

Aktsipetrov, O. A., N. N. Akhmediev, I. M. Baranova, E. D. Mishnina, and V. R. Novak, *Sov. Tech. Lett.* **11**, 249 (1985).

Barzoukas, M., M. Blanchard-Desce, D. Josse, J.-M. Lehn, and J. Zyss, *Chem. Phys.* **133**, 323 (1989).

Berkovic, G., Y. R. Shen, and M. Schadt, *Mol. Cryst. Liq. Cryst. B* **150**, 607 (1987).

Buckley, A., E. Choe, R. DeMartino, T. Leslie, G. Nelson, J. Stamatoff, D. Stuetz, and H. Yoan, *Polym. Mater. Sci. Eng.* **54**, 502 (1986).

Calvert, P. D., and B. D. Moyle, *MRS Symp. Proc.* **109**, 357 (1988).

Carr, N., and M. J. Goodwin, *Makromol. Chem. Rapid Commun.* **8**, 487 (1987).

Choe, E. W., and G. Khanarian, U.S. Patent 4,732,783 (1988).

Choe, E. W., A. Buckley, and A. Garito, U.S. Patent 4,667, 042 (1987).

Choe, E. W., A. Buckley, and A. Garito, U.S. Patent 4,773, 743 (1988).

Cross, G. H., I. R. Girling, I. R. Peterson, N. A. Cade, and J. D. Earls, *J. Opt. Soc. Am. B* **4**, 962 (1987).

Cross, G. H., I. R. Peterson, I. R. Girling, N. A. Cade, M. J. Goodwin, N. Carr, R. S. Sethi, R. Marsden, G. W. Gray, D. Lacey, A. M. McRoberts, R. M. Scrowston, and K. J. Toyne, *Thin Solid Films* **156**, 39 (1988).

Daigo, H., N. Okamoto, and H. Fujimura, *Opt. Commun.* **69**, 177 (1988).

DeMartino, R. N., U.S. Patent 4,717,508 (1988a).

DeMartino, R. N., U.S. Pattent 4,720,355 (1988b).

DeMartino, R. N., E. W. Choe, G. Khanarian, D. Hass, T. Leslie, G. Nelson, J. Stamatoff, D. Steutz, C. C. Teng, and H. Yoon, in P. N. Prasad and D. R. Ulrich (Eds.), *Nonlinear Optical and Electroactive Polymers*, Plenum, New York, 1988, 169.

Dulcic, A., and C. Flytzanis, *Opt. Commun.* **25**, 402 (1978).

Dulcic, A., and C. Sauteret, *J. Chem. Phys.* **69**, 3453 (1978).

Dulcic, A., C. Flytzanis, C. L. Tang, D. Pepin, M. Fitzon, and Y. Hoppiliard, *J. Chem. Phys.* **74**, 1559 (1981).

Everard, K. B., and L. E. Sutton, *J. Chem. Soc.* 2818 (1951).

Girling, I. R., P. V. Kolinsky, N. A. Cade, J. D. Earls, and I. R. Peterson, *Opt. Commun.* **55**, 289 (1985).

Girling, I. R., N. A. Cade, P. V. Kolinsky, R. J. Jones, I. R. Paterson, M. M. Ahmad, D. B. Neal, M. C. Petty, G. G. Roberts, and W. J. Feast, *J. Opt. Soc. Am. B* **4**, 950 (1987).

Günter, P., C. Bosshard, K. Sutter, H. Arend, G. Chapuis, R. J. Twieg, and D. Dobrowlaski, *Appl. Phys. Lett.* **50**, 486 (1987).

Halbout, J. M., S. Blit, W. Donaldson, and C. L. Tang, *IEEE J. Quant. Elec.* **QE-15**, 1176 (1979).

Hall, R. C., G. C. Lindsay, B. Anderson, S. T. Kowel, B. G. Higgins, and P. Stroeve, *Mater. Res. Soc. Proc.* **109**, 351 (1988).

Hampsh, H. L., J. Yang, G. K. Wong, and J. M. Torkelson, *Macromolecules* **21**, 526 (1988).

Hampsh, H. L., J. Yang, G. K. Wong, and J. M. Torkelson, *Polym. Commun.* **30**, 40 (1989).

Hayden, L. M., and S. Kowel, *Opt. Commun.* **61**, 351 (1987).

Heinz, T. F., H. W. K. Tom, and Y. R. Shen, *Phys. Rev. A* **28**, 1883 (1983).

Hulin, D., A. Migus, A. Antonetti, S. Ledoux, J. Badan, J. L. Oudar, and J. Zyss, *Appl. Phys. Lett.* **49**, 761 (1986).

Katz, H. E., K. D. Singer, J. E. Sohn, C. W. Dirk, L. A. King, and H. M. Gordon, *J. Am. Chem. Soc.* **87**, 6561 (1987).

Katz, H. E., C. W. Dirk, M. L. Schilling, K. D. Singer, and J. E. Sohn, in A. J. Heeger, J. Ornstein, and D. Ulrich (Eds.), *Nonlinear Optical Properties of Polymers*, MRS, Pittsburg, 1988, 127.

Lalama, S. J., K. D. Singer, A. F. Garito, and K. N. Desai, *Appl. Phys. Lett.* **39**, 940 (1981).

Ledoux, I., D. Josse, P. Vidakovic, and J. Zyss, *Opt. Eng.* **25**, 202 (1985).

Ledoux, I., D. Josse, P. Vidakovic, J. Zyss, P. A. Hahn, P. F. Gordon, B. D. Bothwell, S. K. Gupta, S. Allen, P. Robin, E. Chastaing, and J. C. Debois, *Europhys. Lett.* **3**, 803 (1987).

Levine, B. F., and C. G. Bethea, *J. Chem. Phys.* **63**, 2666 (1975).

Levine, B. F., C. G. Bethea, E. Wasserman, and L. Leenders, *J. Chem. Phys.* **68**, 5042 (1978).

Levine, B. F., C. G. Bethea, C. D. Thurmond, R. T. Lynch, and J. L. Bernstein, *J. Appl. Phys.* **50**, 2523 (1979).

Loulergue, J. C., M. Dumont, Y. Levy, P. Robin, J. P. Pocholle, and M. Popuchon, *Thin Solid Films* **160**, 399 (1988).

Marowsky, G., and R. Steinhoff, *Opt. Lett.* **13**, 707 (1988).

Marowsky, G., A. Gieralski, R. Steinhoff, D. Dorsch, R. Eldenshnile, and B. Rieger, *J. Opt. Soc. Am. B* **4**, 956 (1987).

Marowsky, G., L. F. Chi, D. Möbius, R. Steinboff, Y. R. Shen, D. Dorsch, and R. Rieger, *Chem. Phys. Lett.* **147**, 420 (1988a).

Marowsky, G., R. Steinhoff, L. F. Chi, J. Huther, and G. Wagniére, *Phys. Rev. B* **38**, 6274 (1988b).

Meredith, G. R., in D. J. Williams (Ed.), *Nonlinear Optical Properties of Organic and Polymeric Materials*, ACS Symp. Ser. No. 233, American Chemical Society, Washington, DC, 1983, p. 32.

Meredith, G. R., V. A. Krongauz, and D. J. Williams, *Chem. Phys. Lett.* **87**, 289 (1982a).

Meredith, G. R., J. G. VanDusen, and D. J. Williams, *Macromolecules* **15**, 1385 (1982b).

Meredith, G. R., D. J. Williams, S. N. Fishman, E. S. Goldbert, and V. A. Krongauz, in D. J. Williams (Ed.), *Nonlinear Optical Properties of Organic and Polymeric Materials*, ACS Symp. Ser. No. 233, American Chemical Society, Washington, DC, 1983a, p. 135f.

Meredith, G. R., D. J. Williams, S. N. Fishman, E. S. Goldbert, and V. Z. Krongauz, *J. Phys. Chem.* **87**, 1697 (1983b).

Miyazaki, T., Watanabe, T, and S. Muyata, *Jpn. J. Appl. Phys.* **27**, 9, L1724 (1988).

Moaz, R., and J. Sagiv, *J. Colloid Interface Sci.* **100**, 415 (1984).

Morley, J. D., V. J. Docherty, and D. Pugh, *J. Chem. Soc. Perkin Trans.* **2**, 1357 (1987a).

Morley, J. D., V. J. Docherty, and D. Pugh, *J. Chem. Soc. Perkin Trans.* **2**, 1351 (1987b).

Neal, D. R., M. C. Petty, G. R. Roberts, M. M. Ahmad, W. J. Feast, I. R. Girling, N. A. Cade, P. V. Kolinsky, and I. R. Paterson, *Electron. Lett.* **22**, 460 (1986).

Nicoud, J. F., and Twieg, R. J., in D. S. Chemla and J. Zyss (Eds.), *Nonlinear Optical Properties of Organic Molecules and Crystals*, Vol. 1, Academic, Orlando, FL, 1987a, p. 238.

Nicound, J. F., and R. J. Twieg, in D. S. Chemla and J. Zyss (Eds.), *Nonlinear Optical Properties of Molecules of Crystals*, Vol. 2, Academic, Orlando, FL, 1987b, p. 255; ibid, 1987c, p. 221.

Oudar, J. L., *J. Chem. Phys.* **67**, 446 (1977).

Oudar, J. L., and D. S. Chemla, *Opt. Commun.* **13**, 164 (1975).

Oudar, J. L., and D. S. Chemla, *J. Chem. Phys.* **66**, 2664 (1977).

Oudar, J. L., and H. LePerson, *Opt. Commun.* **15**, 258 (1975).

Oudar, J. L., and J. Zyss, *Phys. Rev. A* **26**, 2016 (1982).

Pantelis, P., J. R. Hill, S. N. Oliver, and G. J. Davies, *Br. Telecom. Technol. J.* **6**, 5 (1988).

Popovitz-Biro, R., K. Hill, E. M. Landau, M. Lahav, L. Leiserowitz, J. Sagiv, H. Hsiung, G. R. Meredith, and H. Vanherzeele, *J. Am. Chem. Soc.* **110**, 2672 (1988).

Rasing, Th., G. Berkovic, V. K. Shen, S. G. Grub, and M. W. Kim, *Chem. Phys. Lett.* **130**, 2 (1986).

Sagiv, J., *J. Am. Chem. Soc.* **102**, 92 (1980).

Schildkraudt, J. S., T. L. Penner, C. S. Willand, and A. Ulman, *Opt. Lett.* **13**, 134 (1988).

Selfridge, R. H., S. T. Kowel, P. Stroeve, J. Y. S. Lam, and B. Higgins, *Thin Solid Films* **160**, 471 (1988).

Singer, K. D., Sohn, J. E., and S. J. Lalama, *Appl. Phys. Lett.* **49**, 248 (1986).

Singer, K. D., M. G. Kuzyk, W. R. Holland, J. E. Sohn, S. J. Lalama, R. B. Comizzoli, H. E. Katz, and M. L. Schilling, *Appl. Phys. Lett.* **53**, 1800 (1988).

Sohn, J. E., K. D. Singer, M. G. Kuzyk, W. R. Holland, H. E. Katz, C. W. Dirk, M. L. Schilling, and R. B. Comizzoli, in J. Messier, F. Kajzar, P. Prasad, and D. Ulrich (Eds.), *Nonlinear Optical Effects in Organic Polymers, Nato ASI Series*, Vol. 162, Kluwer, Dordrecht, 1989, p. 291.

Stroeve, P., M. P. Srinivason, B. G. Higgins, and S. T. Kowel, *Thin Solid Films* **146**, 209 (1987).

Swalen, J. D., *J. Mol. Electron.* **2**, 155 (1986).

Tam, W., D. F. Eaton, J. C. Calabrese, I. D. Williams, Y. Wang, and A. Anderson, *Chem. Mater.* **1**, 128 (1989).

Tillman, N. A., A. Ulman, J. S. Schildkraudt, and T. L. Penner, *J. Am. Chem. Soc.* **110**, 6136 (1988).

Tomaru, S., S. Zembatsu, M. Kawachi, and M. Kobayashi, *J. Chem. Soc. Chem. Commun.* 1207 (1984).

Twieg, R. J., and C. W. Dirk, *J. Chem. Phys.* **85**, 3539 (1986).

Twieg, R. J., and K. Jain, in D. J. Williams (Ed.), *Nonlinear Optical Properties of Organic and Polymeric Materials*, ACS Symp. Ser. No. 233, American Chemical Society, Washington, DC, 1983, p. 57.

Ulman, A., D. J. Williams, T. L. Penner, D. R. Robello, J. S. Schildkraut, M. Scozzafava, and C. S. Willand, U.S. Patent 4,792, 208 (1988).

Umegaki, S., A. Hiramatsu, Y. Tsukikawa, and S. Tanaka, in G. Khanarian (Ed.), *Molecular and Polymeric Optoelectronic Materials, SPIE*, Vol. 682, Billingham, WA, 1987, p. 187.

Vidakovic, P. V., M. Coquillay, and F. Salin, *J. Opt. Soc. Am. B* **4**, 998 (1987).

Wang, Y., *Chem. Phys. Lett.* **126**, 209 (1986).

Wang, Y., and D. F. Eaton, *Chem. Phys. Lett.* **120**, 441 (1985).

Watanabe, T., K. Yoshinaga, D. Fichou, and S. Miyata, *J. Chem. Soc. Chem. Commun.* 250 (1988).

Weissbuch, I., M. Lahav, L. Leiserwitz, G. R. Meredith, and H. Vanherzule, *Chem. Mater.* **1**, 114 (1989).

Willand, C. S., and D. J. Williams, *Ber. Buns. Gs. Phys. Chem.* **91**, 1304 (1987).

Xutang, T., J. Minhua, X. Dong, and S. Zongshu, *Kexue Tongbao* **33**, 651 (1988).

Ye, C., N. Minami, T. J. Marks, J. Yang, and G. K. Wong, *Macromolecules* **21**, 2899 (1988).

Yoshimura, T., *J. Appl. Phys.* **62**, 2028 (1987).

Ziolo, R., W. H. H. Gunther, G. R. Meredith, and D. J. Williams, *Acta Crystallogr. Sect. B* **38**, 341 (1982).

Zyss, J., D. S. Chemla, and J. F. Nicoud, *J. Chem. Phys.* **74**, 4800 (1981).

Zyss, J., J. F. Nicoud, and M. Coquillay, *J. Chem. Phys.* **81**, 4160 (1984).

Zyss, J., J. L. Oudar, *Phys. Rev. A* **26**, 2028 (1982).

8

THIRD-ORDER NONLINEAR OPTICAL PROCESSES

8.1 THE VARIOUS THIRD-ORDER PROCESSES AND RESULTING POLARIZATIONS

The origin of third-order nonlinear processes was discussed in Chapter 2 by using a power series expansion of the induced polarization. Third-order nonlinear optical interactions yield a rich variety of phenomena that can provide useful fundamental information on the structure and dynamics of molecules and polymers and can be utilized for device concepts. Third-order processes can, in general, be viewed as resulting from four-wave mixing. There are numerous third-order processes that can be distinguished on the basis of frequencies of the output and input waves, as well as by the nature of any material resonance encountered. Consequently, different $\chi^{(3)}$ terms are used to describe the various processes.

Third-harmonic generation is a specific example of the general phenomenon of frequency mixing. Analogous to the second-order prametric mixing processes, third-order interactions can give rise to sum- and difference-frequency mixing. As we shall see in Chapter 12, which deals with devices, the intensity dependence of the refractive index provides a basis for all optical processing of information, which could result in tremendous gain in speed. In addition to this important technological thrust, the intensity-dependent refractive index also gives rise to a large number of phenomena such as self-focusing or defocusing, four-wave mixing, optical bistability, and the optical Kerr gate effect. Furthermore, the third-order nonlinear response coupled with electronic and vibrational resonances yields phenomena such as two-photon absorption and coherent Raman effects. Most processes described in this chapter fall into one of two major categories: frequency mixing or intensity dependence of refractive index. In some processes this distinction is not clear and processes can be viewed as falling into both classes. An effort will be made to address these subtle points. In this chapter the various processes are discussed to familiarize the reader with them. Then selected nonlinear optical processes that are frequently used for measurements of $\chi^{(3)}$ or are of interest from a device perspective (see Chapter 12) are discussed.

8.1.1 Third-Harmonic Generation

Harmonic generation is a special case of frequency mixing that describes the interaction of three input beams of different frequencies to create a nonlinear polarization at a new frequency and the redistribution of power among the beams. Examples of frequency mixing processes are third-harmonic generation, and sum- and difference-frequency generation. The resulting components of polarizations and their relation to the various electric field components and susceptibility functions can be expressed as follows (Reintjes 1984):

$$P_i(3\omega) = \tfrac{1}{4}\chi^{(3)}_{ijkl}(-3\omega; \omega, \omega, \omega)E_j(\omega)E_k(\omega)E_l(\omega)$$

$$P_i(\omega_4 = \omega_1 + \omega_2 + \omega_3) = (\tfrac{6}{4})\chi^{(3)}_{ijkl}(-\omega_4; \omega_1, \omega_2, \omega_3)E_j(\omega_1)E_k(\omega_2)E_l(\omega_3)$$

$$P_i(\omega_4 = \omega_1 + \omega_2 - \omega_3) = (\tfrac{6}{4})\chi^{(3)}_{ijkl}(-\omega_4; \omega_1, \omega_2, -\omega_3)E_j(\omega_1)E_k(\omega_2)E_l^*(\omega_3)$$

$$P_i(\omega_4 = \omega_1 - \omega_2 - \omega_3) = (\tfrac{6}{4})\chi^{(3)}_{ijkl}(-\omega_4; \omega_1, -\omega_2, -\omega_3)E_j(\omega_1)E_k^*(\omega_2)E_l^*(\omega_3)$$

$$(8.1)$$

The notation used for $\chi^{(3)}$ is analogous to what has been defined for $\chi^{(2)}$ in Chapter 5. The numerical factors in front of $\chi^{(3)}$ have two contributions: the degeneracy factor corresponding to the number of distinct permutation possibilities of the incident field E in the numerator, and factor in the denominator from the definition of the complex field amplitude

$$E(\omega, t) = \tfrac{1}{2}[E(\omega)e^{+i(\omega t - kz)} + E^*(\omega)e^{-i(\omega t - kz)}] \qquad (8.2)$$

The input frequency component $-\omega$ in the $\chi^{(3)}$ representation denotes that its field is the complex conjugate E^*. It should be noted that third-harmonic generation derived from $\chi^{(3)}$ is different from third-harmonic generation generally done in the laboratory using two nonlinear crystals in series. The process derived from $\chi^{(3)}$ as discussed here is a one-step process. The third-harmonic generation achieved by using two nonlinear crystals is a two-step process in which one crystal doubles the frequency and the second crystal generates the sum of the second harmonic and the fundamental. Therefore, the latter approach uses the $\chi^{(2)}$ processes, which would require two different crystals for phase-matched generation because phase-matching requirements for each step will be different.

8.1.2 Self-Action

Self-action effects describe a special case of processes derived from the intensity dependence of refractive index in which the nonlinear polarization is created in the medium at the same frequency as the incident beam (Shen 1984, Reintjes 1984). There is only one incident beam that provides the three input photons for interaction. It is called a self-action effect since the nonlinear polarization created by an incident beam affects the propagation or other properties of the same beam through an intensity-dependent refractive index. Some examples of this type of effects are as follows:

1. *Self-focusing*, which occurs as a combined result of positive n_2, the nonlinear refractive index coefficient defined in Chapter 2, and a spatial variation of the laser intensity in which the beam is more intense in the center than at the edges. The result is that the refractive index of the nonlinear medium in the center of the beam is larger than that at the edges and the medium acts as a positive lens focusing the beam.

2. *Self-defocusing*, which results when n_2 is negative. In this case the spatial variation of the laser intensity creates a negative lens resulting in defocusing of the beam.

3. *Self-phase modulation*, which is associated with the temporal behavior of the induced refractive index change, $\Delta n(t) = n_2 I(t)$, and is important only when the optical pulse is of the order of a picosecond or less. The result is a broadening of the frequency profile.

One can understand the frequency broadening derived from self-phase modulation by considering the complex form (8.2) according to which the phase ϕ of a wave is given as

$$\phi(t, z) = \omega t - kz = \omega t - \frac{n \omega z}{c} = \omega t - \frac{[n_0 + n_2 I(t)] \omega z}{c} \tag{8.3}$$

The temporal variation $d\phi/dt$ of the phase gives the frequency ω. In the presence of fast temporal intensity variation, the refractive index n, being dependent on intensity, becomes a function of time. This time dependence of n leads to added temporal variation of phase ϕ through equation 8.3 and, consequently, generates different frequency components, giving rise to broadening of the frequency profile. The nonlinear polarization resulting from the self-action interaction can be written as (Reintjes 1984)

$$P_i(\omega) = \tfrac{3}{4} \text{Re}[\chi_{iikk}^{(3)}(-\omega; \omega, -\omega, \omega)] E_i(\omega) |E_k(\omega)|^2 \tag{8.4}$$

where Re represents the real part of $\chi^{(3)}$ for a general case where it may be a complex quantity.

8.1.3 Two-Photon Absorption

This process occurs when the material has an electronic excited level at twice the frequency ω of the input beam. For this process the resulting polarization is given by (Reintjes 1984)

$$P_i(\omega) = \tfrac{3}{4} \operatorname{Im}\left[\chi^{(3)}_{iikk}(-\omega; \omega, \omega, -\omega)\right] E_i(\omega) |E_k(\omega)|^2 \tag{8.5}$$

Comparing equations 8.4 and 8.5 one can see that the self-action utilizes the real part of $\chi^{(3)}$, while the two-photon absorption involves the imaginary part of $\chi^{(3)}$. A process involving the imaginary part of susceptibility implies damping of the wave in the medium resulting from the exchange of energy between the optical field and the medium. In contrast, neither third-harmonic generation nor the self-action effect require any exchange of energy between the optical field and the nonlinear medium. These processes may be resonantly enhanced if the input and/or output frequencies are near an electronic or vibrational resonance.

8.1.4 Degenerate Four-Wave Mixing

Degenerate four-wave mixing is analogous to self-action in that the input and output waves are at the same frequency (hence, the term degenerate) (Fisher 1983, Reintjes 1984, Shen 1984). Thus, this process is also derived from the intensity dependence of the refractive index. While in the self-action process there is only on beam, the degenerate four-wave mixing process, in general, describes the case where three beams (distinguishable by their direction of propagation and/or polarization, but all of the same frequency) interact to generate a fourth beam of the same frequency. The expression for the induced nonlinear polarization corresponding to the generated field can be written as (Reintjes 1984)

$$P_i(\omega) = \tfrac{3}{4} \chi^{(3)}_{ijkl}(-\omega; \omega, \omega, -\omega) E_j(\omega) E_k(\omega) E_l^*(\omega) \tag{8.6}$$

Comparing (8.4) and (8.6) one can see that while only the real part of $\chi^{(3)}$ contributes to self-action, the degenerate four-wave mixing process derives its contibution from both the real and the imaginary parts of $\chi^{(3)}$. This will become more evident in Section 8.4 where we discuss degenerate four-wave mixing in detail and use a grating picture to illustrate the various processes contributing to this effect.

8.1.5 Coherent Raman Effects

Coherent Raman effects are examples of nondegenerate four-wave mixing, in which the input waves of different frequencies interact in the medium to produce excitations through molecular vibrational resonance (Levenson 1982). It is also an example of frequency mixing because it produces a coherent output at a new frequency. These third-order nonlinear responses occur when the difference in the frequencies of the two input beams at frequencies ω_1 and ω_2 matches ω_R, the frequency of a Raman active vibrational mode:

$$\omega_1 - \omega_2 = \omega_R \tag{8.7}$$

The third input beam of frequency, either ω_1 or ω_2, serves as a probe for the vibrational resonance and is either derived from the same input beam or provided as a spatially separate third beam. The coherent output is the fourth wave. Three different coherent Raman processes are summarized as follows:

1. *Raman-induced Kerr effect*, in which the electric vector polarization of one beam is affected by the presence of another beam when the resonance condition of equation 8.7 is satisfied. In this case, the resulting nonlinear polarization at frequency ω_1 can be written as (Reintjes 1984)

$$P_i(\omega_1) = \tfrac{3}{4} \operatorname{Re}\left[\chi_{iikk}^{(3)}(-\omega_1; \omega_1, \omega_2, -\omega_2)\right]E_i(\omega_1)|E_k(\omega_2)|^2 \tag{8.8}$$

In the four-wave mixing picture, the three input waves are at ω_1, ω_2, and ω_2, the latter two being derived from the same input beam but as complex conjugates. These three waves interact to produce the phase-shifted output wave (electric vector polarization changed) of frequency ω_1. This process is called the Raman-induced Kerr effect because the polarization of a probe beam of frequency ω_1 is affected by the presence of a pump beam of frequency ω_2, but the process is enhanced by meeting the Raman resonance condition $\omega_1 - \omega_2 = \omega_R$. Third-order non-linear processes resulting in polarization rotation are generally ascribed as Kerr processes.

2. *Coherent Stokes–Raman scattering* (CSRS), in which a coherent output wave is produced as Stokes–Raman shifted from the input frequency ω_2 at the new frequency $2\omega_2 - \omega_1$ (with $\omega_1 > \omega_2$ and $\omega_1 - \omega_2 = \omega_R$). The nonlinear polarization describing this process is given as (Reintjes 1984)

$$P_i(\omega_S) = \left(\tfrac{3}{2}\right) \operatorname{Im}\left[\chi_{ijkl}^{(3)}(-\omega_S; \omega_2, \omega_2, -\omega_1)\right]E_j(\omega_2)E_k(\omega_2)E_l^*(\omega_1) \tag{8.9}$$

where the Raman-shifted frequency $\omega_S = 2\omega_2 - \omega_1$.

This process can be explained by the energy diagram shown in Figure 8.1. Energy is transferred from the input field ω_1 to the medium by stimulated Stokes scattering due to the resonance condition and the presence of both ω_1 and ω_2. The resulting nonlinear polarization radiates an output at $\omega_S = 2\omega_2 - \omega_1$ because of the presence of wave ω_2. This process can also be viewed as a four-wave parametric mixing interaction that is resonantly enhanced by a real intermediate Raman transition. In an actual experiment one can use two different geometries. The first involves three beams, in which only two distinct input beams ω_1 and ω_2 are present producing a phase-matched output

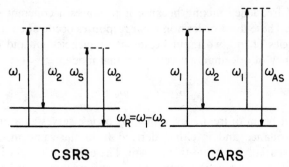

Figure 8.1 Energy-level diagrams generally used to illustrate the CARS and CSRS processes.

(spatially distinguishable) beam at a new frequency ω_S. In this process both waves at ω_2 are derived from the same beam. This type of coherent Raman scattering process is sometimes also called three-wave mixing, even though strictly speaking it is a four-wave mixing process. The second is a four-beam experiment where the two waves at ω_2 are spatially distinct beams. Therefore, there are three spatially distinguishable input beams that produce a phase-matched beam (in a specific direction) at a new output frequency $2\omega_2 - \omega_1$. Since $\omega_1 > \omega_2$, the output wave frequency is on the longer wavelength side (red shifted) of the input wave ω_2 (and therefore ω_1).

3. *Coherent anti-Stokes–Raman scattering (CARS)*, in which a coherent output wave is generated at the frequency $2\omega_1 - \omega_2$ (again with $\omega_1 > \omega_2$ and $\omega_1 - \omega_2 = \omega_R$). The nonlinear polarization for this process is given as (Reintjes 1984).

$$P_i(\omega_{AS}) = (\tfrac{3}{2}) \operatorname{Im} [\chi_{ijkl}^{(3)}(-\omega_{AS}; \omega_1, \omega_1, -\omega_2)] E_j(\omega_1) E_k(\omega_1) E_l^*(\omega_2) \quad (8.10)$$

This process corresponds to the anti-Stokes part of the process discussed above. The corresponding energy diagram is also shown in Figure 8.1. The output produced is on the shorter wavelength side of the input frequency ω_1 (and therefore ω_2 since $\omega_1 > \omega_2$). To avoid interference due to fluorescence that may arise from intrinsic absorption or that due to impurities, one generally studies CARS spectra rather than CSRS. This is because the CARS frequency is on the shorter wavelength side while fluorescence frequency is shifted to the longer wavelength side of the excitation frequency.

In the following sections, some selected third-order processes are discussed in more detail.

8.2 THIRD-HARMONIC GENERATION

The theoretical description of third-harmonic generation follows a treatment similar to that outlined in Chapter 5. The starting point is the wave equation for propagation of a plane wave electric field in a nonlinear medium (equation 5.1), but now with the inclusion of the $\chi^{(3)}$ term. As in the case of second-harmonic generation, treated in Chapter 5, the solution of resulting coupled amplitude equations, under the assumption of small signal or low conversion, leads to the following expression for the efficiency of the generated third-harmonic (Reintjes 1984):

$$\eta = \frac{576\pi^6}{n(3\omega)n^3(\omega)\lambda^2 c^2}|\chi^{(3)}(-3\omega;\omega,\omega,\omega)|^2 I_\omega^2 l^2 \frac{\sin^2(\Delta kl)2)}{(\Delta kl/2)^2} \qquad (8.11)$$

This equation is analogous to Equation 5.7 for second-harmonic generation. In the derivation of the above equation, it has beem assumed that the incident field remains undepleted in the interaction with the medium. The term l represents the interaction length of the nonlinear medium, and $n_{3\omega}$ and n_ω are the refractive indices at 3ω and ω. Δk is the wave-vector mismatch between the third-harmonic and fundamental waves and is given by

$$\Delta k = \frac{6\pi(n_{3\omega} - n_\omega)}{\lambda} \qquad (8.12)$$

Equation 8.11 predicts a dependence of the third-harmonic intensity on the wave-vector mismatch for a medium of constant interaction length l, which in displayed in Figure 8.2. $I_{3\omega}$ shows a symmetric damped-oscillatory behavior about $\Delta k = 0$. The damped oscillations result when, due to phase mismatch, the harmonic field gets out of phase with the polarization that drives it as the waves propagate through the medium. The maximum value for $I_{3\omega}$ is obtained when $\Delta k = 0$. This condition occurs when the phase velocities for the fundamental and the third-harmonic waves are equal. However, as was discussed in Chapter 5, dispersion in the refractive indices (i.e., $n_{3\omega}$ and n_ω are, in general, different) yields different phase velocities for the fundamental and the third-harmonic waves. Therefore, for finite values of phase mismatch, $I_{3\omega}$ decreases from its value at $\Delta k = 0$ and undergoes damped oscillations. A quantity

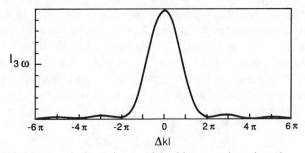

Figure 8.2 Third-harmonic intensity $I_{3\omega}$ in arbitrary units, plotted as a function of Δkl, for a finite wave-vector mismatch Δk.

that serves as a useful guide for a tolerable wave-vector mismatch in a medium while still allowing an appreciable amount of third-harmonic generation is $\Delta k_{1/2}$, the full width of the phase-matching peak at half of the maximum value of $I_{3\omega}$. This quantity is given as (Reintjes 1984)

$$\Delta k_{1/2} = \frac{5.6}{l} \tag{8.13}$$

In anisotropic media, such as birefringent crystals, it is possible to take advantage of the natural birefringence to phase match as is done for efficient second-harmonic generation (Chapter 5, Section 5.2). For example, one can select a direction of propagation for which

$$n_e(3\omega) = n_o(\omega) \tag{8.14}$$

where e and o refer to extraordinary and ordinary rays. These concepts have been discussed in detail in Chapter 5. In practice, however, phase matching for third-harmonic generation in a crystal is very difficult to achieve because of the widely different frequency values of the fundamental and the third-harmonic waves and insufficient birefringence. We are not aware of any crystal where phase-matched third-harmonic generation has been demonstrated.

In pure isotropic media such as gases, liquids, or amorphous polymers, phase matching cannot be met because of dispersion effects on refractive index. However, one can, in principle, achieve phase matching by mixing two isotropic media A and B (such as two gases) (Shen 1984) so that

$$n_A(\omega) + n_B(\omega) = n_A(3\omega) + n_B(3\omega) \tag{8.15}$$

As discussed in Chapter 2 using a harmonic oscillator model, the normal dispersion relation yields $n(\omega) < n(3\omega)$. For equation 8.15 to hold, one of the materials would have to exhibit anamalous dispersion, which describes the condition $n_A(\omega) > n_A(3\omega)$, and compensate for $n_B(\omega) < n_B(3\omega)$. This type of anamalous dispersion has been encountered in alkali metal vapors (Shen 1984, Bloom et al. 1975a, b).

In the absence of phase-matching, one can optimize third-harmonic generation by adjusting the interaction length. For a fixed value of Δk, the harmonic intensity oscillates as the interaction length is changed (either by rotation of the sample, as in the Maker fringe method, or by translation of a wedged-shape sample, as in the wedge fringe method discussed in the next chapter). In this regard, a useful concept, as discussed in Chapter 6 for second-harmonic generation, is that of the coherence length l_c which is defined as (Reintjes 1984):

$$l_c = \frac{\pi}{\Delta k} \tag{8.16}$$

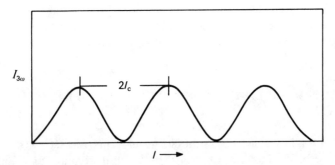

Figure 8.3 Wedge fringe pattern for third-harmonic signal $I_{3\omega}$, obtained as a sample is translated to change the pathlength l.

When the interaction length is changed, the oscillatory behavior of $I_{3\omega}$ is described by a fringing curve. In the case of a wedge-shaped sample, the change of interaction length by translation of the sample produces a wedge fringe pattern, shown in Figure 8.3. It can be seen from equations 8.11 and 8.16 that the interaction length difference between two consecutive maxima for $I_{3\omega}$ is $2l_c$. Therefore, an experimental measurement of this fringe curve also permits determination of the coherence length l_c. As discussed in Chapter 5 and 6 for the case of second-harmonic generation, one can build a physical picture of the coherence length l_c in terms of the power flow between the fundamental and the harmonic waves. As the interaction length in the nonlinear medium increases from 0 to l_c, the power is transferred continuously from the fundamental to the third-harmonic wave. For lengths greater than l_c and less than $2l_c$, the power flow direction now reverses, leading to a reduction in the power of the harmonic wave. At $l = 2l_c$, the direction of the power flow reverses again, giving rise to the oscillatory behavior represented by Figure 8.3.

Harmonic generation (or frequency mixing in general) is a coherent process in the sense that the phase of the outout wave generated at the new frequency has a definite relationship with the phase of the input waves. It is derived from nonlinear interactions which are viewed as instantaneous. Such coherent nonlinearities are derived only from those electronic interactions that do not depend on the population of the excited state. We shall see later that dynamic nonlinearities that involve electronic or molecular excitations are incoherent nonlinear responses of a medium and depend on the population density of excitation in the medium. Dynamic nonlinearities are not probed by third-harmonic generation. However, if the fundamental and/or the third-harmonic frequency is near one-photon, two-photon, or three-photon absorption bands of the material, the electronic microscopic nonlinearity $\gamma(-3\omega; \omega, \omega, \omega)$ and, consequently, the bulk susceptibility $\chi^{(3)}(-3\omega; \omega, \omega, \omega)$ are resonantly enhanced due to dispersion effects. This resonance-enhancement behavior can clearly be seen from the sum-over-states description of γ as discussed in Chapter 2. It is not dependent on the population density. The orientational nonlinearity discussed in Chapter 2 also does not contribute to the third-harmonic generation process, because it is not an instantaneous response and contributes to incoherent nonlinear behavior of the medium.

8.3   ELECTRIC FIELD-INDUCED SECOND-HARMONIC GENERATION

Electric field-induced second-harmonic generation (EFISH) is discussed in Chapter 6. EFISH has contributions both from second- and third-order nonlinearities. However, for centrosymmetric structures or structures with small second-order nonlinearity, the EFISH generation is derived from the third-order susceptibility, γ. The EFISH process can be visualized as a special case of four-wave mixing in which one of the field is a low-frequency or dc-electric field for which $\omega \simeq 0$ compared to the frequency of the optical field. It is, therefore, a special case of sum-frequency generation. The intensity of the second-harmonic beam generated in this case is given as

$$l_{2\omega} \propto |\chi^{(3)}(-2\omega; \omega, \omega, 0)|^2 I_\omega^2 l^2 \frac{\sin^2(\Delta kl/2)}{(\Delta kl/2)^2} \qquad (8.17)$$

The phase mismatch Δk is given as

$$\Delta k = \frac{4\pi(n_{2\omega} - n_\omega)}{\lambda}$$

For rigid solid structures where alignment of molecular dipoles with respect to the electric field is energetically improbable, the EFISH process is derived from γ even though the molecular structure might permit a large β. This is the case with isotropic polymers below their glass transition temperatures.

As for third-harmonic generation, the EFISH process probes only purely electronic third-order nonlinearity, which has an instantaneous response.

8.4   DEGENERATE FOUR-WAVE MIXING

The physical description of degenerate four-wave mixing can, in principle, be treated in a similar manner as that for harmonic generation (or frequency mixing) except that in this process three spatially distinguishable waves of fields $E_1(\omega, t)$, $E_2(\omega, t)$, $E_3(\omega, t)$ of the same frequency ω interact to generate a fourth wave $E_4(\omega, t)$ at the same frequency but with a different propagation direction. The starting point is the nonlinear wave equation (2.42) discussed in Chapter 2 and used in Chapter 5 to derive the expression for second-harmonic generation.

FORWARD WAVE ARRANGEMENT

Nonlinear medium

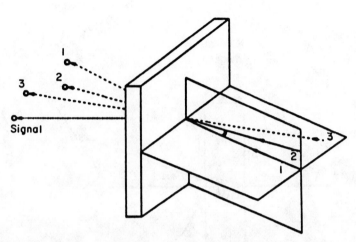

Figure 8.4 Schematic representation of the four beams in the forward-wave arrangements for degenerate four-wave mixing.

This equation will have the following form for the output field $E_4(\omega, t)$:

$$\nabla^2 E_4 = -\frac{\varepsilon}{c^2}\frac{\partial^2 E_4}{\partial t^2} - \frac{4\pi}{c^2}\frac{\partial^2 P}{\partial t^2} \tag{8.18}$$

in which the proper nonlinear polarization is given as

$$P = \chi^{(3)} \vdots E_1 E_2 E_3 \tag{8.19}$$

The complex fields E_1, E_2, and E_3 are defined by equation 8.2. Equations 8.18 and 8.19 describe the coupling of the four fields E_1, E_2, E_3, and E_4.

Equation 8.18 can be solved under the assumption of slowly varying amplitude approximation, which assumes that the lightwave intensity variation along the propagation distance is negligible on the scale of wavelength of light, and negligible pump beam depletion to obtain the expression for field E_4 as a function of propagation distance z through the nonlinear medium. As in the case of harmonic generation, the wave equation also yields the phase-matching requirement $\mathbf{k}_1 + \mathbf{k}_2 + \mathbf{k}_3 + \mathbf{k}_4 = 0$.

Many different beam geometries have been used for degenerate four-wave mixing (DFWM). The two common geometries are shown in Figures 8.4 and 8.5. Figure 8.4 shows the forward-wave geometry arrangement in which all the incident beams $(E_1, E_2,$ and $E_3)$ are propagating in the same direction from right

BACKWARD WAVE ARRANGEMENT

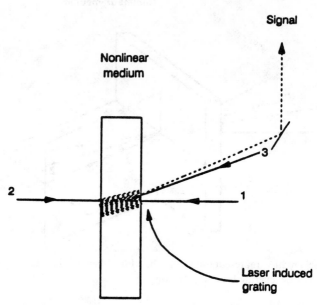

Figure 8.5 Schematics of the four beams in the backward-wave arrangement for degeneate four-wave mixing.

to left. It also shows the phase-matched direction of the output beam (signal) E_4. Figure 8.5 shows a backward-wave geometry in which two beams E_1 and E_2, called the forward and backward beams E_f and E_b, are counterpropagating and the third beam E_3, also called the probe beam E_p, is incident at a small angle θ with respect to E_1. In this geometry $\vec{k}_f = -\vec{k}_b$ so that the phase-matching condition would require $\vec{k}_s = -\vec{k}_p$. In other words, the phase-matched signal will be produced counterpropagating to the probe beam, as shown in Figure 8.5.

The process of DFWM in the backward wave geometry is synonymous with optical phase conjugation (Fisher 1983). In this regard, one considers the interaction of two counterpropagating pump beams I_f and I_b, which set up a phase conjugate mirror. A probe beam I_p incident on it is reflected to retrace itself but can exchange energy with the other beams. The signal I_s is referred to as the conjugate of the probe beam I_p. Therefore, any phase distortion suffered by the probe beam I_p is reconstructed in beam I_s. This process can be used for high-contrast dynamic holography. In the phase-conjugation picture, one often describes the efficiency of the DFWM process in terms of nonlinear reflectivity $R = I_s/I_p$. The solution of the wave equation 8.18 under the assumption of a weak probe beam field E_p, negligible pump beam depletion, and uniform pump amplitude over the interaction pathlength of the probe beam leads to the following expression for the reflectivity R (Yariv 1978):

$$R = \tan^2(k|E_f E_b|l) \tag{8.20}$$

In the above equation E_f and E_b are field amplitudes described by the right side of equation 8.2 and l is the interaction length. The coefficient k is proportional to $\chi^{(3)}(\omega; \omega, -\omega, \omega)/n_0^2$, with n_0 being the linear refractive index at ω. For small values of kl, equation 8.20 can be written as

$$R = k^2 l^2 |E_f E_b|^2 \tag{8.21}$$

Since $I_1 \propto |E_1|^2$, the following expression for the signal intensity I_s is obtained:

$$I_s \propto \frac{(\chi^{(3)})^2}{n_0^4} l^2 I_f I_b I_p \tag{8.22}$$

In the case all the beams I_f, I_b, and I_p are derived from the same parent beam I by splitting it into three parts, equation 8.22 predicts that I_s will have a cubic dependence on the intensity of the input beam only for small values of kl. The signal is generated at the expense of energies of the pump beams.

The DFWM process can also be conceptually understood by using the general picture of a laser-induced grating (Eichler et al. 1986). Here, the interaction of two coherent beams in a material gives rise to intensity modulation. Since the refractive index of a nonlinear medium is dependent on the intensity, the result is the modulation of the refractive index, which is effectively a diffraction grating. If the refractive index of the medium is complex, modulation of the real part of the refractive index gives rise to what is called a phase grating (Eichler et al. 1986, Fayer 1982). Modulation of the imaginary part of the refractive index gives rise to an amplitude grating. In the case of a backward-wave DFWM geometry, there are three different gratings formed by each two-beam pair, but only two of these gratings generate phase-matched conjugate signals (Shen 1984). These gratings can be described by examining the nonlinear polarization (8.19), with the fields given by their complex amplitude forms of equation 8.2. Equation 8.19 can be written in form of the wave vector and frequency description (Shen 1984):

$$
\begin{aligned}
P_s(k_s, \omega) &= \chi^{(3)}(\omega) E_f(k_f) E_b(k_b) E_p^*(k_p) \\
&= A(E_f \cdot E_p^*) E_b + B(E_b \cdot E_p^*) E_f + C(E_f \cdot E_b) E_p^*
\end{aligned}
\tag{8.23}
$$

The three terms on the right side of equation 8.23 represent the contributions due to the three gratings. The first term describes the grating formed by the forward pump beam I_f and the probe beam I_p (the fields in the bracket); the signal is produced by phase-matched Bragg diffraction of the backward beam

I_b. The second term describes the grating formed by the backward pump beam I_b and the probe beam I_p; the signal is produced by phase-matched Bragg diffraction of the forward beam I_f. These two gratings are the static gratings formed by the wave interference. The last term in equation 8.23 describes a standing wave grating formed by the two pump beams I_f and I_b (Shen 1984). It has a temporal oscillation frequency of 2ω because of the pump field product $E_f \cdot E_b$, which can be shown by similar arguments presented in Section 2.4 of Chapter 2 (except now with a complex field) to contain polarization oscillating at 2ω. This grating makes a dominant contribution if the frequency of the optical waves approaches a two-photon resonance of the medium (i.e., if the medium absorbs at 2ω). The coefficients A, B, and C contain $\chi^{(3)}$ and angular dependence on the crossing angle θ. The relative importance of these terms depends on the nature of the nonlinear interaction and the choice of polarization. The various grating contributions can be analyzed by selecting the polarizations of the pump and probe beams. In general, when all the beams have parallel polarizations one measures the term $\chi^{(3)}_{1111}$ to which all the various gratings discussed above contribute. In the case where the forward pump beam and the probe beam have same polarization but the backward beam has transverse polarization, only the grating formed by the beams of same polarization (in this case that formed by I_f and I_p) contributes.

8.5 RESONANT NONLINEARITY

The role of electronic resonances in third-order nonlinear processes is complex and needs to be discussed at some length. As pointed out in various places in this book, electronic resonances enhance purely electronic nonlinearities. For example, third-harmonic generation is enhanced by one-, two-, and three-photon resonances, while effects such as DFWM derived from the intensity dependence of the refractive index are enhanced by one- and two-photon resonances. However, the nonlinearity described by the intensity dependence of the refractive index is affected by material absorption at the frequency of the optical field in another way. The material absorption creates excitations such as excited states, excitons, electron–hole pairs, the concentration of which will depend on the light intensity. The refractive index of the material is dependent on the concentration of the optically generated excitations. Consequently, the refractive index will be dependent on intensity. This population effect also describes a $\chi^{(3)}$ process within the general definition where $\chi^{(3)}$ is a measure of the intensity dependence of the refractive index. However, such a process cannot contribute to third-harmonic generation and thus constitutes a fundamentally different type of effect. This resonant $\chi^{(3)}$ is sometimes also called dynamic nonlinearity or incoherent nonlinearity because it depends on the excited-state population and its response time is determined by the dynamics of the excitation (population) decay.

Under resonant conditions, one can also observe a large dynamic nonlinearity derived from an optical absorption process, which at sufficiently high intensity becomes light-intensity dependent. An optical absorption is described by the imaginary part of the refractive index. However, the Kramers–Kroenig equation (8.24) relates the real part of the refractive index with absorption coefficient α:

$$n(\omega) = \frac{c}{\pi} \text{p.v.} \int_0^\infty \frac{d\omega'\alpha(\omega')}{\omega'^2 - \omega^2} \tag{8.24}$$

where p.v. denotes the principal value of the integral. Therefore, if for a material the absorption to an excited state is intensity dependent, the change in absorption $\Delta\alpha$ as a function of intensity gives rise to a change in the refractive index Δn by the Kramers–Kroenig relation as follows (Finlayson et al. 1989):

$$\Delta n(\omega) = \frac{c}{\pi} \text{p.v.} \int_0^\infty \frac{d\omega'\Delta\alpha(\omega')}{\omega'^2 - \omega^2} \tag{8.25}$$

Equation 8.25 is often used to calculate Δn and $\chi^{(3)}$ from a pump–probe experiment in which the change in the absorption of a weak probe beam by the action of a strong pump beam is measured (Finlayson et al. 1989). Therefore, any process leading to intensity-dependent absorption gives rise to a dynamic third-order optical nonlinearity. We now discuss some of the mechanisms that give rise to the intensity dependence of absorption.

1. *Saturable Absorption*   Materials possessing strong electronic absorptions can behave as saturable absorbers as long as the rate of populating the excited state exceeds the rate of return to the ground state. The physical origin lies in the fact that as the excited states are populated, there is corresponding depletion of the ground-state population, which leads to a decrease in the absorption involving transition from the ground state. Consequently, the absorption saturates at higher intensities and the medium can even bleach (become transparent) at sufficiently high intensities. The result is intensity-dependent absorption.

2. *Phase-Space Filling*   This is a special case of saturation of ground-state absorption originally proposed for multiple-quantum well semiconductors (Schmitt-Rink et al. 1985). Greene et al. (1987) extended the application of this model to linear conjugated polymers, specifically polydiacetylenes. In these systems, the lowest energy excitation produces excitons, which are correlated (bound) electron–hole pairs. These excitons have a certain geometric size corresponding to the average separation between the electron and the hole. At low concentrations (excitation densities) of excitons, the excitons behave like bosons (neutral particles). At higher excitation densities (produced by increased intensity) the excitonic separation approaches the geometric size of excitons. At this excitation density, the excitons no longer behave like bosons, and the fermion nature (chargd particle repulsion) of the electron–hole pair now becomes

manifested in blocking further photogeneration of excitons. This is called the phase-space filling of excitons, which leads to reduction of the oscillator strength (absorption strength). The result is an intensity-dependent absorption, which, according to Kramers–Kroenig relation (8.25), leads to an intensity-dependent refractive index and hence $\chi^{(3)}$. According to Greene et al. (1987), the saturation density N_s for excitons confined in a linear polymer chain is given as

$$N_s = \frac{2}{3\sigma_c \xi_0} \tag{8.26}$$

where σ_c is the chain cross section and ξ_0 is the exciton length. Greene et al. have also derived an expression for n_2, the nonlinear refractive index coefficient, due to phase-space filling, which, for a case where the pulse duration is shorter than the response of the optical nonlinearity, is given as

$$n_2 = (1 - R) \frac{\alpha \tau_p}{h\nu} \frac{n}{2N_s} \tag{8.27}$$

where R is the reflectivity of the polymeric material, α is the absorption coefficient of the excitonic transition, τ_p is the duration of the laser pulse, and n is the real part of the linear refractive index.

3. *Conformational Deformation in Conjugated Polymers with Bond Alternation* Certain linear polymeric structures permit conformational deformation to occur following a band-to-band transition (Su et al. 1979). An example is a polyacetylene polymer, whose structure is represented in Figure 8.6. The backbone structure of this polymer consists of alternate single and double bonds (hence, bond alternation). The electronic states of this polymer have been described in terms of the valence and conduction band model (Skotheim 1986). Theoretical work of Su et al. (1979) suggests that photoexcitation of an

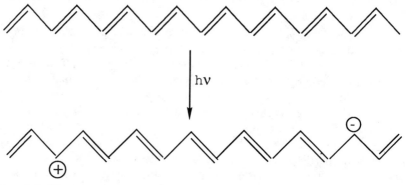

Figure 8.6 Photogeneration of a solution and antisoliton pair represented by the $\oplus$ and $\ominus$ sites, which represent the domains where the bond alternation changes.

electron from the valence band to the conduction band leads to a rapid conformational deformation creating solitonic defect centers, as shown in Figure 8.6. A soliton center defines a domain that separates two structures that are energetically equivalent (degenerate) but have different (opposite in this case) bond alternation (Skotheim 1986). In the case of polyacetylene, a soliton–antisoliton pair is produced and new states corresponding to them are created at midgap between the valence and the conduction bands. There is a redistribution of the oscillator strength from the original valence-to-conduction band transition, to transitions involving these midgap states. This change in absorption again gives rise to a change in the refractive index according to equation (8.25). Transient pump–probe experiments in polyacetylene have provided evidence for rapid conformational deformations and appearance of new spectroscopic transitions involving the midgap states (Shank et al. 1982).

Another example of conformational deformation is provided by polythiophene in which the two different bond alternation schemes, as shown in Figure 8.7, are energetically nonequivalent. Consequently, a conformational deformation in polythiophene produces polaronic defects that separate the two energetically nonequivalent bond alternations (Skotheim 1986). Nonetheless, the photo-excitation creates new polaronic states, leading to a redistribution of the oscillator strength. This again provides a contribution to $\chi^{(3)}$. The physics of these types of new excitations (solitons, polarons) is unique to these polymeric structures.

Dynamic resonant processes make the major contribution to $\chi^{(3)}$ in multiple-quantum well semiconductors, one of the very promising group of nonlinear materials (Haug 1988). However, as pointed out earlier, they do not contribute to third-harmonic generation. They can be probed by processes that depend on the intensity dependent refractive index, such as degenerate four-wave mixing

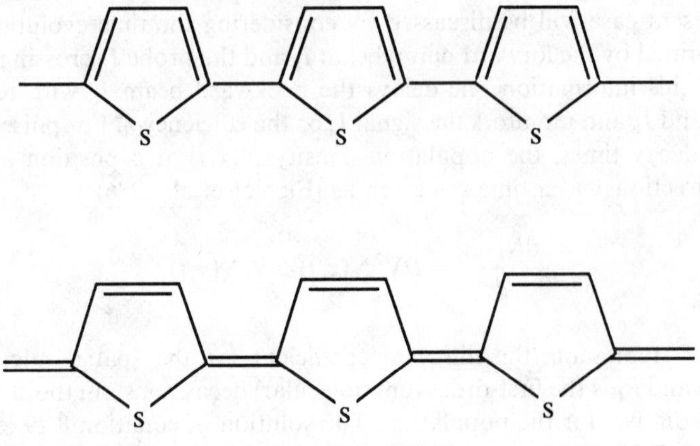

Figure 8.7 The low-energy aromatic (*top*) and high-energy quinonoid (*bottom*) forms with different bond alternation for polythiophene.

discussed above. Alternatively, they can also be probed by transient absorption experiments using the pump-probe technique mentioned above.

For probing dynamic resonant processes we discuss degenerate four-wave mixing in the backward-wave geometry. In the case where all the three beams are derived from the same parent input beam (i.e., I_f, I_b, and I_p are all proportional to the input beam intensity I), the grating efficiency $\eta = I_s/I_p$ can be described as (Eichler et al. 1986)

$$\eta^{(\lambda)} \propto \left(\frac{\partial n}{\partial N} \right)^2 g(\lambda)^2 I^{2\beta}(\lambda) f^2(t, \theta) \tag{8.28}$$

where N relates to the excitation population density and $g(\lambda)$ is the quantum yield (photogeneration efficiency) at wavelength λ. The term f is a function of the crossing angle θ of I_f and I_p and t is the time delay between the forward and the backward pulses. The exponent β depends on the photophysics of excitation. If the excitation is generated by a one-photon absorption, $\beta = 1$. The reflectivity is then dependent on I^2 and the signal I_s exhibits a cubic power dependence on the parent input beam intensity. If, on the other hand, a two-photon absorption creates the population grating, the signal I_s will exhibit a fifth-power dependence on the input intensity ($\beta = 2$). Therefore, for resonant nonlinear processes, the determination of β from the dependence of the DFWM signal I_s on the input pulse intensity I gives information on the nature of the photophysical process creating material excitation. The time evolution of the grating provides information on the buildup and decay of the excitation. The decay of the grating due to spatial migration of excitation will be dependent on the grating spacing and, therefore, on the crossing angle θ. Hence, an angular dependence study of the grating decay provides information on the dynamics of excitation migration.

At this point, a discussion of the time response of a dynamic nonlinearity as measured by degenerate four-wave mixing is presented. The time response in the present case will be discussed by considering the time evolution of the grating formed by the forward pump beam I_f and the probe I_p crossing at angle θ. To get this information one delays the backward beam I_b with respect to beams I_f and I_b and monitors the signal I_s or the efficiency η. For pulses shorter than the decay times, the population density $N(x, t)$ at a position along the grating direction x at a time t is given as (Eichler et al. 1986)

$$\frac{\partial N(x, t)}{dt} = D\nabla^2 N(x, t) - k_1 N(x, t) \tag{8.29}$$

where D represents the diffusion coefficient for the spatial migration of excitation and k_1 is the first-order (unimolecular) decay constant (both radiative and nonradiative) for the population. The solution of equation 8.29 leads to

$$f(t, \theta) = \exp[-2\Gamma(\theta)t] \tag{8.30}$$

in which

$$\Gamma(\theta) = \frac{16\pi^2 D \sin^2 \theta/2}{\lambda^2} + k_1 \tag{8.31}$$

Equation 8.30 predicts that the grating efficiency given by equation 8.28 will show an exponential decay with a rate of 2Γ, which, in general, will depend on the crossing angle θ and the excited-state decay constant k_1. For organic structures, the diffusion constant for excitation migration is generally small; the first term, therefore, is small compared to k_1. In such a situation the signal I_s will decay with a rate constant of $2k_1$, twice as fast as the excited-state population. This also makes sense intuitively, since the signal is proportional to the square of the population density.

In some materials, a bimolecular mechanism may dominate the decay. In other words, two excited-state species decay together. An example of this type of process is exciton–exciton annihilation. In the presence of a bimolecular decay, one has to include a term $-k_2 N^2(x, t)$ on the right side of the kinetic equation 8.29 to describe the time dynamics. The result is a nonexponential, power-dependent (population-dependent), excited-state decay, which is reflected in the response of the nonlinear behavior (Samoc and Prasad 1989). Therefore, an important point to remember is that the dynamic nonlinear response of a system may be intensity dependent, being faster at higher intensity and slower at lower intensity if bimolecular decay mechanism is manifested (Samoc and Prasad 1989).

One important point to notice is that the resonant $\chi^{(3)}$ value is not well defined; under certain conditions its magnitude may be intensity dependent and pulse-width (photon fluence) dependent (Casstevens et al. 1990). As discussed above, depending on the resonant condition (one-photon or two-photon excitation), the signal may show a I^3 or I^5 dpendence on the incident intensity. Furthermore, in the case of saturation of a transition, the power dependence is often less than cubic. In such cases, if one uses the definition $\Delta n = n_2 I$ to determine n_2 (consequently $\chi^{(3)}$), the value of n_2 can be intensity dependent. Therefore, one often uses a concept of an effective $\chi^{(3)}$ to describe the dynamic nonlinearity, with an understanding that this effective $\chi^{(3)}$ may be intensity dependent (Casstevens et al. 1990). Also, for the case where the decay of the excited-state population is comparable to or longer than the pulse width, use of longer pulses will lead to more photon fluence and more excited-state population. This will have an effect of enhancing the effective $\chi^{(3)}$ value. Therefore, the effective $\chi^{(3)}$ describing a resonant dynamic nonlinearity can also be pulse-width dependent as has been reported for phthalocyanine (Prasad 1988a).

It was pointed out above that the response time (decay time) of the effective $\chi^{(3)}$ can be power dependent if a bimolecular decay process is involved. Such intensity-dependent decay processes have been reported for organic systems (Samoc and Prasad 1989). To conclude this discussion of dynamic nonlinearity, it should be emphasized that the exact conditions of measurement of the magnitude and the response time of the effective $\chi^{(3)}$ must be specified before drawing conclusions.

8.6   OTHER MECHANISMS CONTRIBUTING TO THE INTENSITY DEPENDENCE OF THE REFRACTIVE INDEX

In addition to the electronic effects discussed above, there are other processes that contribute to the intensity dependence of the refractive index and hence give rise to a signal in the degenerate four-wave mixing experiment. One such process is the orientational contribution in the case of anisotropic fluids, as discussed in Chapter 2.

In the case of resonant processes, local nonradiative relaxations produce local heating, which gives rise to a rise of local temperature and also sets up density waves leading to counterpropagating (standing) ultrasonic waves of wavelength equal to the grating spacing (Eichler et al. 1986, Fayer 1982). One often uses the term thermal nonlinearity to describe these effects. Therefore, the observed nonlinear response in a four-wave mixing experiment is derived from many physical processes. However, the time response is different and can be used to determine their relative contributions (Prasad 1988b). Table 8.1

TABLE 8.1

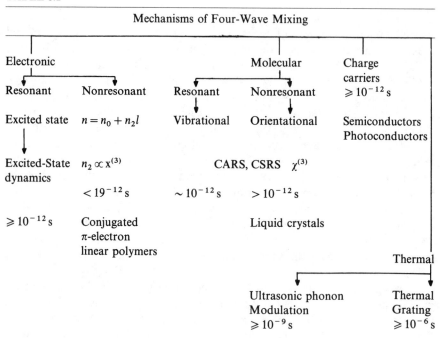

Mechanisms of Four-Wave Mixing				
Electronic			Molecular	Charge carriers $\geqslant 10^{-12}$ s
Resonant	Nonresonant	Resonant	Nonresonant	
Excited state	$n = n_0 + n_2 I$	Vibrational	Orientational	Semiconductors Photoconductors
Excited-State dynamics	$n_2 \propto \chi^{(3)}$	CARS, CSRS $\chi^{(3)}$		
	$< 19^{-12}$ s	$\sim 10^{-12}$ s	$> 10^{-12}$ s	
$\geqslant 10^{-12}$ s	Conjugated π-electron linear polymers		Liquid crystals	
				Thermal
		Ultrasonic phonon Modulation $\geqslant 10^{-9}$ s		Thermal Grating $\geqslant 10^{-6}$ s

summarizes the various resonant and nonresonant processes that contribute to DFWM. It also lists the time domains over which these processes occur. To separate these contributions in a DFWM experiment, ultrafast laser pulses of several picoseconds or less are highly desirable. For optical pulses in the nanosecond range, the resonant nonlinearity is often dominated by thermal effects.

8.7   INTENSITY-DEPENDENT PHASE SHIFT

The intensity-dependent refractive index change $\Delta n = n_2 I$ derived from the nonlinear refractive index n_2 (proportional to $\chi^{(3)}$) can be observed in several kinds of processes. The simplest is self-focusing and defocusing described above. In these processes the beam profile can be monitored as a function of the propagation length through the nonlinear medium. Alternatively, one can simply measure the output power density (intensity per unit area, or power in a given area that is smaller than the diameter of the output beam). Another manifestation of intensity-dpendent refractive index is the power-dependent phase shift. The phase of a wavefront propagating through a medium of length l is given as

$$\phi = kl = \frac{2\pi n v l}{c} = \frac{2\pi n l}{\lambda} \tag{8.32}$$

and the nonlinear phase shift $\Delta\phi$ by

$$\Delta\phi = \frac{2\pi \Delta n l}{\lambda} = \frac{2\pi n_2 I l}{\lambda} = \frac{2\pi n_2 v I l}{c} \tag{8.33}$$

This phase shift can be measured in a Fabry–Perot etalon, which is simply a cavity with parallel partially reflecting mirrors on each end. The transmission of the cavity depends on the phase relationships of the standing waves in the cavity and the intensity-dependent refractive index can change these relationships in a light-intensity-dependent manner. The nonlinear Fabry–Perot etalon is discussed in more detail in the next chapter.

In the case of guided waves, where a wave propagates either in a wave guide or as a surface plasmon (discussed in detail in the next chapter) at a metal–dielectric interface, the change of refractive index results in a shift of the angle at which the bulk wave is coupled as a guided wave. The change in the coupling angle is related to $n_2 I$. For these processes, both the magnitude and the sign of n_2 (and hence $\chi^{(3)}$) play important roles. Therefore, studies of the nonlinearity through these processes also provide determination of the sign of $\chi^{(3)}$. These processes are discussed in detail in the next chapter.

The nonlinear phase shift also causes optically induced birefringence, often referred to as the optical Kerr effect. For plane-polarized light propagating through an isotropic nonlinear medium, the induced birefringence Δn is given as (Ho et al. 1987)

$$\Delta n = \delta n_{\parallel} - \delta n_{\perp} = \frac{12\pi}{n_0}(\chi^{(3)}_{1111} - \chi^{(3)}_{1122})$$ (8.34)

where the designations $\parallel$ and $\perp$ refer to directions parallel and perpendicular to that of the incident light polarizations. This induced birefringence can be probed by another weaker beam of same or different wavelength. This phenomenon forms the basis for an optical Kerr gate, discussed in Chapter 9.

8.8   CASCADING EFFECT

As reported by Meredith and coworkers (1982, 1983) the third-order nonlinear response of noncentrosymmetric molecules can be affected by cascading effects, which yield an overestimated "true" second hyperpolarizability value. Cascading effects occur when lower-order optical nonlinearities contribute in a multistep or cascaded manner to higher-order nonlinear phenomena. These effects arise because, in addition to the electric fields applied to the sample, microscopic electric fields are generated by lower-order nonlinear polarizations. Interactions between these electric field components leads to cascaded polarization terms which are superimposed upon the process under investigation. In other words, a cascaded first hyperpolarizability (second-order coefficient) may interact with the applied electric field to generate a third-order nonlinear response. As reported by Meredith and Buchalter (1983), more than 50% of the third-order nonlinear coefficient for p-nitroaniline measured by the third-harmonic generation (THG) experiment in the solution phase is due to the cascading effect. This contribution is even larger for 2-methyl nitroaniline in the THG measurement.

To evaluate the cascading contribution for degenerate four-wave mixing experiments in solution, one can use the same approach presented by Meredith. For more details on cascading effects, interested readers are referred to the original work of Meredith and coworkers (1982, 1983). Taking into account the cascading term, one can write the following expression for γ of a noncentro-symmetric molecule:

$$\langle \gamma_{sol} \rangle = \langle \gamma_{true} \rangle + \langle \gamma_{casc} \rangle$$ (8.35)

Here, γ_{sol} is the measured second hyperpolarizability of the solute, and γ_{true} is the "true" electronic second hyperpolarizability of the compound under consideration. The cascading process for DFWM experiment involves two terms: a 2ω electric field cascading with a $-\omega$ electric field component and a static electric field component cascading with a ω electric field. When the degeneracy factor is taken into account, $\langle \gamma_{casc} \rangle$ is given by

$$\langle \gamma_{casc} \rangle = (\tfrac{2}{3})[\langle \beta^{\lambda}(-\omega; -\omega, 2\omega) \cdot C_1^{\lambda} \cdot \beta^{\lambda}(-2\omega; \omega, \omega) \rangle$$
$$+ 2\langle \beta^{\lambda}(-\omega; \omega, 0) \cdot C_2^{\lambda} \cdot \beta^{\lambda}(0; \omega, -\omega) \rangle]$$ (8.36)

The superscript λ stands for the different chemical species of the solution; β is the first hyperpolarizability; C_1 and C_2 are the cascading tensors, which are given as follows (Meredith and Buchalter 1983):

$$C_1^\lambda = \frac{(8\pi/9v^\lambda)[(n_{2\omega}^\lambda)^2 - 1][(n_{2\omega}^\lambda)^2 + 2]}{(n_{2\omega}^\lambda)^2 + 2(n_{2\omega})^2} U \tag{8.37}$$

$$C_2^\lambda = \frac{(8\pi/9v^\lambda)(\varepsilon - 1)(\varepsilon^\lambda + 2)}{\varepsilon^\lambda + 2\varepsilon} U \tag{8.38}$$

where v^λ, $n_{2\omega}^\lambda$, and ε^λ are the volume, the refractive index, and the dielectric constant of the chemical species λ, respectively; $n_{2\omega}$ and ε are the refractive index and the zero-frequency dielectric constant of the solvent. U is a unit diagonal tensor. Equations 8.36–8.38 are different from those reported by Meredith because here the expressions have been derived for DFWM, while Meredith and coworkers (1983) report it for the third-harmonic generation. Equation 8.38 includes the reorientation contribution to cascading (Meredith 1982).

REFERENCES

Bloom, D. M., G. W. Bekkers, J. F. Young, and S. E. Harris, *Appl. Phys. Lett.* **26**, 687 (1975a).

Bloom, D. M., J. F. Young, and S. E. Harris, *Appl. Phys. Lett.* **27**, 390 (1975b).

Casstevens, M. K., M. Samoc, J. Pfleger, and P. N. Prasad, *J. Chem. Phys.* **92**, 2019 (1990).

Eichler, H., P. Gunther, and D. W. Pohl, *Laser Induced Dynamic Gratings*, Springer-Verlag, Berlin, 1986.

Fayer, M. D., *Annu. Rev. Phys. Chem.* **33**, 63 (1982).

Finlayson, N., W. C. Banyai, C. T. Seaton, G. I. Stegeman, M. O'Neill, T. J. Cullen, and C. N. Ironside, *J. Opt. Soc. Am. B* **6**, 675 (1989).

Fisher, R. A. (Ed.), *Optical Phase Conjugation*, Academic, New York, 1983.

Greene, B. I., J. Orenstein, R. R. Millard, and L. R. Williams, *Phys. Rev. Lett.* **58**, 2750 (1987).

Haug, H. (Ed.), *Optical Nonlinearities and Instabilities in Semiconductors*, Academic, San Diego, 1988.

Ho, P. P., N. L. Yang, T. Jimbo, Q. Z. Wang, and R. R. Alfano, *J. Opt. Soc. Am. B* **4**, 1025 (1987).

Levenson, M. D., *Introduction to Nonlinear Laser Spectroscopy*, Academic, New York, 1982.

Meredith, G. R., *Chem. Phys. Lett.* **92**, 165 (1982).

Meredith, G. R., and B. Buchalter, *J. Chem. Phys.* **78**, 1938 (1983).

Prasad, P. N., in A. J. Heeger, J. Orenstein, and D. R. Ulrich (Eds.), *Nonlinear Optical Properties of Polymers*, Materials Research Society Symposium, Vol. 109, MRS, Pittsburgh, 1988a, p. 271.

Prasad, P. N., in P. N. Prasad and D. R. Ulrich (Eds.), *Nonlinear Optical and Electroactive Polymers*, Plenum, New York, 1988b, p. 41.

Reintjes, J F., *Nonlinear Optical Parametric Process in Liquids and Gases*, Academic, New York, 1984.

Samoc, M., and P. N. Prasad, *J. Chem. Phys.* **91**, 6643 (1989).

Schmitt-Rink, S., D. S. Chemla, and D. A. B. Miller, *Phys. Rev. B* **32**, 6601 (1985).

Shank, C. V., R. Yen, R. L. Fork, J. Orenstein, and G. L. Baker, *Phys. Rev. Lett.* **49**, 1660 (1982).

Shen, Y. R., *The Principles of Nonlinear Optics*, Wiley-Interscience, New York, 1984.

Skotheim, T. A. (Ed.), *Handbook of Conducting Polymers*, Vols. 1 and 2, Dekker, New York, 1986.

Su, W. P., J. R. Schrieffer, and A. J. Heeger, *Phys. Rev. Lett.* **42**, 1698 (1979).

Yariv, A., *IEEE J. Quantum Electron.* **QE-14**, 650 (1978).

9

MEASUREMENT TECHNIQUES FOR THIRD-ORDER NONLINEAR OPTICAL EFFECTS

9.1 A GENERAL DISCUSSION OF THE MEASUREMENT OF $\chi^{(3)}$

As discussed in the previous chapter, the third-order nonlinear optical interactions give rise to a large number of phenomena. Any of these phenomena can be monitored to obtain information on third-order optical nonlinearity. Experimental probes generally used to measure $\chi^{(3)}$ are based on the following effects:

1. Third-harmonic generation
2. Electric field-induced second-harmonic generation
3. Degenerate four-wave mixing
4. Optical Kerr gate
5. Self-focusing

In addition, processes involving an intensity-dependent phase shift due to intensity-dependent refractive index can also be used to measure $\chi^{(3)}$. The examples of these processes are found in nonlinear optical waveguides, Fabry–Perot etalons, and surface plasmon optics. The intensity-dependent phase shift changes the resonance condition, which defines the transmission characteristics of the wave through the waveguide, Fabry–Perot, or the surface-plasmon coupling, as the intensity of the input (or pump beam) is changed.

Although one loosely describes the strength of optical nonlinearity of a medium by its value, in reality there are a number of relevant parameters that

199

describe both the second- and third-order optical nonlinearities. Many of these parameters have been discussed in earlier chapters; they are collectively discussed here in relation to $\chi^{(3)}$ measurements. Some of these parameters, such as the response time and the sign, are more significant in relation to third-order optical nonlinearity, which is the main reason for their discussion in this Chapter.

1. *The $\chi^{(3)}$ Tensor*   As we discussed in an earlier chapter, $\chi^{(3)}$ is a fourth-rank tensor, which, even in isotropic media such as liquids, solutions, random solids, or amorphous polymers has three independent components $\chi^{(3)}_{1111}$, $\chi^{(3)}_{1212}$, and $\chi^{(3)}_{1122}$. They are defined by the relative polarizations of the four waves. For an isotropic medium far from any resonance and for purely electronic (instantaneous) nonlinearities $\chi^{(3)}_{1122} = \chi^{(3)}_{1212} = \frac{1}{3}\chi^{(3)}_{1111}$. This relationship even further reduces the number of independent components of the $\chi^{(3)}$ tensor for an isotropic medium. Most frequently one measures $\chi^{(3)}_{1111}$ where all the four waves have the same polarization and refers to it as the $\chi^{(3)}$ value for the medium.

2. *Response Time of the Nonlinearity*   The response time of the nonlinearity relates to its mechanism. Therefore, its determination is of considerable value in establishing the mechanism of optical nonlinearity. In addition, the response time is a valuable parameter for device applications. One of the major attractive features for the interest in optical signal processing is the gain of speed compared to electronics. Therefore, optical nonlinearities with ultrafast speeds that can beat the currently achievable electronic speed will be sought after. The nonresonant electronic nonlinearity, which involves only virtual electronic states as intermediate levels for interaction, has the fastest response time, limited only by the laser pulse width. However, some resonant electronic nonlinearity can also have extremely fast response time when the excited-state relaxation is ultrafast. Therefore, one needs to use the best time resolution available, preferably subpicoseconds, to study the time response of the nonlinearity when investigating its mechanism.

3. *Wavelength Dispersion of $\chi^{(3)}$*   The $\chi^{(3)}$ value is dependent on the frequencies of the interacting waves. When these frequencies are far away from electronic or vibrational resonances of the nonlinear material, this wavelength (frequency) dependence is weak. However, near a resonance, the dependence may be highly pronounced. Therefore, strictly speaking, one should specify the $\chi^{(3)}$ dispersion as $\chi^{(3)}(-\omega_4; \omega_1, \omega_2, \omega_3)$ while quoting a value. This feature also cautions one to be careful in comparing the $\chi^{(3)}$ values obtained by the various techniques. As discussed in the earlier chapter, one measures $\chi^{(3)}(-3\omega; \omega, \omega, \omega)$ by third-harmonic generation, and $\chi^{(3)}(-\omega; \omega, -\omega, \omega)$ by degenerate four-wave mixing. The two values are not expected to be identical because of the dispersion effect. Still, a qualitative correlation of the two values serves a useful purpose in identifying whether one is measuring a nonresonant purely electronic nonlinearity.

4. *Sign of $\chi^{(3)}$*   The nonlinear susceptibility $\chi^{(3)}$ also has a sign, which is an important fundamental property relating to the microscopic nature of optical

nonlinearity. Furthermore, it is an important device parameter that determines whether self-focusing or self-defocusing will occur in a nonlinear medium through which an intense pulse is propagating.

5. *Real or Complex*   The $\chi^{(3)}$ value may not just be a real number. It can also be a complex number. This situation occurs when any frequency of the interacting waves approaches that of a one-photon, two-photon, or three-photon electronic resonance (the latter only for third-harmonic generation). One-photon resonances are easier to recognize because they can be seen in the linear absorption spectra or can be recognized even by the color of the material. Two-photon or three-photon resonances are weak and do not appear in the absorption spectra obtained with ordinary lamp sources (as used in the commercial spectrophotometers). Therefore, even though a wavelength appears to be out of the absorption region, it may be at a multiphoton resonance, making the $\chi^{(3)}$ value at this wavelength a complex quantity.

6. *Temperature Dependence of $\chi^{(3)}$*   There are several physical processes that may make $\chi^{(3)}$ temperature dependent. First, if the material shows any structural transition at a temperature, it would manifest itself in a change of $\chi^{(3)}$. This effect can be more pronounced when the structural change also involves a change in the effective π-electron conjugation. This type of behavior has been seen in solution cast thin films of a soluble polydiacetylene, commonly known as poly-4-BCMU, which is reported to undergo a conformational transition involving a change in the efffective π-electron conjugation (Rao et al. 1986). Second, if the photon-induced processes such as electron–phonon interaction play an important role in determining optical nonlinearity, $\chi^{(3)}$ may be temperature dependent. The temperature dependence of $\chi^{(3)}$ is also important from the viewpoint of devices. A device made to operate under the atmospheric conditions may not be operational in space, if $\chi^{(3)}$ has a strong temperature dependence. Therefore, this characteristic of $\chi^{(3)}$ is important to be investgated.

It is often difficult to get complete information on all the relevant parameters of third-order nonlinearity using one single technique. However, one can use a combination of techniques to probe the various aspects of the $\chi^{(3)}$ behavior. Now we describe some of the techniques used to measure $\chi^{(3)}$.

9.2 SELECTIVE MEASUREMENT TECHNIQUES

9.2.1 Third-Harmonic Generation

A typical experimental arrangement for third-harmonic generation (THG) is shown in figure 9.1. Generally one utilizes a Q-switched pulse Nd:Yag laser which provides nanosecond pulses at low repetition rates (10–30 Hz). In the third-harmonic generation method one does not usually measure the time response; therefore, the pulse width of the laser, is not as crucial. However, if longer pulses (higher photon flux) are found to cause sample decomposition due to absorption, then it may be advisable to use a cw-Q-switched and

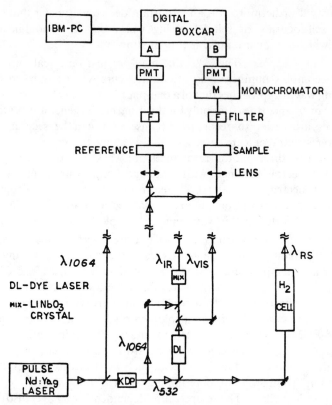

Figure 9.1   Experimental arrangement for third-harmonic generation. PMT, photo-multiplier tube; KDP, doubling crystal. A LiNbO₃ nonlinear crystal is used for mixing to generate the difference frequency in the near IR.

mode-locked Nd:Yag laser where the strongest pulses are selected through an electrooptic pulse selector. Usually, the organic systems have limited transparency towards the UV spectral range. The selection of wavelength should first be made so that the third-harmonic signal does not fall in the UV region of high absorption. For this reason, either the fundamental output of the Nd:Yag laser is Raman–Stokes shifted in a H_2 gas cell to a longer wavelength in the near IR, or a dye is pumped and mixed with the green (or fundamental) from the Yag to generate the difference frequency. After proper selection of wavelength and polarization, the laser beam is split into two parts, one being used to generate the third harmonic in the sample and the other to generate third harmonic in a reference. For the THG technique, glass is generally taken as the reference. For the non-phase-matched THG one uses the Maker fringe or wedge fringe method in which the path length of the sample (and the reference) is varied and the third-harmonic signal is monitored as a function of the interaction length l to obtain the fringes, as discussed in Chapter 8. From the fringes one determines the coherence length for both the sample and the reference. The ratio of the third-harmonic signals for the sample and the reference for the same input intensity and the same interaction length is given by

$$\frac{I(3\omega)_{\text{sample}}}{I(3\omega)_{\text{reference}}} = \left(\frac{\chi^{(3)}_{\text{sample}}}{\chi^{(3)}_{\text{reference}}}\right)^2 \left(\frac{l_c^{\text{reference}}}{l_c^{\text{sample}}}\right)^2 \tag{9.1}$$

From this expression one can determine $\chi^{(3)}$ of the sample by obtaining from the experiment the third-harmonic signals, $I(3\omega)_{\text{sample}}$ and $I(3\omega)_{\text{reference}}$, and coherence lengths, l_c^{sample} and $l_c^{\text{reference}}$. In the original method of Maker, the sample is in a rectangular form (slab in the case of a solid sample, or a cell for a liquid sample). The sample is rotated to vary the path length as represented in Figure 9.2. The observed Maker fringe pattern is fitted using a suitable computer program to obtain the coherence length. Alternatively, one can use a wedge-shaped sample, also shown in Figure 9.2. In this case, the sample is simply translated to vary the path length, which yields the wedge fringe.

Since all media including air show third-order nonlinear optical effect, one has to be extremely careful in third-harmonic measurement using a non-phase-matched condition. This point has been emphasized by Meredith et al. (1983a, b) and Kajzar and Messier (1985, 1987a, b). In third-harmonic measurements, contributions from air and the walls of the cell (the latter in the case of liquid samples) may even be dominant, especially with samples having low $\chi^{(3)}$. The resulting harmonic field is equal to the sum of contributions from consecutive nonlinear media (Kajzar and Messier 1985):

$$E^{3\omega}(r, t) = \sum_j E_j^{3\omega}(r_j, t)T_j \tag{9.2}$$

where $E_j^{3\omega}$ is the field in medium j and T_j is the corresponding whole transmission factor. The resulting harmonic intensity, which is given by the absolute square of the field represented by equation 9.2, will show complex interference patterns. For a liquid cell in open atmosphere one has to sum equation 9.2 over five media: two air media, two optical windows, and the liquid length. For a polymer

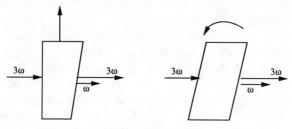

Figure 9.2 Sample geometries for wedge fringe (*left*) and maker fringe (*right*) studies of third-harmonic generation.

film supported on a glass substrate, there are four media to be included in the sum on the right side of equation 9.2.

Meredith et al. (1983a, b) demonstrated that contributions due to air can be minimized by working in a vacuum or with very tight focusing. In such a case the coherent interference of fields (as described by equation 9.2) due to the two windows with that generated in the liquid under study (for a liquid in a cell) can be handled in one of the following ways (Stevenson and Meredith 1986):

1. An analysis of the complex Maker fringe or wedge fringe arising from the coherent interference of the field in all three media (Meredith et al. 1983a, Kajzar and Messier 1985).

2. Rotation of windows to eliminate their third-harmonic field contributions (Thalhammer and Penzkofer 1983).

3. Use a longer liquid cell so that the fundamental beam intensity at the back window is sufficiently weak. In this case the liquid nonlinearity then can be calibrated relative to the front window nonlinearity by fitting of the fringe curve and comparing with the measured fringe pattern from an identical block of window glass (Meredith et al. 1983b). Since this method requires a long path length (over 6 cm), the liquid sample must be transparent (nonabsorbing at both the fundamental and the third-harmonic).

4. Making both the liquid chamber and the front window sufficiently thick compared to beam focal parameters. In this case the harmonic fields generated at the front surface and at the back window are insignificant. The $\chi^{(3)}$ is determined relative to a reference liquid in an identical second chamber in the same cell (Stevenson and Meredith 1986).

The third-harmonic generation method has the advantage that it probes purely electronic nonlinearity. Therefore, orientational and thermal effects as well as other dynamic nonlinearities derived from excitations under resonance conditions are eliminated. The involved instrumentation is simple, since one can use a nanosecond laser source for this type of study. However, for most device applications utilizing third-order processes, to be discussed in chapter 12, one makes use of an intensity-dependent refractive index described by $\chi^{(3)}$ $(-\omega; \omega, -\omega, \omega)$. As a matter of fact, any dynamic nonlinearity (excited-state process) that gives rise to the intensity dependence of a refractive index is important from this point of view. The THG method does not provide any information on dynamic nonlinearity and measures $\chi^{(3)}$ $(-3\omega; \omega, \omega, \omega)$. As discussed above, if the third-order nonlinearity has a strong frequency dispersion, $\chi^{(3)}$ $(-3\omega; \omega, \omega, \omega)$ and $\chi^{(3)}$ $(-\omega; \omega, -\omega, \omega)$ will be different. An important class of third-order nonlinear optical material is multiple-quantum well semiconductors, which show very large resonant dynamic nonlinearity $(\chi^{(3)})$ that cannot be probed by THG. The THG method does not provide any information on the time response of optical nonlinearity. Another disadvantage of the method is that one has to consider resonances at ω, 2ω, and 3ω as opposed to methods that are based on the intensity dependence of the refractive index, in which case only resonances at ω and 2ω manifest.

9.2.2   Electric Field-Induced Second-Harmonic Generation

The method is similar to the one described for third-harmonic generation. This method has also been discussed in Chapter 6 in relation to the measurement of $\chi^{(2)}$ and the molecular hyperpolarizability β. By this method one measures $\chi_{\text{eff}}^{(3)}$, which has two different components, in general: one component derived from the third-order nonlinearity, $\chi^{(3)}$, the other component coming from $\chi^{(2)}$ due to alignment of dipoles. In the solid phase, where the reorientation of the molecules is energetically not favorable (much below the glass temperature of the polymer), the pure $\chi^{(3)}$ contribution is manifested. Therefore, this method measure $\chi^{(3)}(-2\omega; \omega, \omega, 0)$. Again, a Maker fringe or wedge fringe technique can be used to obtain the EFISH signal and the coherence length for both the sample and the reference. To measure $\chi_{1111}^{(3)}(-2\omega; \omega, \omega, 0)$ for solid samples, the applied electric field has to be parallel to the electric field vector of the laser pulse and, therefore, perpendicular to the direction of propagation of the sample. Hence, it may be advisable to use a wedge-shaped sample with electrodes deposited on the top and the bottom. To measure $\chi_{1111}^{(3)}(-2\omega; \omega, \omega, 0)$ for liquid samples, one should use a wedge-shaped cell with electrodes at the top and the bottom. This cell design was discussed in Chapter 6.

The merit of EFISH method is that it is simple in instrumentation. The method may be suitable for samples with $\beta = 0$. As in the case of THG, the EFISH method does not measure dynamic nonlinearity and does not provide any information on the time response of nonlinearity.

9.2.3   Degenerate Four-Wave Mixing

Degenerate four-wave mixing (DFWM) is a convenient method of measuring $\chi^{(3)}$, which includes both electronic and dynamic nonlinearities, and of obtaining its time response. In the counterpropagating backward-wave geometry described in Section 8.4, the phase-matching condition is automatically satisfied. Since the nonlinear response is phase matched, this feature makes this method highly sensitive. For transparent or weakly absorbing samples, this geometry is most convenient. As discussed in Section 8.4, there are many processes that contribute to the DFWM signal. Therefore, care is needed to analyze the four-wave mixing results. Since the nonresonant nonlinearities in organic systems are generally weak, even a weak absorption in the sample can give dominant thermal nonlinearity (thermally induced refractive index change), especially with a long pulse laser source. For investigating electronic nonlinearities, therefore, one should avoid using nanosecond laser pulses and perform DFWM experiments only with laser sources that provide pulses $\leqslant 10$ ps. The peak powers generally needed for investigating nonresonant nonlinearities of organic systems are in the range $10\,MW/cm^2$ to $1\,GW/cm^2$. For this reason, high peak power picosecond (subpicosecond) laser pulses such as those achieved with pulse amplification are required. Also, the temporal and spatial qualities of pulses are very important. If the laser pulses are not close to the transform limit (the frequency width and the pulse width related by the uncertainty principle), artifacts in the temporal response behavior of the nonlinearity can be expected. Since the nonresonant DFWM signal depends on the cubic power of the parent beam intensity (equation 8.22), the pulse-to-pulse power fluctuation should be minimized when using the data collection in averaging mode.

A representative experimental arrangement for DFWM is shown in Figure 9.3 (Prasad et al. 1988). It utilizes a cw mode-locked Nd:Yag laser, the pulses from which are about 80 ps and are compressed in a fibre pulse compressor to about 5 ps and then frequency doubled. The frequency-doubled green power is stabilized by a feedback mechanism to a Bragg cell which controls the transmission of the fundamental IR power of the Nd:Yag laser. The stabilized green pulses are used to synchronously pump a dye laser, the output pulses from which are above 350 fs wide but do not have sufficient peak power for the measurement of optical nonlinearities in organic systems. For conjugated polymeric systems with relatively large $\chi^{(3)}$, one may be able to use cavity dumping of the dye laser to increase the peak power by one to two orders of magnitude. But for most cases, amplification of pulses are required. For this

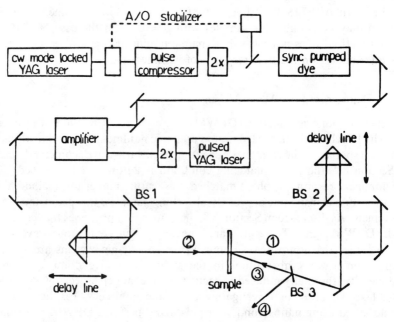

Figure 9.3 Schematics of a laser system for subpicosecond degenerate four–wave mixing studies. A/O, acoustic optics; BS, beam splitter.

purpose the experiment arrangement of Figure 9.3 uses a three-stage amplifier pumped by a low repetition rate (10–30 Hz) pulse Nd:Yag laser, whereby the peak power is amplified by five orders of magnitude. The pulses are split, each passing through a separately adjustable delay line, and meet at the sample in a backward-wave geometry. Beams 1 and 3 are synchronized in time by their cross-correlation. Then the DFWM signal is studied as a function of time delay of the backward beam 2. The DFWM signal (the maximum value) obtained from the sample is compared with that obtained from CS_2 as the reference by using the same power levels for each. The $\chi^{(3)}$ value then is obtained by using the following equation for the case of a nonabsorbing sample:

$$\frac{\chi^{(3)}_{sample}}{\chi^{(3)}_{CS_2}} = \left(\frac{n^0_{sample}}{n^0_{CS_2}}\right)^2 \frac{l_{CS_2}}{l_{sample}} \sqrt{\frac{I_{sample}}{I_{CS_2}}} \tag{9.3}$$

n^0_{sample} and $n^0_{CS_2}$ are the linear refractive indices of the sample and CS_2; l_{CS_2} and l_{sample} are the path lengths of the two media. I_{sample} and I_{CS_2} are the respective DFWM signals from the sample and CS_2; The use of the above equation assumes that the interaction lengths of CS_2 and the sample are the same as the path lengths and in each case a cubic dependence of the DFWM signal on the input power density (equation 8.22) is valid. To assure the first requisite, one should use the smallest angle possible between beams 1 and 3, avoid tight focusing, and use the smallest sample thickness with which an easily detectable signal can be observed. One should also make, in each case, a power-dependence study of the DFWM signal to assure that under the conditions of power densities used, the cubic relation holds both for the sample and the reference. Great disparity in the path lengths of the sample and the reference or in the DFWM signal levels should be avoided. For thin films (micrometers), special CS_2 cells with dimensions of $< 100\,\mu m$ should be used. For solid samples and dilute solutions that have low DFWM signals, it is advisable to use a different reference with low values of $\chi^{(3)}$. For the latter case, CCl_4 is suggested as a better reference material. Widely varying values of $\chi^{(3)}$ have been reported for CS_2, making it difficult for a new researcher to decide which value to use. Based on studies of a number of organic liquids and consistencies of their relative $\chi^{(3)}$ values, we suggest a value of 5×10^{-13} esu for CS_2 when using pulses $\leqslant 4$ ps.

For absorbing samples, a correction due to absorption losses is needed which yields the following modification of equation (9.3):

$$\frac{\chi^{(3)}_{sample}}{\chi^{(3)}_{CS_2}} = \left(\frac{n^0_{sample}}{n^0_{CS_2}}\right)^2 \frac{l_{CS_2}}{l_{sample}} \sqrt{\frac{I_{sample}}{I_{CS_2}}} \frac{\alpha l_{sample}}{\exp(-\alpha l_{sample}/2)(1 - \exp(-\alpha l_{sample}))} \tag{9.4}$$

In the above equation α is the linear absorption coefficient. The CS_2 sample has been assumed to be nonabsorbing at the wavelength of study.

It can be seen from equations 9.3 and 9.4 that the DFWM method of determination of $\chi^{(3)}$ requires the knowledge of the linear refractive index of the sample. For a liquid sample one can simply use an Abe refractometer. For solid samples, a crude method is to use index matching liquids. For thin films, a convenient method is the m-line technique proposed by Ding and Garmire (1983, Singh and Prasad 1988) which uses a quasi-waveguide arrangement.

Unlike a true waveguide, in a quasi-waveguide the refractive index (n) of the guiding film is less than that of the substrate (coupling prism in this case). The light in a quasi-waveguide structure is guided by the air–film interface, while the substrate–film interface is leaky, because the light propagating in a zigzag path suffers only partial reflection at this interface. In a waveguide, interference effects permit only modes of certain discrete values of propagation constants (see Chapter 11). In-plane scattering of light into other directions of the same mode or other modes gives rise to a series of lines, called m-lines, which represent the mode spectra of the waveguide. These m-lines carry information on the substrate and film (guiding) properties. If N_i and N_j are the effective refractive indices for the ith and jth TE modes, respectively, m the order of mode, and λ the wavelength of light used, the refractive index of the film is given by

$$n^2 = N_i^2 + \frac{(m_i + 1)^2}{(m_j + 1)^2 - (m_i + 1)^2}(N_i^2 - N_j^2) \qquad (9.5)$$

For the geometry used in this case, the effective index N of a given mode is related to the synchronous angle θ with respect to the surface normal for the given mode and the refractive index n_p of the coupling element:

$$N = n_p \sin \theta \qquad (9.6)$$

An experimental setup used for determining refractive indices of films is shown in Figure 9.4. The quasi-waveguide structure is formed by coating a thin film ($\sim 1\,\mu m$) of the material on the clean flat surface of a high refractive index (e.g.,

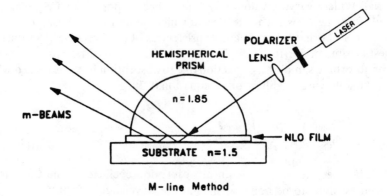

M-line Method

Figure 9.4 Experimental arrangement for the m-line method to determine the refractive index of a film. NLO film represents the nonlinear medium.

$n_p = 1.783$ at 632.8 nm) glass hemispherical prism. The film can be deposited by evaporating the substance (in case of a small molecule or oligomer) in a vacuum chamber evacuated to 10^{-5} torr. For high molecular weight oligomers or polymers, spin-coating or doctor-blading methods can be used.

The quasi-waveguide structure thus formed is mounted on a rotary stage and light from a TE polarized laser beam (e.g., an He–Ne laser for $\lambda = 632.8$ nm) is focused into the base of the hemispherical coupler. By turning the rotary stage, the m-lines can be observed on the screen when a particular TE mode of the quasi-waveguide is excited. Coupling angles are measured for each selectively excited TE mode and with the use of equations 9.5 and 9.6 the refractive index of the film can be evaluated.

For frequency dispersion studies of $\chi^{(3)}$ in the DFWM experimental arrangement (Figure 9.3), one can change the dye in the oscillator and/or amplifier. Alternatively, one can generate a continuum by focusing the subpicosecond amplified pulses on a water cell, select another wavelength, and amplify it.

The most important merit of the method is that one can get valuable information about the various nonlinear processes that contribute to the intensity dependence of the refractive index. The information obtained on the time response of optical nonlinearity is important in identifying the mechanism of optical nonlinearity. The method is highly sensitive because the signal is generated as a phase-matched nonlinear optical response. The drawback of the method is that it is highly sophisticated, which requires picosecond time resolution and careful control of experimental conditions and data analysis.

9.2.4 Optical Kerr Gate

This method utilizes the optical Kerr effect, discussed in Chapter 8, whereby an intense linearly polarized light pulse traveling through an optically isotropic $\chi^{(3)}$ medium induces optical birefringence. This optically induced birefringence can be probed by a linearly polarized weaker probe pulse to obtain the birefringence $\delta n = \delta n_{\parallel} - \delta n_{\perp}$, which provides a measurement of $\chi^{(3)}$. Furthermore, evolution of the birefringence and hence the response time of $\chi^{(3)}$ can be probed by delaying the probe beam with respect to the intense pump beam. Because the optically induced anisotropy (δn) is small, the method is better suited for isotropic materials such as liquids or amorphous (unoriented) polymers.

This method was described in detail by Ho (1984). We briefly discuss it following Ho's approach. A suitable picosecond experimental arrangement for the optical Kerr gate experiment is shown in Figure 9.5. A picosecond pulse of appropriate wavelength is split in two parts: a strong (100 MW/cm^2) orienting pump pulse and a weak probe pulse. The weak pulse undergoes a variable delay.

Linear polarizations of both beams are selected through polarizers (P). Although the two beams can be used in a collinear geometry, generally one prefers to use them crossing at an angle θ (usually $\theta = 45°$) to avoid any extra

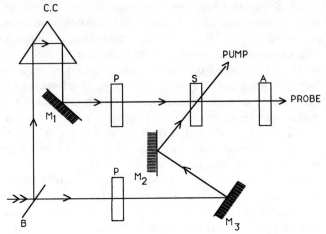

Figure 9.5 Schematics of Kerr gate experimental arrangement. A, analyzer; P, polarizer; S, sample; M, mirrors; B, beam splitter; C.C., corner cube for optical delay.

optical element in the beam. To obtain information on the optically induced birefringence, the probe beam passes through a cross polarizer (A). The intensity of the probe beam transmitted through the analyzer is measured with a slow photodetector system as a function of the delay time between the probe and the orienting pulses.

The signal $I_t(\tau)$ transmitted by the analyzer for a time dealy τ between the pump and probe beam is given as

$$I_t(\tau) = \int_{-\infty}^{\infty} \langle E_{\text{probe}}^2(t - \tau_{\text{D}}) \rangle \sin^2 \left[\frac{\delta\phi(t)}{2} \right] dt \qquad (9.7)$$

where E_{probe} is the electric field of the probe beam and $\delta\phi$ is the phase retardation of the probe beam due to optically induced birefringence. It is given as

$$\delta\phi(t) = \frac{2\pi l}{\lambda} \delta n(t) \qquad (9.8)$$

where l is the sample path length, λ is the wavelength of the probe beam, and $\delta n(t)$ is the induced refractive index change which can be written as

$$\delta n(t) = n_2^f \langle E_1^2(t) \rangle + \sum_i \frac{n_{2i}^s}{\tau_i} \int_{-\infty}^{t} \langle E_1^2(t') \rangle \exp\left(-\frac{t - t'}{\tau_i} \right) dt' \qquad (9.9)$$

In the above equation n_2^f is the intensity dependent refractive index derived from fast responding electronic $\chi^{(3)}$; the second term consists of various slow responding (dynamic) nonlinearity with response time τ_i. The electric field of the pump pulse is represented by E_1. In the case of nonresonant electronic nonlinearity,

$$\delta n = n'_2 I_{\text{pump}} \tag{9.10}$$

which will define the peak value of the phase retradation $\delta\phi$ (equation 9.8) to be

$$\delta\phi = \frac{2\pi l}{\lambda} n'_2 I_{\text{pump}} \tag{9.11}$$

From a measurement of the peak value of transmitted probe signal I, one obtains $\delta\phi$ and hence n'_2. The n'_2 measurement by the optical Kerr gate method is related to $\chi^{(3)}$ by the following expression

$$n'_2 = \frac{12\pi}{n_0}(\chi^{(3)}_{1111} - \chi^{(3)}_{1122}) \tag{9.12}$$

In general, the optical Kerr gate method provides direct information on the $\chi^{(3)}_{1111}$ component, usually measured by DFWM, only when $\chi^{(3)}_{1122}$ is relatively small. However, at a wavelength far off from resonance and for isotropic medium, a purely electronic nonlinearity leads to $\chi^{(3)}_{1122} = \frac{1}{3}\chi^{(3)}_{1111}$. In such a case, the measurement of n'_2 by this technique will yield a direct determination of $\chi^{(3)}_{1111}$.

As discussed above, the optically induced birefringence is small, which puts stringent conditions on the experimental conditions. One requires a high power density, a very stable laser source (minimum pulse to pulse fluctuation), a clean polarization, a long interaction length, and an isotropic medium. Consequently, the method is better suited for liquid samples.

9.2.5 Self-Focusing Methods

As described in Chapter 8, self-focusing (or defocusing) processes are derived from the combined action of an intensity-dependent refractive index and a spatial variation of the laser intensity in which the beam is more intense in the center than at the edges. The self-focusing effect has also been used for measurement of optical nonlinearity. Two methods utilizing self-focusing have been used for the measurement of $\chi^{(3)}$. The first one, called the power-limiting method, was developed by Soileau et al. (1983). This method utilizes the experimental configuration shown in Figure 9.6. The beam path at low input power is represented by the solid line. The laser beam is focused into the nonlinear medium under investigation by a lense L1. The transmitted power is imaged by a lens L2 through an aperture onto a detector D. As the input power is increased, the transmitted power measured at D initially increases linearly till a critical value P_C for self-focusing is reached. At this point the nonlinear

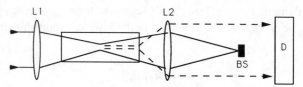

Figure 9.6 experimental arrangement for self-focusing. L_1 and L_2, lenses; BS, aperture; D, detector.

medium acts as a lens which changes the focusing conditions onto detector D. The beam path under this condition is shown by the dotted line. As a result, deviation from the linear relationship between the incident power and the transmitted power is observed. Beyond the critical power P_C, many other nonlinear processes (nonlinear absorption, laser induced breakdown, beam self-trapping) give rise to a leveling off of the transmitted power. The result is that this arrangement at $P > P_C$ acts as a power-limiting device. Marburger (1977) used the solution of the nonlinear wave equation for a focused Gaussian beam to relate P_C to n_2, the nonlinear refractive index coefficient. According to him this relationship in esu is given as

$$P_C = \frac{3.72C\lambda^2}{32\pi^2 n_2} \tag{9.13}$$

Therefore, from a determination of P_C by the above experiment, one can determine n_2. This method was used by Frazier et al. (1987) to measure the nonlinearity of palladium polyyne film. More recently, Winter et al. (1988) used this method to measure the third-order nonlinearity of ferrocene in the molten form and in solutions.

Recently, Sheik-bahae et al. (1989) used a z-scan method based on self-focusing to measure the sign as well as the magnitude of $\chi^{(3)}$. The z-scan experimental arrangement they used is shown in Figure 9.7. A Gaussian beam is tightly focused by a lens and passed through the nonlinear medium. The

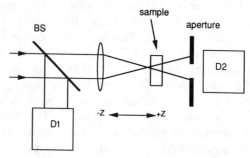

Figure 9.7 Schematics of the z-scan method. BS, beam splitter; D_1 and D_2, detectors.

output power is measured by detector D_2 after passing through a finite aperture. This transmitted power is obtained as a function of the sample position on the z axis measured with respect to the focal plane. From a quantitative analysis of this z-scan transmitted power profile, they obtained both the magnitude and the sign of the nonlinearty. Here, only a qualitative description of the z-scan transmitted power profile will be presented. Let us consider a case where a nonlinear material with a nonlinear refractive index $n_2 < 0$ (case of self-defocusing) and thickness less than the depth of focus is located in front of the focus ($-z$). As the sample is moved toward the focus, the increased power density gives rise to a negative lensing effect with a net result of collimating the beam and consequently increasing the power throughput through the aperture. With the sample moved to the $+z$ side of the focus, the negative lensing effect diverges the beam and, therefore, reduces the power throughput passing through the aperture. At $z = 0$, the effect is analogous to placing a thin lens at the focus that will cause a minimal far-field pattern change. The resultng curve for $n_2 < 0$ can be expected to be of the form represented in Figure 9.8. For a self-focusing nonlinearity ($n_2 > 0$), a profile opposite in sign to that represented in figure 9.8 will be observed. In other words, reduced transmission for $-z$ and enchanced transmission for $+z$ will be observed for $n_2 > 0$.

These self-focusing methods have the appeal that they are simple and, in principle, offer information on both the sign and the magnitude of nonlinearity. The method, however, puts the stringent requirement on the beam quality that it must be Gaussian (TEM_{00}). Because it requires a high power density and long interaction length, the method is not useful for thin polymeric films. Frazier

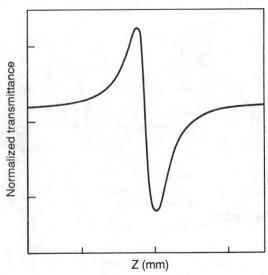

Figure 9.8 The normalized transmittance profile obtained as a function of z-scan in a medium exhibiting self-defocusing.

et al. (1987) used a polymer film of thickness $120\,\mu$m. Sheik-bahae et al. (1989) used a 1-mm-thick CS_2 cell and a 2.5-mm-thick BaF_2 crystal in their investigation using the z-scan method. The method does not provide any information on the time response of the nonlinearity. Furthermore, because of the use of longer path lengths, even weakly absorbing samples will provide large refractive index changes derived from the thermal effect. This effect will get more serious when a nanosecond pulse (with a large photon fluence) is used. A more serious problem with a solid sample may be the dielectric breakdown and sample damage at power densities needed for these self-focusing methods. The method may have some merit in qualitative screening of transparent liquid samples.

9.2.6   Surface Plasmon Nonlinear Optics

This method is especially suited for investigation of optical nonlinearities in ultrathin films, such as Langmuir–Blodgett monolayer and multilayer systems. First, we briefly discuss surface plasmons and the method used to excite them. Surface plasmons are electromagnetic waves that propagate along the interface between a metal and a dielectric material such as organic films (Wallis and Stegeman 1986). Since the surface plasmons propagate in the frequency and wave-vector ranges for which no propagation is allowed in either of the two

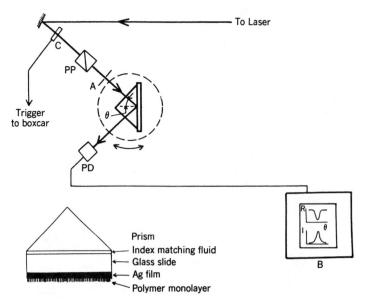

Figure 9.9   Schematics of the experimental arrangement for a surface plasmon experiment. θ, the coupling angle; C, chopper; PP, prism polarizer; A, aperature; PD, photodiode; B, boxcar integrator. The inset shows the layered system and prism assembly for the Kretschmann geometry.

media, no direct excitation of surface plasmons is possible. The most commonly used method to generate a surface plasmon wave is attenuated total reflection (ATR).

The Kretschmann configuration of ATR is widely used to excite surface plasmons (Wallis and Stegeman 1986). This configuration is shown in Figure 9.9. A microscopic slide is coated with a thin film of metal (usually a 400- to 500-Å-thick silver film by vacuum deposition). Then the organic film (such as polymer monolayer) to be investigated is coated on the metal surface. The microscopic slide is now coupled to a prism through an index-matching fluid. A laser beam is incident at the prism at an angle larger than the critical angle. The total attenuated reflection of the laser beam is monitored. At a certain θ_{sp}, the electromagnetic wave couples to the interface as a surface plasmon. At this angle the ATR signal drops. The angle is determined by the relationship

$$k_{sp} = kn_p \sin \theta_{sp} \qquad (9.14)$$

where k_{sp} is the wave vector of the surface plasmon, k is the wave vector of the bulk electromagnetic wave, and n_p is the refractive index of the prism. The surface plasmon wave vector k_{sp} is given by

$$k_{sp} = \frac{\omega}{c} \sqrt{\frac{\varepsilon_m \varepsilon_d}{\varepsilon_m + \varepsilon_d}} \qquad (9.15)$$

where ω is the optical frequency, c the speed of light, and ε_m and ε_d are the relative dielectric constants of the metal and the dielectric, respectively. In the case of a bare silver film, ε_d is the dielectric constant of air and the dip in reflectivity occurs at one angle. In the case of silver coated with an ultrathin organic film this angle shifts. In this experiment one measures the angle for the reflectivity minimum, the minimum value of reflectivity, and the width of the resonance curves. These observables are used for a computer fit of the resonance curve using a least-squares fitting procedure with the Fresnel reflection formulas, which yields three parameters: the real and imaginary parts of the refractive index and the film thickness. The $\chi^{(3)}$ values can be obtained from an intensity-dependence study of the refractive index. The experiment utilizes the study of angular shift (change in θ_{sp}) and broadening of the surface plasmon resonance as a function of intensity of the incident laser beam. Intensity dependence of the angle θ_{sp} gives the magnitude of the intensity-dependent refractive index coefficient n_2 and, consequently, $\chi^{(3)}$. An important advantage of this method is that one can also obtain the sign of n_2 (and hence $\chi^{(3)}$) from the direction of shift of θ_{sp}.

Combining equations 9.14 and 9.15 one can see that the change $\delta\theta$ in the surface plasmon resonance angle (the angle corresponding to minimum reflectivity; for simplicity the subscript sp is dropped) caused by changes $\delta\varepsilon_m$ and $\delta\varepsilon_d$ in the dielectric constants of the metal and film, respectively, is given by (Nunzi and Ricard 1984)

$$\cot \theta \, \delta\theta = \frac{1}{2\varepsilon_m \varepsilon_d (\varepsilon_m + \varepsilon_d)} (\varepsilon_m^2 \delta\varepsilon_d + \varepsilon_d^2 \delta\varepsilon_m) \tag{9.16}$$

Since $|\varepsilon_m| \gg |\varepsilon_d|$, the change in θ is much more sensitive to a change in ε_d than to a change in ε_m. Therefore, this method appears to be ideally suited to obtain $\delta\varepsilon_d$ as a function of laser intensity and hence to obtain n_2. In actual experiment, however, utmost care is needed to interpret the data and obtain meaningful results. The reason is that for the visible wavelength region(~632 nm), the surface plasmon propagation is of the order of 10 μm in which the wave is completely damped. Therefore, even though the coated polymer film may be optically clear at this wavelength, the damping in the metal leads to heating. The $\delta\varepsilon_d$ and $\delta\varepsilon_m$ values in such a situation have two contributions: one due to intrinsic $\chi^{(3)}$ and the other due to thermal effect ($\partial n/\partial T$ or $\partial\varepsilon/\partial T$). Further complication may arise if $\partial\varepsilon_d/\partial T$ and $\partial\varepsilon_m/\partial T$ have different signs. In such a case, depending on what contribution dominates, one may see a positive shift ($\delta\theta$ positive) or a negative shift as the laser beam intensity is increased. Experimental study of poly-4-BCMU monolayer films using this method has been discussed by Prasad (1988) and clearly reveals the complication due to thermal effects. This method is not recommended for measurement of weak optical nonlinearity, especially with nanosecond pulses. With very short pulses (< 10 ps) at low repetition rate, the method may yield meaningful results.

9.2.7 Nonlinear Fabry–Perot Method

This methd is based on the intensity-dependent phase shift resulting from the intensity-dependent refractive index. A Fabry–Perot etalon or interferometer is an optical resonator formed by two parallel mirrors of high reflectivity ($R > 99\%$). An excellent description of the Fabry–Perot etalon can be found in the book *Optical Electronics* by Yariv (1985). We briefly discuss here the principles of a Fabry–Perot etalon, which is schematically shown in Figure 9.10. It consists of two mirrors separated by a medium of refractive index n and thickness l. A beam entering the etalon at an angle θ' undergoes successive reflections from these two mirrors. The phase delay between two transmitted waves I_1 and I_2, which differ in pathlength by one round-tip within the cavity, is given by

$$\delta = \frac{4\pi n l \cos\theta}{\lambda} \tag{9.17}$$

For a normal incidence, which is generally used,

$$\delta = \frac{4\pi n l}{\lambda}$$

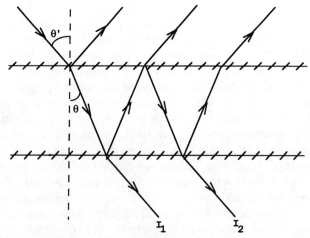

Figure 9.10   Schematics of successive reflections of rays in a Fabry–Perot etalon.

The total fraction of the incident intensity transmitted is determined by the sum of all the intensities I and is given by

$$\frac{I_t}{I_i} = \frac{(1-R)^2}{(1-R)^2 + 4R \sin^2(\delta/2)} \tag{9.18}$$

This expression assumes no losses in the cavity. If there are losses due to scattering or absorption in the medium, a correction must be introduced.

The transmission has a maximum whenever δ is an integral multiple of 2π. This condition defines the resonances of a Fabry–Perot cavity. Therefore, for cavity resonances,

$$\delta = \frac{4\pi nl}{\lambda} = 2m\pi \tag{9.19}$$

or

$$\frac{nl}{\lambda} = \frac{m}{2} \tag{9.20}$$

If one scans the length of the cavity (separation between the two mirrors) one would see a series of maxima. The separation between the two successive maxima is defined by

$$\frac{n(l_1 - l_2)}{\lambda} = \frac{1}{2} \tag{9.21}$$

Therefore, from a measurement of the cavity lengths corresponding to the two successive maxima one can obtain the refractive index of the medium. If the refractive index of the medium is intensity dependent, then the resonance condition will be different at low power and high power. In other words, from a power-dependence study of the resonance conditions (9.21), one can determine n_2 and hence $\chi^{(3)}$. This method, like any method that utilizes intensity-dependent phase shift, also provides the sign of nonlinearity by observing the direction in which the successive maxima move while scanning the cavity length.

For solution measurements, one can use a cell in which a wall can be a mirror. For solid samples, a thin film of the nonlinear material can directly be deposited on one of the mirrors. For the resolution needed to determine n_2, the separation between the mirrors must be very precisely varied. Therefore, a high-resolution piezoelectric scan control of the cavity is required. This method is very sensitive to any residual (or weak absorption) in the sample, which gets amplified by successive reflections in the cavity. Therefore, thermally induced nonlinearity at high intensity levels will dominate. Like in the surface plasmon optics method, one should use short pulses at a low repetition rate to avoid complications due to any residual absorption. The method is not suitable for an ultrathin film because of the small Δn produced, which will be hard to measure accurately. Uchiki and Kobayashi (1988) have used a method that utilizes the quadratic electrooptic coefficient of a film in a Fabry–Perot geometry, which, according to them, provides good sensitivity. The quadratic electrooptic effect is derived from the nonlinear susceptibility $\chi^{(3)}(-\omega;\omega,0,0)$ whereby the refractive index of a material depends on the square of the applied dc or low-frequency ac field. They used this method to measure $\chi^{(3)}$ of 4-diethylamino-4'-nitrostilbene (DEANS) doped in a polycarbonate film. The Fabry–Perot cavity contained a polymer thin film doped with DEANS. The cavity mirrors were thin aluminum films that also served as electrodes for the applied electric field. A dc electric field-induced change of the refractive index was observed by monitoring the change in transmission behavior of the Fabry–Perot cavity.

9.3 MEASUREMENT OF MICROSCOPIC NONLINEARITIES, γ

The measurement of $\chi^{(3)}$ of solutions can be used to determine the microscopic nonlinearities γ (the molecular second hyperpolarizabilities) of a solute, provided γ of the solvent is known. This measurement also provides information on the sign of γ (and hence $\chi^{(3)}$) of the molecules (solute) if one knows the sign of γ for the solvent. Under favorable conditions one can also use solution measurements to determine whether γ (and hence $\chi^{(3)}$) is a complex quantity. The method utilizes two basic assumptions: (1) The nonlinearities of the solute and the solvent molecules are additive, and (2) Lorentz approximation can be used for the local field correction. Under these two assumptions one can write the $\chi^{(3)}$

of the solution to be equal to

$$\chi^{(3)} = f^4[N_{\text{solute}}\langle\gamma\rangle_{\text{solute}} + N_{\text{solvent}}\langle\gamma\rangle_{\text{solvent}}] \tag{9.22}$$

In equation 9.22 $\langle\gamma\rangle$ represents the orientationally averaged second hyperpolarizability defined as

$$\langle\gamma\rangle = \tfrac{1}{5}(\gamma_{xxxx} + \gamma_{yyyy} + \gamma_{zzzz} + 2\gamma_{xxyy} + 2\gamma_{xxzz} + 2\gamma_{yyzz}) \tag{9.23}$$

N is the number density in the units of number of molecules per cm^3. f is the Lorentz correction factor for the local field given by

$$f = \frac{n^2 + 2}{3} \tag{9.24}$$

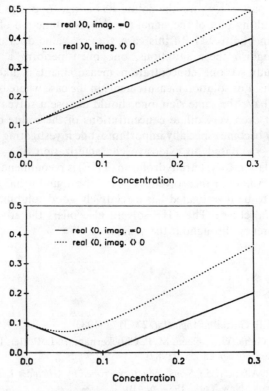

Figure 9.11 Concentration dependence of $\chi^{(3)}$ of a solution. *Top:*—, the real part of $\chi^{(3)}$ for the solute is positive and the imaginary part is zero;..., the real part is positive but the imaginary part is nonzero. *Bottom:*—the real part of $\chi^{(3)}$ for the solute is negative with the imaginary part zero;..., the real part is negative with the imaginary part nonzero.

If the solute is in dilute concentration, equation 9.22 can be written as

$$\chi^{(3)} = f^4 [N_{solute} \langle \gamma \rangle_{solute}] + \chi^{(3)}_{solvent} \tag{9.25}$$

If the $\langle \gamma \rangle$ values for both the solute and the solvent have the same sign, equation 9.25 predicts a linear dependence of $\chi^{(3)}$ with the concentration of the solution as shown in Figure 9.11. By a least-squares fit of this concentration dependence, one can readily obtain $\langle \gamma \rangle$ of the solute molecule. If the signs of the nonlinearities are opposite but both are real quantities, a concentration dependence study would yield a curve, also shown in Figure 9.11. The value of the resultant $\chi^{(3)}$ of the solution decreases and goes to zero at some concentration. In the case when $\langle \gamma \rangle$ of the solute is complex, equation 9.25 yields the signal given by

$$I \propto |\chi^{(3)}|^2 = |f^4 [N_{solute} \langle \gamma^{Re}_{solute} \rangle] + \chi^{(3)}_{solvent}|^2 + |f^4 N_{solute} \langle \gamma^{Im}_{solute} \rangle|^2 \tag{9.26}$$

When the real part of $\langle \gamma \rangle$ for the solute has a sign opposite to that for the solvent, the resulting plot of the signal (or $\chi^{(3)}$) as a function of concentration is also shown in Figure 9.11. In this case, the $\chi^{(3)}$ value does not go through zero. To distinguish these situations, one must perform a concentration-dependence study. A one concentration measurement is likely to lead to erroneous results. For solution measurements in the case where both the solvent and the solute have the same sign, one should choose a solvent that has very low $\chi^{(3)}$ so that even very dilute concentrations of the solute can be studied. This factor may become especially important while investigating molecules with relatively low $\langle \gamma \rangle$. Based on extensive characterization of various common solvents in our laboratory, tetrahydrofuran (THF) is recommended as a suitable choice. The $\chi^{(3)}$ value for this solvent is $\sim 10^{-14}$ esu and it has a positive sign. If the material to be investigated has a relatively good solubility in THF, this solvent is a good choice. The THF solvent also offers the advantage of wide optical transparency throughout the visible.

REFERENCES

Ding, T. M., and E. Garmire, *Appl. Opt.* **22**, 3177 (1983).

Frazier, C. C., S. Guha, W. P. Chen, M. P. Chockerham, P. L. Porter, E. A. Chauchard, and C. H. Lee, *Polymer* **28**, 553 (1987).

Ho, P. P., in R. R. Alfano (Ed.), *Semiconductors Probed by Ultrafast Laser Spectroscopy*, Vol. 2, Academic, New York, 1984, p. 410.

Kajzar, F., and J. Messier, *Phys. Rev. A* **32**, 2352 (1985).

Kajzar, F., and J. Messier, in D. S. Chemla and J. Zyss (Eds.), *Nonlinear Optical Properties of Organic Molecules and Crystals*, Academic, Orlando, Fl, 1987a, p. 51.

Kajzar, F., and J. Messier, *Rev. Sci. Instrum.* **56**, 2081 (1987b).

Marburger, J. H., in J. H. Sandom and S. Stenholm (Eds.), *Progress of Quantum Electronics*, Pergamon, New York, 1977 p. 35.

Meredith, G. R., B. Buchalter, and C. Hanzlik, *J. Chem. Phys.* **78**, 1533 (1983a); **78**, 1543 (1983b).

Nunzi, J. M., and D. Ricard, *Appl. Phys. B* **35**, 209 (1984).

Prasad, P. N., in P. N. Prasad and D. R. Ulrich (Eds.), *Nonlinear Optical and Electroactive Polymers*, Plenum, New York, 1988.

Prasad, P. N., J. Swiatkiewicz, and J. Pfleger, *Mol. Cryst. Liq. Cryst.* **160**, 53 (1988).

Rao, D. N., P. Chopra, S. K. Ghoshal, J. Swiatkiewicz, and P. N. Prasad, *J. Chem. Phys.* **84**, 7049 (1986).

Sheik-bahae, M., A. A. Said, and E. W. VanStryland, *Opt. Lett.* **14**, 955 (1989).

Singh, B. P., and P. N. Prasad, *J. Opt. Soc. Am. B* **5**, 453 (1988).

Soileau, M. J., W. E. Williams, and E. W. VanStryland, *IEEE J. Quant. Electron.* **QE-19**, 731 (1983).

Stevenson, S. H., and G. R. Meredith, *SPIE Proc.* **682**, 147 (1986).

Thalhammer, M., and A. Penzkofer, *Appl. Phys. B* **32**, 137 (1983).

Uchiki, H., and T. Kobayashi, *Mat. Res. Soc. Symp. Proc.* **109**, 373 (1988).

Wallis, R. F., and G. I. Stegeman (Eds.), *Electromagnetic Surface Excitations* Springer-Verlag, Berlin, 1986.

Winter, C. S., S. N. Oliver, and J. D. Rush, *Opt. Commun.* **69**, 45 (1988).

Yariv, A., *Optical Electronics*, Holt, Rinehart and Winston, New York, 1985.

10

A SURVEY OF THIRD-ORDER NONLINEAR OPTICAL MATERIALS

10.1 PERSPECTIVE

Compared to the extensive amount of research conducted on synthesis and characterization of new molecular structures for second-order nonlinear optical applications (Chapter 7), the study of third-order nonlinear processes on molecular materials has received relatively limited attention. Recently, the scope of synthesis and characterization of third-order materials has expanded considerably. The impetus for this increased activity has been a quest for fundamental understanding of the structure–property relationship for third-order optical nonlinearity and the strong technological interest in all-optical signal processing provided by the third-order processes. As a result, the data base for third-order nonlinear materials is rapidly expanding, although it is still highly limited compared to that for second-order materials.

It is interesting to look at the third-order nonlinear processes in organic structures from a historical perspective. Although studies of third-order optical nonlinearities go back as far as the 1960, during which several measurements of optical Kerr effect were performed on organic liquids, this field showed only modest activity. The first systematic investigation of third-order nonlinearity on conjugated systems was conducted by Hermann et al. (1973). They measured the γ value for *trans* β-carotene, a polyene with 11 double bonds (discussed below in detail). This work generated considerable interest in conjugated structures and spurred theoretical analysis by Rustagi and Ducuing (1974) who recognized the importance of π conjugation in determining the third-order nonlinear optical properties. Sauteret et al. (1976) reported the first investigation of a conjugated polymer where they studied third-harmonic generation in polydiacetylene produced by solid-state polymerization of a monomer crystal. Their work clearly revealed a strong dependence of the $\chi^{(3)}$ value on the π-electron conjugation. The focus of much materials research during the late 1970s and early 1980s remained on the second-order nonlinear organic materials, particularly molecular design and crystal growth.

As discussed in Chapter 7, much of the effort in design of the second-order material has been placed on synthesizing molecules with low-lying charge-transfer states and preparing noncentrosymmetric condensed phases for expressing the bulk nonlinearity ($\chi^{(2)}$). Therefore, ordered crystalline states and poled polymers formed important classes of $\chi^{(2)}$ materials. As discussed in Section 10.2, the third-order optical nonlinearity does not require noncentrosymmetric molecular structures, intramolecular charge transfer, or any bulk order. The lack of symmetry control may widen the range of possible structures for consideration for third-order nonlinear optical applications, although electronic and other material requirements for practical applications may be extremely difficult to achieve.

Synthesis can play a very important role in the development of third-order materials by using the tremendous flexibility a molecular material offers to modify its chemical structure. Unfortunately, our theoretical understanding of third-order nonlinearity is very limited and guidelines for molecular structural requirements for enhancing the microscopic coefficient, γ, can be only crudely stated. From this point of view, systematic studies of γ measurements on sequentially built and systematically derivatized chemical structures can provide useful insight for identifying structural features for enhancement of third-order nonlinearity. Therefore, molecular design and syntheses of novel structures are expected to play an important role in the development of third-order nonlinear materials. A widely investigated class of material in this regard is conjugated polymers. While earlier work focused mainly on the various polydiacetylenes, only recently have efforts expanded to include other classes of conjugated polymers.

In this chapter the structural requirements for enhanced third-order optical nonlinearity are reviewed. Then measurements of third-order optical nonlinearities are surveyed. This survey is limited to liquid and solid phases and covers some selected important structures. It is, therefore, by no means a thorough review of all reported measurements.

Third-order nonlinear optical interactions give rise to a large variety of processes, as discussed in Chapter 8. In comparison, second-order processes are rather limited in number, the principal manifestations being frequency mixing and the electrooptic effect. In Chapter 9, it was pointed out that one can use many different processes for characterization of third-order optical non-linearities. This provides tremendous flexibility in selecting a method to measure $\chi^{(3)}$. The data surveyed here are a good representation of this diversity in measurement techniques. Results presented here are for (1) third-harmonic generation, (2) degenerate four-wave mixing, (3) Kerr effect, and (4) EFISH. This diversity in the measurements, however, also poses problems in correlation between $\chi^{(3)}$ and γ of a material measured by different techniques, since different methods probe different $\chi^{(3)}$ terms. Further complication is created by the use of different reference standards used in various techniques. The complexity involved in the interpretation of the third-order measurements combined with the factors mentioned above may compromise the reliability of some reported measurements. In this chapter measurements are listed as reported in the literature and it is left to the reader to ascertain the underlying assumptions and conditions of the experiment.

10.2   STRUCTURAL REQUIREMENTS FOR THIRD-ORDER OPTICAL NONLINEARITY

Electronic structural requirements for third-order nonlinear organic systems are different from those for second-order materials. Although the understanding of structure–property relationship for third-order effect is highly limited, the microscopic theoretical models discussed in Chapter 3 predict large nonresonant third-order optical nonlinearity associated with delocalized π-electron systems. These molecular structures do not have to be acentric because γ is a fourth-rank tensor. Nevertheless, they must possess anharmonicity. Conjugated polymers with alternate single and multiple bonds in their backbone structure provide a molecular frame for extensive conjugation and have emerged as the most widely studied group of $\chi^{(3)}$ organic materials. Examples of conjugated polymers are polydiacetylenes, poly-p-phenylenevinylene, and polythiophenes, which are discussed below in detail.

The optical nonlinearity is strongly dependent on the extent of π-electron delocalization from one repeat unit to another in the polymer (or oligomer) structure. As might be expected this effective delocalization is not equally manifested in all conjugated polymers but depends on the details of repeat unit electronic structure and order. For example, in a sequentially built structure, discussed below, the π-delocalization effect on γ is found to be more effective for the thiophene oligomers than it is for the benzene oligomers.

The largest component of the γ tensor is in the conjugation direction. Therefore, even though no particular bulk symmetry is required for nonzero $\chi^{(3)}$, a medium in which all conjugated polymeric chains align in the same direction should have a larger $\chi^{(3)}$ value along the chain direction relative to that in an amorphous or disordered form of the same polymer. Studies of $\chi^{(3)}$ in ordered or stretch-oriented polymers, as discussed below, confirm this prediction. Finally, the polymeric chains should pack as closely as possible to maximize the hyperpolarizability density and hence $\chi^{(3)}$.

Since third-order processes do not require bulk order, all materials, including liquids, liquid crystals, and solids, exhibit $\chi^{(3)}$. In the solid state, one can use various bulk forms such as crystalline materials, amorphous polymers, glass structures, oriented polymers, polymer blends as well as mono- and multilayer Langmuir–Blodgett films. Amorphous polymeric or glassy structures are particularly useful media for third-order nonlinear processes because they can be readily processed into device structures requiring film or fiber formats.

It should be noted that extensive π conjugation is often associated with enhanced conductivity in organic systems. Conducting polymers, which have been a topic of considerable interest in the past (Skotheim 1986), are of considerable interest in the context of nonlinear optics. On the other hand, no direct correlation between high conductivity and $\chi^{(3)}$ has been observed. As discussed below, polyacetylene and polythiophene, which in the doped state exhibit very high electrical conductivity, also exhibit relatively large third-order nonlinear optical effects in the undoped (nonconducting state). It should be remembered that conductivity is a bulk property that is heavily influenced by intrachain as well as interchain carrier transports. In contrast, the origin of third-order nonlinearity in conjugated polymers is primarily microscopic, determined by the structure of the polymer chain. Therefore, a conjugated polymer may be a very good $\chi^{(3)}$ material but not necessarily a good conductor. Polydiacetylene is a good example; it exhibits a large nonresonant $\chi^{(3)}$ value but is a wide band-gap semiconductor and therefore a poor conductor.

10.3 LIQUID MATERIALS AND SOLUTIONS

A large number of materials have been investigated in the liquid phase and in solutions. As discussed in Chapter 9, Section 9.3, liquid and solution phase measurements yield an orientationally averaged value of the nonlinear coefficient. Earlier work focused on saturated alkanes with general formula C_nH_{2n+2} and cycloalkanes with general formula C_nH_{2n}. Hermann and Ducuing (1974) investigated the third-order hyperpolarizabilities of long-chain alkanes ($n = 5$–9, 14, 15) and cycloalkanes ($n = 5$–8) using third-harmonic generation with fundamentals at 1.89 and 2.47 μm. The values for γ were reported to be in the range 1.8–5.2 $\times$ 10^{-36} esu. Meredith et al. (1983a) reported the $\chi^{(3)}$ and γ values for neat liquids of several halogen-substituted alkanes (such as CCl_4 and $CHCl_3$), again using third-harmonic generation with the fundamental wave length of 1.91 μm. They also reported γ values of the order of magnitude of 10^{-36} esu. Kajzar and Messier (1985, 1987b) investigated neat liquids of several alkanes and their derivatives with the general formula

$$CH_3–(CH_2)_{n-2}–CH_2X \qquad \text{with } X = H, Cl, Br, I$$

The emphasis of this work was to test the bond additivity model of optical nonlinearity discussed in Chapter 2. In their earlier work (Kajzar and Messier 1985) they used the fundamental wavelengths of 1.064 and 1.907 μm, while in their latter work (1987b) they used only 1.064 μm. The later work also addressed the question of local electric field and the concepts of group hyperpolarizabilities. Recently, Samoc et al. (1989) have investigated the role of heavy atoms in various halomethanes by using the EFISH method as well as DFWM. The γ values

TABLE 10.1 Reported γ Values for Some Liquids

Compounds	Structure	Measurement Techniques	Wavelength Used (μm)	$\gamma \times 10^{36}$ (esu)	Reference
Benzene	(benzene ring)	THG	1.907	4.07	Kajzar and Messier (1985)
		THG	1.064	4.40	Hermann (1973)
		THG	1.89	2.4	Hermann (1973)
		THG	1.91	3.85	Meredith et al. (1983b)
		EFISH	1.318	2.06	Levine and Bethea (1975)
Allo-ocimene	Three double bonds	THG	1.89	9.7 ± 1.7	Hermann and Ducuing (1974)
trans-Retinol (vitamin A) (melt)	Five double bonds	THG	1.89	46 ± 12	Hermann and Ducuing (1974)
trans-Retinal (vitamin A aldehyde) (melt)	Six double bonds	THG	1.89	90 ± 20	Hermann and Ducuing (1974)
Nitrobenzene	(benzene–NO_2)	THG	1.91	5.37	Meredith et al. (1983b)
Iodobenzene	(benzene–I)	THG	1.91	8.19	Meredith et al. (1983b)
Pyridine	(pyridine ring)	THG	1.907	3.46	Kajzar and Messier (1985)
Thiophene	(thiophene ring, S)	DFWM	0.602	4.1	Zhao et al. (1988)

for these simple molecules were reported to be in the range of 10^{-36}–10^{-35} esu. They also found that iodo substitution significantly enhanced the γ value.

Aromatic hydrocarbons and their derivatives have been extensively investigated by various methods. Some of the important earlier reports quoted here have used third-harmonic generation (Hermann 1973, Meredith et al. 1983b), EFISH (Oudar and LePerson 1975, Levine and Bethea 1975), and four-wave mixing (Levenson and Bloembergen 1974). Benzene is the most widely investigated aromatic compound. The reported value of the microscopic nonlinearity $\langle \gamma \rangle$ for this molecule varies from 2.04 to 5.4 $\times$ 10^{-36} esu. Recent results obtained by subpicosecond degenerate four-wave mixing for benzene (Zhao et al. 1989) gives a $\langle \gamma \rangle$ value of 6 $\times$ 10^{-36}. This method, however, can have molecular orientational contributions, which can be minimized if sufficiently short pulses are used. Table 10.1 lists the reported values of some of the liquids with π-conjugated structures. The column Resonant/Nonresonant refers only to the one-photon resonance condition at the wavelength of study. Therefore, a wavelength labeled N (nonresonant) in Table 10.1 may still be near a two-photon resonance.

Solution measurements have been carried out on many types of conjugated structures, most of them being organic dyes and pigments. Meredith and Buchalter (1983) used third-harmonic generation measurement on paranitroaniline and its methyl derivatives (MNA) to investigate two different local field models: Debye and Onsager models. For p-nitroaniline in methanol, the Debye model yields $\langle \gamma \rangle = 27.8 \times 10^{-36}$ esu, while the value obtained from the Onsager model is 21.0×10^{-36} esu. Zhao et al. (1988) used the solution measurement on thiophene oligomers to obtain γ and then correlate it with the measurement of $\chi^{(3)}$ in the solid phase to test the validity of the local field approximation. For this particular case they found good agreement.

para-Nitroaniline   MNA

Hermann (1974) investigated various cyanine dyes in dimethyl sulfoxide solution using third-harmonic generation with the fundamental wavelength at 1.89 μm. The structure of a representative dye of this class is

In this study the number of alternate double bonds, N, was varied from 3 to 7. The value of $\langle \gamma \rangle$ reported varies from 5×10^{-35} esu for $N = 3$ to 2×10^{-31} esu for $N = 7$. This work was the first to establish a strong dependence of γ on the number of double bonds and, hence, the delocalization length. Hermann (1974) also reports that all dyes except one show a negative sign for γ, which is characteristic of a resonant nonlinearity.

Another conjugated structure investigated by many workers (Hermann et al. 1973, 1974, Maloney and Blau 1987, Meredith and Stevenson 1989) is β-carotene:

The value reported by Hermann et al. (1973) obtained from third-harmonic generation measurement in benzene solution with a wavelength of $\lambda_F = 1.89 \, \mu m$ is 4.8×10^{-33} esu. Maloney and Blau (1987) used degenerate four-wave mixing with 160-ps pulses from a Nd:Yag laser at a wavelength of $1.064 \, \mu m$, in an ethanol solution of β-carotene. The value they reported is $\sim 7.2 \times 10^{-31}$ esu. Meredith and Stevenson (1989) report a value of $\sim 1.1 \times 10^{-32}$ esu for β-carotene using a fundamental wavelength of $1.908 \, \mu m$ for third-harmonic generation. The large difference in the values obtained by THG and by DFWM may be attributable to the fact that DFWM has contributions other than purely electronic processes. This topic is discussed in detail in Chapter 8.

Stevenson et al. (1988) prepared and investigated the polarizabilities and hyperpolarizabilities of a large number of thiazole, benzthiazole cyanines, and azacyanines by third-harmonic generation. To understand the structure–property relationship, they analyzed the results in terms of structural anharmonicity factors discussed by Mehendale and Rustagi (1979). Their work suggests that enhancement of γ in a linear conjugated structure can be achieved by structural modifications that increase the anharmonicity factor. A symmetric aza substitution in the cyanine-like structure produced this effect. The values of γ reported by Stevenson et al. for the various dyes range from 5×10^{-37} esu to 4.9×10^{-34}. A representative of this class of structures is the following azabenthiazole with $\gamma = (4.90 \pm 0.40) \times 10^{-36}$ esu:

Azabenzthiazole

Zhao et al. (1988, 1989) have conducted a systematic study of third-order nonlinearities in sequentially built and systematically derivatized structures using solution measurements. These studies also had the objective to understand structure–property relationships. Thiophene and benzene oligomers with the structures

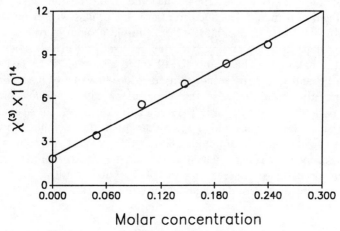

Thiophene oligomers

X, Y = H, NO$_2$, I

n = 1–6

Benzene oligomers

n = 1–3

were investigated by degenerate four-wave mixing. Various microscopic theories discussed in Chapter 3 predict a positive sign for γ for nonresonant cases. In agreement with the theoretical predictions, Zhao et al. found a positive value of γ for all the oligomers they investigated. A typical concentration dependence plot (for THF solution) along with a linear least-squares fit is shown in Figure 10.1. The dependence of the $\langle \gamma \rangle$ value on the number of repeat units was analyzed to determine the effect of π-electron delocalization. This plot for the thiophene oligomers is shown in Figure 10.2. This result clearly demonstrates that third-order nonlinearity is highly dependent on the π-electron conjugation length. When the number of repeat unit n increases, the effective conjugation length increases resulting in significant enhancement of $\langle \gamma \rangle$. Zhao et al. used the following equation to fit $\langle \gamma \rangle$ to a power law in n:

$$\langle \gamma \rangle = A + B(n - \delta)^C \qquad (10.1)$$

Figure 10.1 Concentration dependence of $\chi^{(3)}$ of the thiophene trimer in THF solution. The straight line is the linear least-squares fit.

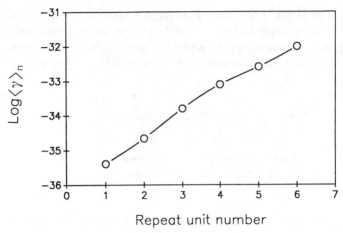

Figure 10.2   The dependence of $\langle \gamma \rangle$ on the number of repeat units n for the thiophene oligomers. The $\langle \gamma \rangle$ value is in the logarithmic scale.

Their analysis yields an exponent of ~ 4, which is to be compared with a value of 5 predicted by the free electron model (Rustagi and Ducuing 1974) and 3.5 predicted by ab initio calculations (Chopra et al. 1989). The experimental result reported by Zhao et al. (1989) also suggests that substitution by nitro and iodo groups enhances the third-order optical nonlinearity of the thiophene units.

Another interesting group of structures investigated (Kminek et al. 1989) in the solution phase are cumulenes with the general formula in which R, R', R'', and R''' are aryl or butyl groups. They represent extended π-electron structures and are rich in π-electron density. However, they are not typical conjugated structures, which represent alternate single and multiple bonds. Cumulenes have rod-like backbones (C=C chains) which possess two alternating, mutually perpendicular π bonds (Fischer 1964). The band gap corresponding to the lowest π–π^* transition is highly dependent on the value of n as well as on the nature of the substituents R, R', R'', and R'''. For example, aromatic group substitution in the cumulene structure of a given n value, especially when n is an odd number, reduces the band gap. This points to enhanced π conjugation when n is odd. This can be understood by seeing that, for the odd value of n, the orbitals at the ends of the molecule are in the plane of the cumulene molecular frame, which has the effect of extending the conjugation through the aromatic rings on both sides. Cumulenes with large values of n are unstable. An ab initio quantum calculation (Chopra et al. 1989) of the second hyperpolarizability of cumulenes reports another unusual feature in that the component of γ along the chain direction is negative in sign, whereas it is positive for conjugated structures involving alternate single and multiple bonds.

$$\begin{array}{c} R \\ \diagdown \\ \diagup \\ R' \end{array} C(\!\!=\!\!C)_n \begin{array}{c} R'' \\ \diagup \\ \diagdown \\ R''' \end{array}$$

Recently, Kminek et al. (1989) have synthesized various derivatized cumulenes with $n = 3$ and $n = 5$ and measured the orientationally averaged $\langle \gamma \rangle$ value in THF solution using subpicosecond degenerate four-wave mixing at 602 nm. For some of the derivatives (e.g., $R = R' = R'' = R''' = \langle \text{O} \rangle$ and $n = 5$), the wavelength 602 nm is within the absorption band; therefore, the measured value of γ is under resonant conditions. For others (e.g., $R = R'' = \langle \text{O} \rangle - OCH_3$, $R' = R''' = \langle \text{O} \rangle$, and $n = 3$; $R = R''' = \langle \text{O} \rangle$, $R' = R''' = C(CH_3)_3$ and $n = 5$), the absorption at this wavelength is negligible and hence the measurement at 602 nm should yield a nonresonant value. The $\langle \gamma \rangle$ values for the various cumulenes studied by Kminek et al. are in the range 10^{-32}–10^{-36} esu. The value of $\langle \gamma \rangle$ for 1,1,6,6-tetraphenyl hexapentaene with the structure

is -1×10^{-32} esu.

10.4 CONJUGATED POLYMERS

As discussed above, a polymeric backbone provides a structural framework for extensive π-electron delocalization. Consequently, conjugated polymers have been the subject of intense interest for the study of third-order nonlinear optical processes. In this section, we discuss some selective examples of conjugated polymers that have been investigated.

10.4.1 Polydiacetylenes

Polydiacetylenes are the most widely investigated class of conjugated polymers for third-order nonlinear optical effects. The general structure of this class of polymers is

Polydiacetylenes are prepared by the solid-state polymerization of the corresponding monomer (Schott and Wegner 1987):

$$R_2\text{–C}\equiv\text{C–C}\equiv\text{C–}R_1$$

This type of polymerization is an example of a topochemical polymerization where bonds rearrange but movement of nuclei is minimal. It can be induced

TABLE 10.2 $\chi^{(3)}$ **Measurement on Various Polydiacetylenes**

Substituents R_1, R_2	Common Name	Measurement Technique	Wavelength Used (μm)	Material Forms	R/NR	$\chi^{(3)}$ (esu)	Reference
$R_1 = R_2 = (CH_2)_4 OCONHC_6H_5$	TCDU	THG	2.62	Crystal	NR	$3.7 \pm 1.4 \times 10^{-11}$ Parallel to chain $<4 \times 10^{-13}$ perpendicular to chain	Sauteret et al. (1976)
PTS		THG	2.62	Crystal	NR	$1.6 \pm 1.0 \times 10^{-10}$ parallel to chain $<2.0 \times 10^{-12}$ $\perp$ to chain	Sauteret et al. (1976)
	PTS	DFWM	0.651 0.700	Crystal	R NR	9.0×10^{-9} 5×10^{-10}	Carter et al. (1987)
PTS-12		EFISH THG		Solution in DMF, CHCl$_3$	NR NR	5×10^{-10} $\gamma = 1.9 \times 10^{-34}$	Kajzar and Messier (1987)
Poly-4-BCMU		DFWM	0.602	Solution cast film in red form	NR	4.0×10^{-10}	Rao et al. (1986a)
				Film in yellow form	NR	2.5×10^{-11}	Rao et al. (1986a)
		DFWM	1.17 eV	Solution gel		1.1×10^{-11}	Nunzi et al. (1989)

Structure (PTS): $-CH_2-O-$ attached to benzene ring with para $-SO_2-$ ... CH_3

Structure (PTS-12): $-(CH_2)_4-O-SO_2-$ attached to benzene ring with para CH_3

Structure (Poly-4-BCMU): $-(CH_2)_4-OC-N-CH_2-C-OC_4H_9$ with $\parallel O$, $\parallel O$, $-H$

232

Polymer	R groups	Technique	$\hbar\omega$	Sample form	R/NR	Value	Reference
Poly-4-BCMU		Kerr	$\hbar\omega_1$ = broad band $\hbar\omega_2$ = 1.99 eV	PMMA doped with poly-4-BCMU solution cast film	R	$\leqslant 3 \times 10^{-10}$	P. P. Ho et al. (1986, 1987)
Poly-4-BCMU		THG	1.06	LB film	NR	3×10^{-11} red form	Berkovic et al. (1987)
	$R_1 = R_2 = C_6H_4NHCOC_{17}H_{35}$	THG	1.90	Cast film		1.4×10^{-11}	Kurihara et al. (1987)
	$R_1 = R_2 = (CH_2)_3OCON(CH_3)_2$	THG	1.90	Cast film		5.2×10^{-12}	Tomaru et al. (1987)
	$R_1 = R_2 = CH_2OCONHC_4H_9$	THG	1.90	Vacuum-deposited monomer polymerized by UV light (blue form)		1.4×10^{-11}	
	$R_1 = R_2 = (CH_2)_4OCONHC_3H_7$	THG	1.90	Oriented by unidirectional rubbing		3.80×10^{-10}	Koda et al. (1988)
	$R_1 = CH_3-(CH_2)_{15}$ $R_2 = -(CH_2)_8-COOH$	Waveguided coupling silver plasmon	0.650 / 0.750	LB films 500 Å	R / NR	4.0×10^{-10} / 4.0×10^{-11}	Carter et al. (1987)
Poly-3-BCMU		DFWM				9×10^{-10}	Nunzi et al. (1989)
p-DCH	$R_1 = R_2 = -CH_2-N-C_{12}H_8$	THG	1.35–1.45	Monomer vacuum deposited on alkali halide and polymerized		$(1 \pm 0.1) \times 10^{-10}$	
		EFISH	1.35–1.45			$(6.4 \pm 0.4) \times 10^{-11}$	LeMoigne et al. (1988)

thermally, or by UV light or by γ irradiation whereby a single crystal of the monomer is converted into a single crystal of the polymer (Schott and Wegner 1987). A great variety of polydiacetylenes can be made, depending on the nature of the substituents R_1 and R_2. By appropriate choice of the substituents (such as when R_1 is a long alkyl chain and R_2 is a carboxylic group) one can produce amphiphilic structures that form Langmuir–Blodgett films. Polymerization, even in the monolayer, has been accomplished to produce a polymeric film that can subsequently be transferred as monolayer or as multiple layers by the Langmuir–Blodgett technique. Depending on the nature of the R_1 and R_2 groups, the polymer can be insoluble or soluble in common organic solvents. The polymer formed in solid state is in the blue form ($\lambda_{max} \sim 650$ nm). For some soluble polymers, recasting from the solution can yield the original blue form. However, for many polymers, recasting from solution leads to another form which is red ($\lambda_{max} = 530$ nm). The red shift is generally associated with a less conjugated conformation. These polymers are also thermochromic since, when the blue or red form is heated, it converts to an even further less conjugated yellow form ($\lambda_{max} = 460$ nm). Therefore, polydiacetylenes provide unique opportunities to study conformational effects (ordered crystalline polymers, solution cast disordered form, etc.), effect of conjugation, and the role of substituents on third-order nonlinearities.

The $\chi^{(3)}$ values for a large number of polydiacetylenes have been investigated using solutions (for soluble polymers), solution cast films, crystals, and Langmuir–Blodgett films. The measurement techniques used have been third-harmonic generation, degenerate four-wave mixing, surface plasmon waveguide coupling, electric field-induced second-harmonic generation, and Kerr gate experiments. Table 10.2 lists some of the results for various polydiacetylenes.

Two specific types of polydiacetylenes will be discussed in somewhat more detail. These are commonly abbreviated as PTS and poly-BCMU. For PTS the groups R_1 and R_2 are p-toluene sulfonate, whose structure is shown in Table 10.2. The PTS crystal was investigated by Sauteret et al. (1976) using third-harmonic generation. They showed that the crystal exhibited a large anisotropy in the value of $\chi^{(3)}$. Along the direction parallel to the chain, the value of $\chi^{(3)}$ is more than an order of magnitude larger than that perpendicular to the chain. This is what one expects from the π-electron contribution, as suggested by various models discussed in Chapter 3.

Carter et al. (1987) investigated the $\chi^{(3)}$ behavior of polydiacetylenes using degenerate four-wave mixing and optical waveguide methods as well as the effect of dispersion on $\chi^{(3)}$. Their result shows a clear enhancement in the $\chi^{(3)}$ value, by more than one order of magnitude, as the one-photon resonance is approached.

Poly-BCMU is another group of polydiacetylenes that has been widely investigated. The general structure of the substituents $R_1 = R_2$ is

$$R_1 = R_2 = -(CH_2)_m - O - \underset{\underset{O}{\|}}{C} - \underset{\underset{H}{|}}{N} - CH_2 - \underset{\underset{O}{\|}}{C} - OC_4H_9$$

Depending on the value of m, names such as poly-3-BCMU $(m = 3)$, poly-4-BCMU $(m = 4)$, and poly-9-BCMU $(m = 9)$ are used. These polymers are soluble and can be prepared in the form of Langmuir–Blodgett films. The work by Biegajski et al. (1986) shows that a monolayer film of poly-4-BCMU is in the yellow form, but when compressed it undergoes a monolayer to bilayer transition, which also results in a change of conformation from the yellow (less conjugated) form to the red (more conjugated) form. Berkovic et al. (1987) reported the first study of $\chi^{(3)}$ from a monolayer film (12 Å in the case of poly-4-BCMU) using third-harmonic generation. They studied the monolayer-to-bilayer conformational transition in poly-4-BCMU using third-harmonic generation and found a large increase in $\chi^{(3)}$ as the yellow form converted to the relatively more conjugated red form, confirming the π-conjugation length dependence of the process.

A solution cast film of poly-4-BCMU when heated also undergoes the transition from the red form to the yellow form. The study of $\chi^{(3)}$ behavior of this film by degenerate four-wave mixing (Rao et al. 1986a) also revealed a drastic reduction of $\chi^{(3)}$ as the film converted from the red form to the yellow form.

10.4.2   Poly-p-phenylene Vinylenes and Analogues

The polyparaphenylene vinylene represents a class of polymers with the general formula (Gagnon et al. 1987)

$$X = Y = R, OR, H$$

in which R is an alkyl group. Also, one can make the thienylene analogue with the following structure:

This group of polymers offers many unique features. First, good optical-quality films of the polymer can be prepared through a soluble sulfonium salt precursor. Some of these precursors are even water soluble. This feature allows one to make a device structure with the precursor polymer and then convert it to the final polyparaphenylene (or thienylene) vinylene structure by a heat-treatment procedure discussed by Gagnon et al. (1987). The parent polymer, poly-p-

phenylene vinylene $(X = Y = H)$, commonly abbreviated as PPV, can be synthesized to yield a high molecular weight polymer of relatively high purity and narrow molecular weight distribution, which helps in the control of structural homogeneity for avoiding domain structures. PPV and its 2, 5-dimethoxy derivatives $(X = Y = OCH_3)$ have been processed by Foster–Miller Inc. of Waltham, Massachusetts, as good optical-quality, stretch-oriented, free-standing films, providing an opportunity to study the polymer chain orientation effect on $\chi^{(3)}$. In a uniaxially stretch-oriented film, one would expect the conjugated chains to align along the draw (stretch) direction. These films have very good mechanical strength and high optical damage threshold under picosecond and femtosecond pulse illumination. In stretched films, doping produces high electrical conductivity along the draw direction indicating a high effective π conjugation in this polymer.

The band gap for PPV is $\sim 430\,nm$. Kaino et al. (1987) performed a third-harmonic generation measurement on an as-cast (unoriented) film of PPV at a fundamental wavelength of $1.85\,\mu m$. They reported a value of $\sim 10^{-11}\,esu$ at this wavelength. A more recent third-harmonic generation study by Bubeck et al. (1989) reports a value of $\chi^{(3)} = (1.5 \pm 0.6) \times 10^{-10}\,esu$ at a fundamental wavelength of $1.064\,\mu m$. Of course, the third-harmonic $(355\,nm)$ in the latter study will be in resonance.

Singh et al. (1988) performed a subpicosecond degenerate four-wave mixing study of a 10:1 uniaxially stretch-oriented, free-standing film of PPV at wavelengths of 602 and 580 nm. They observe a very fast rise time, limited by the rise time of the optical pulse (laser pulse width 350 fs). The decay of the signal is initially fast, again within the autocorrelation width. Therefore, a sub-picosecond response of $\chi^{(3)}$ is observed, as expected under nonresonant conditions. However, Singh et al. (1988) also observed the presence of a weak tail with a longer time constant, which they attributed to the presence of absorbing impurities. Therefore, even the apparent nonresonant regime (below the band gap) is not completely free of absorption. The $\chi^{(3)}$ value along the draw direction is $1.2 \times 10^{-9}\,esu$ at 602 nm for a 6:1 uniaxial film (Swiatkiewicz, Prasad, and Karasz, unpublished result).

Singh et al. (1988) also studied the anisotropy of $\chi^{(3)}$. According to their study, the change in $\chi^{(3)}$ value (actually the square root of DFWM signal) as a function of film rotation with respect to the incident electric field vector yields a polar plot, shown in Figure 10.3. The highest value of $\chi^{(3)}$ is obtained when the electric vectors of all the four waves are parallel to the draw direction. The minimum value for $\chi^{(3)}$ is for the orientation when all the electric vectors are perpendicular to the draw direction . The $(\chi^{(3)})_{\parallel}/(\chi^{(3)})_{\perp}$ ratio is 39, indicating a very high degree of orientational anisotropy. The x-ray study of such stretch-oriented PPV films shows a high degree of polymer chain alignment along the draw direction. This result also provides confirmation that the largest component of the $\chi^{(3)}$ (and hence the microscopic nonlinearity γ) tensor is along the chain direction, as is expected from theoretical calculations of microscopic nonlinearities in π-conjugated polymeric (or oligomeric) structures.

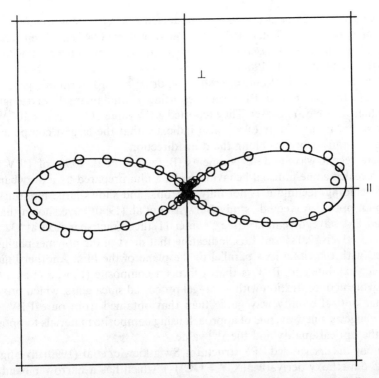

Figure 10.3 Polar plot of square root of DFWM signal for a 10:1 uniaxially stretch-oriented poly(paraphenylene vinylene)polymer. The directions parallel and perpendicular to the draw directions are labeled ∥ and ⊥.

To explain the observed polar plot, Singh et al. used a simple transformation of the fourth-rank tensor $\chi^{(3)}$ from the film-based to the laboratory-based coordinate system. This relationship is discussed in Chapter 4. Since only in-plane anisotropic measurements are made, this transformation reduces to

$$\chi^{(3)}_{1111,L} = \chi^{(3)}_{1111,F}\cos^4\theta + \{\chi^{(3)}_{1122,F} + \chi^{(3)}_{1221,F} + \chi^{(3)}_{2211,F} + \chi^{(3)}_{2121,F}$$
$$+ \chi^{(3)}_{1212,F} + \cdots\}\cos^2\theta\sin^2\theta + \chi^{(3)}_{2222,F}\sin^4\theta \qquad (10.2)$$

In the above equation, $\chi^{(3)}_{1111,L}$ refers to the value of $\chi^{(3)}$ in the laboratory coordinate system, when all the beam polarizations are vertical. $\chi^{(3)}_{1111,F}$ is the component along the draw direction, and $\chi^{(3)}_{2222,F}$ is the component perpendicular to the draw direction. The other components are the in-plane off-diagonal terms. The angle θ is the angle between the electric polarization vector (vertical in our measurement) and the draw direction. The solid line in the figure represents the theoretical fit using equation 10.2. The experimental data points are adequately described by the theoretical curve.

Recently, Bubeck et al. (1989) also conducted a degenerate four-wave mixing study of an unoriented film of PPV using 400-fs pulses at 647 nm. Their reported value of $\chi^{(3)}$ $(-\omega; \omega, -\omega, \omega) \sim 10^{-10}$ esu is in excellent agreement with that reported by Singh et al. (1988).

McBranch et al. (1989) have recently reported a third-harmonic generation study of stretch-oriented PPV samples using a fundamental wavelength of 1.06 μm from a Nd:Yag laser. They reported a $\chi^{(3)}$ value of $(2 \pm 1.5) \times 10^{-11}$ esu. Their study of anisotropy of $\chi^{(3)}$ also indicates that the largest component of the fourth-rank tensor is along the draw direction.

More recent waveguide experiments (Burzynski et al. 1990) of PPV films have revealed some unusual behavior. Here, a film prepared by a drawbar (the doctor-blading) technique is not fully isotropic and shows large birefringence. From an optical waveguide study Burzynski et al. (1990) have shown that for TE and TM waveguide modes (see Chapter 11) the refractive indices at 632.8 nm are, respectively, 2.085 and 1.63, indicating that most of the polymer chains are aligned with the chain axis parallel to the plane of the film. Another unusual behavior exhibited by PPV is that it forms a composite (Prasad et al. 1989a) in very high concentration with a sol–gel processed silica glass, which provides a better optical quality waveguide than that obtained from pure PPV. This result suggests a new avenue of approach using composite materials to optimize both the optical quality and the $\chi^{(3)}$ value.

Among the derivatized PPV structures, Swiatkiewicz et al. (1990) investigated the 2,5-dimethoxy derivative (X, Y = OCH$_3$), which has a narrower band gap (580 nm) compared to PPV. This reduction of band gap is attributed to the electron-donating character of the methoxy groups. Swiatkiewicz et al. (1990) conducted characterization of linear optical properties and performed subpicosecond degenerate four-wave mixing studies of a 6:1 uniaxially stretch-oriented film of dimethoxy PPV at 602 nm. Their result reveals a strong dichroic behavior, a large birefringence, and a very high degree of anisotropy of $\chi^{(3)}$. The $\chi^{(3)}$ value along the draw direction is 4×10^{-9} esu, but it is a resonant value since the decay of the four-wave mixing signal is clearly longer than the optical pulse width.

Kaino et al. (1989) have recently studied the poly-p-theinylene vinylene (the thiophene analog of PPV) by third-harmonic generation and reported a higher value of $\chi^{(3)}$ than that for PPV.

10.4.3 Polyacetylene

Polyacetylene is an example of a conjugated polymer that can support a conformational defect in form of solitons which can be photogenerated. The physical picture associated with this type of process is discussed in Chapter 8, Section 8.5.

Femtosecond time-resolved transient absorption experiments (Shank et al. 1982, Vardeny et al. 1982) and more recent subpicosecond pump–probe experiments (Rothberg et al. 1986, Etemad et al. 1988) have shown that conformational distortions occur within femtoseconds to produce new (midgap) states. Consequently, a redistribution of oscillator strength to photoinduced absorption involving midgap states occurs. As discussed in Section 8.5 the Kramers–Kronig relation then predicts that a redistribution of oscillator strength (change in absorbance) will produce a corresponding change in the refractive index, which in turn contributes to the dynamic $\chi^{(3)}$.

The $\chi^{(3)}$ behavior of *trans*-polyacetylene has been investigated by third-harmonic generation, pump–probe transient absorption, and degenerate four-wave mixing. Kajzar et al. (1987) studied the dispersion of $\chi^{(3)}$ in polyacetylene using third-harmonic generation. They reported for the range of 1.17–1.5 eV, a nonresonant $\chi^{(3)}$ value of $\sim 10^{-10}$ esu. At ~ 2.0 eV the $\chi^{(3)}$ value is $\sim 10^{-9}$ esu. More recently, Fann et al. (1989) extended the wavelength of study to lower energies in the infrared using a free electron laser. According to their analysis this dispersion shows the manifestations of one- and two-photon resonances. Sinclair et al. (1988) have also performed a third-harmonic generation study of *trans*-polyacetylene with a fundamental wavelength of 1.06 μm and report a value of $\chi^{(3)} = 4 \times 10^{-10}$ esu. They also studied the cis isomer of polyacetylene and found that the $\chi^{(3)}$ value for *cis*-polyacetylene is one-tenth of that for *trans*-polyacetylene.

cis-Polyacetylene

The problem with polyacetylene is that it is an insoluble polymer and is extremely air sensitive. Therefore, once the polymer is deposited it cannot be processed. Prasad et al. (1988a) have studied the polyacetylene–polymethylmethacrylate graft copolymer, which is solution processible and not as air sensitive. The value of $\chi^{(3)}$ they reported, using degenerate four-wave mixing at 0.602 μm, is $\sim 10^{-10}$ esu.

A related polymer polyphenylacetylene shows $\chi^{(3)} = 10^{-11}$ esu (Prasad 1987). The reduced value for this polymer may be because of reduced conjugation effect due to the distortions caused by the phenyl rings.

10.4.4 Polythiophene

Polythiophene is another example of a conjugated polymer where a conformational defect can exist in the form of polarons. This topic is also discussed in Section 8.5.

Again, it has been suggested on the basis of transient absorption that photoexcitation of this material across the band gap (band gap = 2 eV) leads to rapid conformational deformation, producing polaronic states in the gap. The result is a redistribution of oscillator strength, which, as discussed above, is expected to contribute to resonant $\chi^{(3)}$.

The $\chi^{(3)}$ behavior of polythiophene has been studied in several forms. Prasad et al. (1988a) studied electrochemically polymerized films by degenerate four-wave mixing. They report a $\chi^{(3)}$ value of $\sim 4 \times 10^{-10}$ esu at 602 nm using 350-fs pulses. Even though it is a resonant nonlinearity, they report a subpicosecond response.

Unlike polyacetylene, polythiophene is air stable and, furthermore, by long alkyl substitution at the 3-position, the resulting polymer

R = long-chain alkyl group

is soluble. This substituted polythiphene is reported to form a monolayer, which can be transferred as a Langmuir–Blodgett film (Logsdon et al. 1988). The nonlinearity of polythiophene is large enough to obtain the degenerate four-wave mixing, phase-conjugate signal even from a monolayer (22 Å thick) film. The value of $\chi^{(3)}$ obtained from these films is $\sim 10^{-9}$ esu at 602 nm (Prasad et al. 1988b, Logsdon et al. 1988). As discussed in Chapter 8, there have been suggestions that large resonant nonlinearity in polythiophene (Heeger et al. 1987) may be predominantly from conformational deformation producing polaronic states. Indeed, photoexcitation spectroscopic studies (Moraes et al. 1984, Vardeny et al. 1986) confirm the photogeneration of polaronic states in the band gap. More recent femtosecond transient absorption studies of Stamm et al. (1989, in press) reveal that the rise time for the appearance of the new absorption bands involving the photogenerated polaronic states is ~ 100 fs, as suggested by the Su et al. model (1979). They observe the decay of the polaronic states to be in the picosecond time range. Similar results have been obtained by Vardeny et al. (1989). They observed a complicated decay of transient polaronic state absorption, which involves several time constants corresponding to decays in picoseconds.

Singh et al. (1990) and Pang and Prasad (1990) have recently performed femtosecond time-resolved degenerate four-wave mixing studies on soluble dodecylpolythiophene films. They find that the rise time for the nonlinear response is almost instantaneous, while most of the decay occurs in less than 200 fs. These features are very clear in the investigation of Pang and Prasad (1990) who used 50-fs pulses from an amplified colliding pulse mode-locked laser system. They conclude that this behavior is not consistent with the dominant polaronic contributions in determining optical nonlinearity. They suggest that the initial dynamic nonlinearity in polythiophene is primarily determined by the phase-space filling of photogenerated unrelaxed excitons. This mechanism is discussed in Chapter 8.

Recently, Yang et al. (1989) and Worland et al. (1989) also reported $\chi^{(3)}$ measurements for polythiophenes and their values of $\sim 10^{-9}$ esu are in excellent agreement with those of Logsdon et al. (1988). Stamm et al. (1989, in press) investigated the nonlinear optical behavior of polythiophene using femtosecond pump–probe absorption experiments.

10.4.5 Other Conjugated Polymers

Another group of conjugated polymers that have been investigated (Rao et al. 1986b, Prasad 1988) are aromatic heterocyclic rigid rod polymers such as poly(p-phenylene benzobisthiazole) and poly(p-phenylene benzobisoxazole) with the following structures:

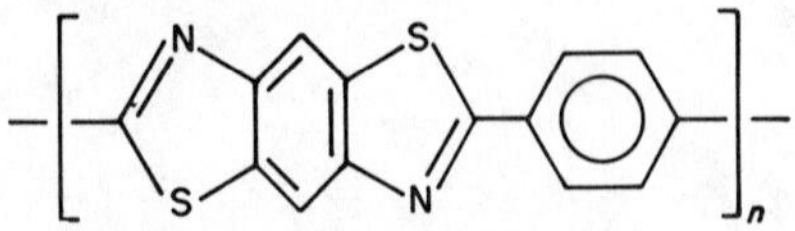

Poly-p-phenylene benzobishthiazole 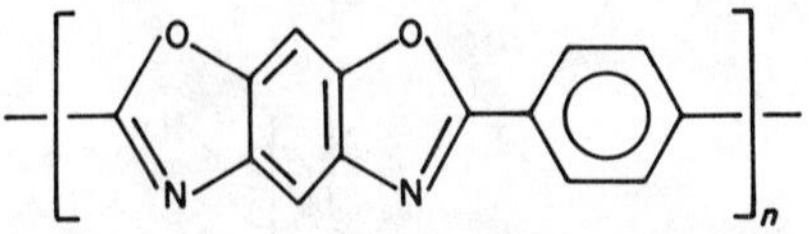   Poly-p-phenylene benzobisxazole

In the past, the common abbreviations used for these polymers were PBT and PBO, respectively, but recently there has been some preference to rename them PBZT and PBZO. These polymers, because of their rigid rod-type molecular conformation exhibit very high mechanical strength and stability. They are processible through liquid-crystalline-phase strong acid solutions such as in polyphosphoric acid or methane sulfonic acid. Furthermore, Foster–Miller, Inc. of Waltham, Massachusetts, has developed techniques to fabricate oriented uniaxial and biaxial films of these materials. Rao et al. (1986b) used subpicosecond degenerate four-wave mixing to measure $\chi^{(3)}$ in a biaxial film of PBT. They report a value of $\chi^{(3)}$ of $\sim 10^{-11}$ esu at both 585 and 605 nm, with the response in subpicoseconds. They also studied the anisotropy of $\chi^{(3)}$ by obtaining the polar plot as a function of angular orientation of the film with respect to the laser polarization. As discussed above for the stretch-oriented PPV polymer, the anisotropic behavior of $\chi^{(3)}$ conforms to its fourth-rank tensor property. The measurement of PBO (Prasad, unpublished results) yields a value of $\chi^{(3)}$ similar in magnitude to that for PBT.

A more recent study of PBT Films (Swiatkiewicz et al. 1990) has revealed that the $\chi^{(3)}$ value for this polymer is dependent on the processing condition. The $\chi^{(3)}$ value of a better optical quality film is considerably higher than that initially reported (Rao et al. 1986b).

Another group of polymers with a very high mechanical stability are polyimides. One specific example of this group, commonly known as LARC-TPI, has the following structure:

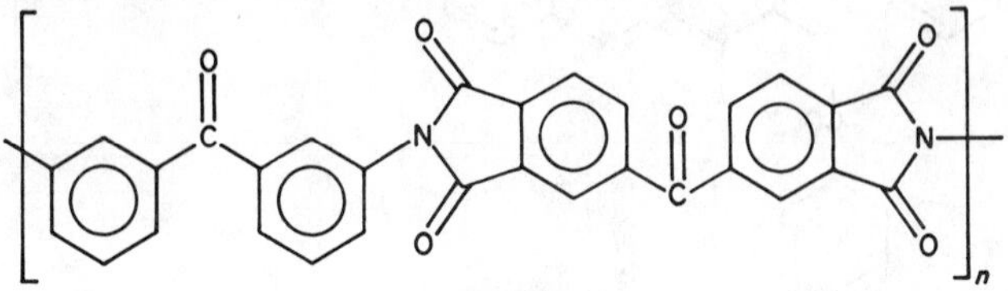

LARC-TPI

This polymer does not have an extensively π-conjugated structure. However, like the PPV polymer discussed above, it is processible through a precursor route and can be stretch-oriented into good optical quality films. Prasad (1988) reports a study of $\chi^{(3)}$ in a biaxial film of LARC-TPI using subpicosecond degenerate four-wave mixing at 605 nm. The observed anisotropy of $\chi^{(3)}$ again is explained by using its fourth-rank tensor property.

Dalton and coworkers (1988, in press) have synthesized and investigated a large group of fused ring ladder polymers. These polymers have the general formula

with X = O, S, NH

in which Φ is a substituted phenyl group or a fused ring aromatic group. Dalton et al. used a synthetic route that produces a soluble prepolymeric structure, which, as for PPV, can be converted into the final ladder polymer. They have used picosecond degenerate four-wave mixing to measure the nonlinearities of these ladder polymers and report values in the range of $\sim 10^{-9}$ esu at 585 nm, at which these materials have significant absorption.

Recently Yu and Dalton (1989) have prepared and investigated a group of polymers that are block copolymers of electroactive and flexible chain segments. These polymers offer the advantage that nonlinear optical properties can be controlled by systematically varying the electroactive segment, while the solubility and solid-state morphology can be controlled by the flexible chain spacers. Two examples of this class of copolymers are

Their preliminary measurements of $\chi^{(3)}$ made at several wavelengths using picosecond degenerate four-wave mixing indicate that the values are $\sim 10^{-8}$ esu at 532 nm, $\sim 4 \times 10^{-9}$ esu at 585 nm, and 10^{-10}–10^{-11} esu at 1064 nm.

10.5 MACROCYCLES

Macrocycles are examples of large conjugated organic structures which can be thought of as two-dimensionally delocalized π-electron systems. One specific example of this class of material is phthalocyanine, with a structure shown in Figure 10.4. The phthalocyanines can be prepared with various substituents on the ring to control solubility and morphology. Furthermore, the central positions in the phthalocyanine moiety can be metal free or substituted by copper, nickel, platinum, palladium, or even silicon. Metal substitution introduces low-lying energy states derived from metal-to-ligand and ligand-to-metal charge transfer, which may contribute to optical nonlinearity. In addition, one can prepare polymeric phthalocyanines, in which the phthalocyanine rings are connected in a shish-kabob-type arrangement through a bridging atom. The phthalocyanines exhibit strong electronic transitions in the visible region between 600 and 800 nm. This is a π–π^* excitation and is often called the Q-band, the absorption coefficient for which is in excess of 10^4 cm^{-1}. A very important feature of phthalocyanines is their high chemical and thermal stability. By appropriate derivatization, many soluble phthalocyanine structures have been synthesized. Some of the soluble phthalocyanines have also been reported to form Langmuir–Blodgett films. Examples are tetrakis (cumylphenoxy)

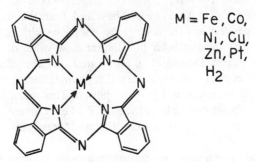

Figure 10.4   General structure of a phthalocyanine macrocycle.

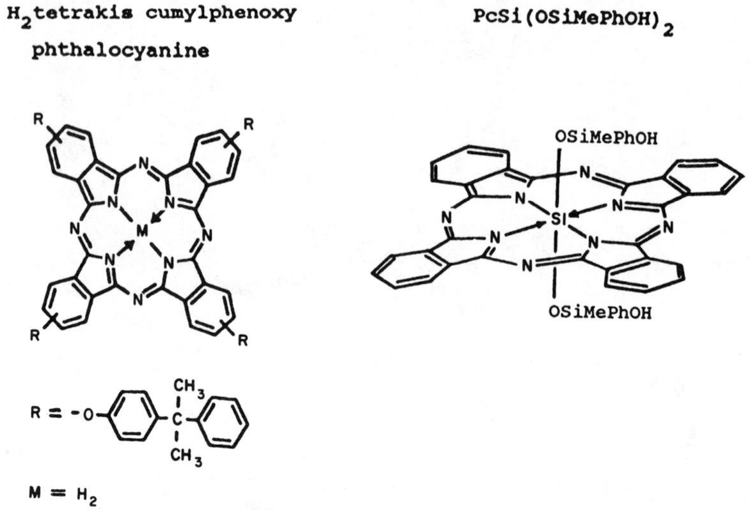

H$_2$tetrakis cumylphenoxy
phthalocyanine

PcSi(OSiMePhOH)$_2$

R = -o-⟨⟩-C(CH$_3$)(CH$_3$)-⟨⟩

M = H$_2$

Figure 10.5 Structures of tetrakis(cumylphenoxy)phthalocyanine and silicon phthalocyanine.

phthalocyanine (Barger et al. 1985, Prasad et al. 1988b) and a silicon phthalocyanine (Prasad et al. 1989b). The structures of these compounds are shown in Figure 10.5.

There have been several reports of the study of nonlinear optical behavior of various phthalocyanines. Z. Z. Ho et al. (1987) used third-harmonic generation with a fundamental wavelength of 1.06 μm and reported $\chi^{(3)}$ for fluoroaluminum and chlorogallium phthalocyanine polycrystalline films to be 5×10^{-11} and 2.5×10^{-11} esu. Prasad et al. (1987, 1988b, 1989b) have investigated the resonant $\chi^{(3)}$ behavior (dynamic nonlinearity as defined in Chapter 8) for several phthalocyanines in the form of vacuum deposited films or Langmuir–Blodgett films, using degenerate four-wave mixing with picosecond and subpicosecond pulses at 605 nm. The resonant $\chi^{(3)}$ value is large enough that the DFWM signal can be observed even from monolayer Langmuir films of tetrakis cumylphenoxy phthalocyanine (Prasad 1988b) and silicon phthalocyanine (Prasad et al. 1989b). They find that the magnitude of the effective $\chi^{(3)}$ is dependent on the laser pulse width, which can be expected for resonant nonlinearity as discussed in Chapter 8. A value of $\sim 10^{-9}$ esu was found with 350-fs pulses. Prasad et al. also found that the effective $\chi^{(3)}$ values for the nickel- and copper-substituted tetrakis (cumylphenoxy)phthalocyanines were similar to that for the metal-free compound. This result is not surprising because at 605 nm, the wavelength for their DFWM experiment, the nonlinearity is dominated by the contribution from the $\pi-\pi^*$ resonance of the organic macrocyclic structure.

Casstevens et al. (1990) used the exciton phase-space filling model discussed in Chapter 8 to explain the resonant nonlinearity of phthalocyanines. They also report that the response time, being in picoseconds, is strongly dependent on the pulse intensity and they attribute this to a bimolecular excited-state decay mechanism involving exciton–exciton annihilation. Using differential transmission spectroscopy with femtosecond pulses at 620 nm, Z. Z. Ho et al. (1988a; 1988b) have studied nonlinear optical response of vacuum-evaporated polycrystalline films of fluoroaluminum phthalocyanine. They also report an intensity-dependent decay derived from exciton–exciton annihilation.

Wu et al. (1989) have investigated the optical nonlinearity of another phthalocyanine-type structure, the silicon naphthalocyanine, by the study of their saturable absorption behavior, and they quote a value for the nonlinear refractive index coefficient n_2 of 1×10^{-4} cm^2/kW at 810 nm. More recently, Shirk et al. (1989) investigated the third-order nonlinearity of the Pt, Pb, and metal-free tetrakis(cumylphenoxy)phthalocyanines using degenerate four-wave mixing with 35-ps pulses at 1.064 μm. They report that at this wavelength, which is far from the one-photon resonance of the π–π* transition, metal substitution strongly enhances the $\chi^{(3)}$ value. The $\chi^{(3)}$ values reported by them are 2×10^{-10} esu for Pt–phthalocyanine, 2×10^{-11} esu for Pb–phthalocyanine, and 4×10^{-12} esu for the metal-free form. They suggest that low-lying charge-transfer states involving the metal atom may be contributing to the observed enhancement of the nonlinearity at this longer wavelength.

10.6 POLYSILANES

Polysilanes with the following structure

$$-\left[-\underset{R_2}{\overset{R_1}{Si}}-\right]_n-$$

are nonconjugated inorganic polymers that contain no π electrons in the polymer backbone. However, they are an interesting group of polymers because they show delocalization of σ electrons along the polymer backbone. Their physical and chemical properties can be tailored by the choice of appropriate substituents R_1 and R_2, which also determine the conformation of the polymer and its solubility. Because of the delocalization of the σ electrons, these polymers show a rich variety of photophysical properties. For this reason, several research groups have shown a considerable interest in investigating the nonlinear optical properties of various polysilanes and a related group of polymers, polygermanes, which are the germanium analogues. These groups of polymeric materials show optical transparency throughout the visible. Kajzar et al. (1986) used third-harmonic generation measurements by transmission at the fundamental wavelength of 1.064 μm to investigate a spin-coated thin film of a polysilane in which $R_1 = CH_3$ and $R_2 = \bigcirc$. The value of $\chi^{(3)}$ they reported is $(1.5 \pm 0.1) \times 10^{-12}$ esu at 1.064 μm. They also report to have determined the phase of $\chi^{(3)}$. They find that the real part of $\chi^{(3)}$ is positive and the imaginary part is considerably smaller. Kajzar et al. suggest that the value of $\chi^{(3)}$ as measured by third-harmonic generation at $1.064\,\mu m$ is enhanced by a three-photon resonance. Baumert et al. (1988) have also reported third-harmonic generation studies on thin films of planar zigzag polysilanes and polygermanes. Their value is 11.3×10^{-12} esu, which was found to vary with thermally induced reversible changes in the polymer backbone conformation.

The dynamic behavior of the third-order nonlinear optical response in the polysilane, again with $R_1 = CH_3$ and $R_2 = \bigcirc$, was investigated by Yang et al. (1988) using both the picosecond Kerr gate and forward-wave degenerate four-wave mixing. In the Kerr gate experiment, the pump and the probe wavelengths were at 1.06 and $0.53\,\mu m$, respectively. Yang et al. (1988) report the electronic nonlinear response in the polysilane thin film, as determined by the Kerr gate experiment, to be faster than 3 ps. Their Kerr gate experiment yields a value of $\chi^{(3)}$ of the order of $(2.0 \pm 0.6) \times 10^{-12}$ esu. Yang et al. (1989) performed the degenerate four-wave mixing studies on the polysilane solution with various wavelengths in the range $0.53-1.06\,\mu m$. The $\chi^{(3)}$ value obtained by degenerate four-wave mixing is of the order of $(1.6 \pm 0.2) \times 10^{-12}$ esu. More recently, McGraw et al. (1989) studied the optical nonlinearity of octyl methylpolysilane ($R_1 = C_8H_{17}$ and $R_2 = CH_3$ in the general formula). Their reported value for the electronic $\chi^{(3)}$ contribution is $(1.8 \pm 0.5) \times 10^{-12}$ esu.

10.7 ORGANOMETALLIC STRUCTURES

Organometallic systems provide many interesting and unique structure and bonding schemes for molecular engineering of new materials. Furthermore, insertion of a metal unit in a conjugated structure significantly influences the π-electron behavior that can have important manifestations in optical nonlinearity. Organometallic structures containing transition metals offer new properties derived from the richness of the various excited states due to low-lying d–d transitions present in these systems. Furthermore, they offer the prospect of the tailorability of metal–organic ligand interactions. For these reasons, organometallic structures have received some attention recently.

One group of organometallic structures is represented by sandwich-type compounds, in which a metal atom acts as a bridge between two aromatic rings. An example is ferrocene, in which two aromatic cyclopentadienyl rings

Ferrocene

interact with a metal atom through their π-electron system. An obvious question is whether there is an effective π conjugation through the bridging metal which can lead to enhanced third-order nonlinearity. Winter et al. (1988) used a self-focusing-type study to measure $\chi^{(3)}$ of ferrocene in the molten liquid form as well as in the form of solution in ethanol using nanosecond pulses at the wavelength of $1.06\,\mu$m. The value of $\langle\gamma\rangle$ reported by them is $\sim 2 \times 10^{-34}$ esu. Ghosal et al. (1990) conducted a study of the nonlinear optical behavior of ferrocene, its various oligomers, and a number of their aryl vinyl derivatives in the THF solution using subpicosecond degenerate four wave mixing at the wavelength of 602 nm. The lowest-lying excitation in these compounds is due to the metal d–d transition. The wavelength of 602 nm is close to this resonance. The value of $\langle\gamma\rangle$ for ferrocene reported by Ghosal et al. is $\sim 10^{-35}$ esu, which is lower than that reported by Winter et al. Ghosal et al. (1990) suggest that this difference may partly arise because their degenerate four-wave mixing study used subpicosecond pulses, which minimized the role of nonelectronic (slowly responding) nonlinearities. A comparison of the γ value for the ferrocene monomer with those of the various oligomers did not show any significant increase for the oligomers. This behavior is different from what has been observed in the study of other sequentially built oligomers of conjugated structures, such as the thiophene oligomers discussed above. Ghosal et al. conclude from this result that the third-order nonlinearity in these structures is determined by the π-electron delocalization through the directly coupled organic units and the delocalization effect is not carried through the metal. Another interesting observation reported by them is that even though the 602-nm wavelength is in resonance with the d–d transition of many of these structures, the behavior of the third-order nonlinearity is as if it is nonresonant. The γ value is found to be positive, with the imaginary part being negligible. Based on this result they suggest that the metal d–d transitions in ferrocene-type structures do not make significant contribution to optical nonlinearity as compared to that derived from the π–π^* transitions.

Another group of organometallic structures investigated for third-order optical nonlinearity is transition metal polyynes, such as

$$M = Pt, Pd$$

Frazier et al. (1987, 1988) used four-wave mixing to investigate the third-order nonlinearity of these groups of organometallic compounds in the THF solution. They report that two-photon absorptions in these metal polyynes make significant contributions to the nonlinearity. They find that the γ value is complex. The largest value is reported for the Pt polymer and it is

1.45×10^{-33} esu. The γ value per repeat unit increases only slightly with the chain length. Recently, Guha et al. (1989) have investigated the third-order nonlinearity of several platinum polyynes using 23-ps laser pulses. They used the optical Kerr effect with the pump beam of wavelength $1.064 \mu m$ and the probe beam of $0.532 \mu m$ to measure the real part of γ. To measure the imaginary part of γ, they used two-photon absorption at $0.532 \mu m$. Dilute solutions of the polymers in THF were used for these measurements. For the Pt–polyynes of the structure represented above, they report the real part of γ to be 8.56×10^{-34} esu and the imaginary part to be 3.57×10^{-33} esu. Therefore, even though th polymer is transparent in the visible, the two-photon resonances makes a significant contribution, giving rise to predominantly imaginary γ.

REFERENCES

Barger, W. R., A. W. Snow, H. Wohltjen, and N. L. Jarvis, *Thin Solid Films* **133**, 197 (1985).

Baumert, J. C., G. C. Bjorklund, D. M. Jundt, M. C. Jurich, H. Looser, R. D. Miller, J. Rabolt, R. Sooriyakumaran, J. D. Swalen, and R. J. Twieg, *Appl. Phys. Lett.* **53**, 1147 (1988).

Berkovic, G., Y. R. Shen, and P. N. Prasad, *J. Chem. Phys.* **87**, 1897 (1987).

Biegajski, J., R. Burzynski, D. A. Cadenhead, and P. N. Prasad, *Macromolecules* **19**, 2457 (1986).

Bubeck, G., A. Kaltbeitzel, R. W. Lenz, D. Neher, J. D. Stenger-Smith, and G. Wegner, in J. Messier, F. Kajzar, P. Prasad, and D. Ulrich, (Eds.), *Nonlinear Optical Effects in Organic Polymers* NATO ASI Series, Kluwer Academic, Dordrecht, 1989, p. 143.

Burzynski, R., P. N. Prasad, and F. E. Karasz, *Polymer* **31**, 627 (1990).

Carter, G. M., Y. J. Chen, M. F. Rubner, D. J. Sandman, M. K. Thakur, and S. K. Tripathy, in D. S. Chemla and J. Zyss (Eds.), *Nonlinear Optical Properties of Organic Molecules and Crystals*, Vol. 2, Academic, New York, 1987, p. 85.

Casstevens, M., M. Samoc, J. Pfleger, and P. N. Prasad, *J. Chem. Phys.* **92**, 2019 (1990).

Chopra, P., L. Carlacci, H. F. King, and P. N. Prasad, *J. Phys. Chem.* **93**, 7120 (1989).

Dalton, L. R., in A. J. Heeger, J. Orenstein, and D. R. Ulrich (Eds.), *Nonlinear Optical Properties of Polymers, Materials Research Society Symposium Proceedings*, Vol. 109, Materials Research Society, Pittsburgh, 1988, p. 301.

Dalton, L. R., R. Vac, and L. P. Yu, in T. Skotheim (Eds.), *Handbook of Electroresponsive Polymers*, Dekker, in press.

Etemad, S., G. L. Baker, L. Rothberg, and F. Kajzar, in A. J. Heeger, J. Orenstein, and D. R. Ulrich (Eds.), *Nonlinear Optical Properties of Polymers, Materials Research Society Symposium Proceedings*, Vol. 109, Materials Research Society, Pittsburgh, 1988, p. 217.

Fann, W. S., J. Madey, S. V. Benson, S. Etemad, G. L. Baker, and F. Kajzar, unpublished results, 1989.

Fischer, H., in S. Patai (Ed.), *The Chemistry of Alkenes*, Interscience, London, 1964, p. 1027.

Frazier, C. C., S. Guha, W. P. Chen, M. P. Cockerham, P. L. Porter, E. A. Chauchard, and C. H. Lele, *Polymer* **28**, 553 (1987).

Frazier, C. C., E. A. Chauchard, M. P. Cockerham, and P. L. Porter, in A. J. Heeger, J. Orenstein, and D. R. Ulrich (Eds.), *Nonlinear Optical Properties of Polymers, Materials Research Society Symposium Proceedings*, Vol. 109, Material Research Society, Pittsburgh, 1988, p. 323.

Gagnon, D. R., J. D. Capistran, F. E. Karasz, R. W. Lenz, and S. Antoun, Polymer **28**, 567 (1987).

Ghosal, S., M. Samoc, P. N. Prasad, and J. J. Tufariello, *J. Phys. Chem.* **94**, 2847 (1990).

Guha, S., C. C. Frazier, P. L. Porter, K. Kang, and S. E. Finberg, *Opt. Lett.* **14**, 952 (1989).

Heeger, A. J., D. Moses, and M. Sinclair, *Synth. Met.* **17**, 347 (1987).

Hermann, J. P., *Opt. Commun.* **9**, 74 (1973).

Hermann, J. P., *Opt. Commun.* **12**, 102 (1974).

Hermann, J. P., and J. Ducuing, *J. Appl. Phys.* **45**, 5100 (1974).

Hermann, J. P., D. Richard, and J. Ducuing, *Appl. Phys. Lett.* **23**, 178 (1973).

Ho, P. P., R. Dorsinville, N. L. Yang, G. Odian, G. Eichmann, T. Jimbo, Q. Z. Wang, G. C. Tang, N. D. Chen, W. K. Zou, Y. Li, and R. R. Alfano, *SPIE Proc.* **682**, 36 (1986).

Ho, P. P., N. L. Yang, T. Jimbo, Q. Z. Wang, and R. R. Alafano, *J. Opt. Soc. Am. B* **4**, 1025 (1987).

Ho, Z. Z., and W. Peyghambarian, *Chem. Phys. Lett.* **148**, 107 (1988a).

Ho, Z. Z., C. Y. Ju, and W. M. Hetherington, *J. Appl. Phys.* **62**, 716 (1987).

Ho, Z. Z., V. Williams, N. Peyghambarian, and W. M. Hetherington, *SPIE Proc.* **971**, 51 (1988b).

Kaino, T., K. Kubodera, S. Tomaru, T. Kurihara, S. Saito, T. Tsutsui, and S. Tokito, *Electron. Lett.* **23**, 1095 (1987).

Kaino, T., et al., private communications, 1989.

Kajzar, F., and J. Messier, *Phys. Rev. A* **32**, 2352 (1985).

Kajzar, F., and J. Messier, in D. S. Chemla and J. Zyss (Eds.), *Nonlinear Optical Properties of Organic Molecules and Crystals*, Vol. 2, Academic, New York, 1987a, p. 51.

Kajzar, F., and J. Messier, *J. Opt. Soc. Am. B* **4**, 1040 (1987b).

Kajzar, F., J. Messier, and C. Rosilio, *J. Appl. Phys.* **60**, 3040 (1986).

Kajzar, F., S. Etemad, G. L. Baker, and J. Messier, *Synth. Met.* **17**, 563 (1987).

Kminek, I., J. Klimovic, and P. N. Prasad, unpublished work, 1989.

Koda, T., K. Ishikawa, T. Kanetake, T. Nishikawa, Y. Tokura, S. Koshihara, K. Takeda, and K. Kubodera, Third Asia Pacific Physics Conference, Hong Kong, June 1988.

Kurihara, T., K. Kubodera, S. Matsumoto, and T. Kaino, *Polymer Preprints*, The Society of Polymer Science, Japan, **36**, 1157 (1987).

LeMoigne, J., A. Thierry, P. A. Chollet, F. Kajzar, and J. Messier, *J. Chem. Phys.* **88**, 6647 (1988).

Levenson, M. D., and N. Bloembergen, *J. Chem. Phys.* **60**, 1323 (1974).

Levine, B. F., and C. G. Bethea, *J. Chem. Phys.* **63**, 2666 (1975).

Logsdon, P., J. Pfleger, and P. N. Prasad, *Synth. Met.* **26**, 369 (1988).

Maloney, C., and W. Blau, *J. Opt. Soc. Am. B* **4**, 1035 (1987).

McBranch, D., M. Sinclair, A. J. Heeger, A. O. Patil, S. Shi, S. Askari, and F. Wudl, *Synth. Met.* **29**, E85 (1989).

McGraw, D. J., A. E. Siegman, G. M. Wallraff, and R. D. Miller, *Appl. Phys. Lett.* **54**, 1713 (1989).

Mehendale, S. C., and K. C. Rustagi, *Opt. Commun.* **28**, 359 (1979).

Meredith, G. R., and B. Buchalter, *J. Chem. Phys.* **78**, 1938 (1983).

Meredith, G. R., and S. H. Stevenson, in J. Messier, F. Kajzar, P. Prasad, and D. Ulrich (Eds.), *Nonlinear Optical Effects in Organic Polymers*, NATO ASI Series, Kluwer Academic, Dordrecht, 1989, p. 105.

Meredith, G. R., B. Buchalter, and C. Hanzlik, *J. Chem. Phys.* **78**, 1533 (1983a).

Meredith, G. R., B. Buchalter, and C. Hanzlik, *J. Chem. Phys.* **78**, 1543 (1983b).

Moraes, F., M. Schaffer, M. Kobayashi, A. J. Heeger, and F. Wudl, *Phys. Rev. B* **30**, 2948 (1984).

Nunzi, J. M., J. L. Ferrier, and R. Chevalier, in J. Messier, F. Kajzar, P. Prasad, and D. Ulrich (Eds.), *Nonlinear Optical Effects in Organic Polymers*, NATO ASI Series, Vol. 12, Kluwer Academic, Dordrecht, 1989, p. 365.

Oudar, J. L., and H. LePerson, *Opt. Commun.* **2**, 258 (1975).

Pang, Y., and P. N. Prasad, *J. Chem. Phys.* **93**, 2201 (1990).

Prasad, P. N., *Thin Solid Films* **152**, 275 (1987).

Prasad, P. N., in P. N. Prasad and D. R. Ulrich (Eds.), *Nonlinear Optical and Electroactive Polymers*, Plenum, New York, 1988, p. 41.

Prasad, P. N., J. Swiatkiewicz, and J. Pfleger, *Mol. Cryst. Liq. Cryst.* **160**, 53 (1988a).

Prasad, P. N., M. K. Casstevens, J. Pfleger, and P. Logsdon, *SPIE Proc.* **878**, 106 (1988b).

Prasad, P. N., F. E. Karasz, Y. Pang and C. J. Wung, U. S. Patent application 312132 (1989a).

Prasad, P. N., M. Casstevens, and M. Samoc, *SPIE Proc.* **1056**, 117 (1989b).

Rao, D. N., P. Chopra, S. K. Ghoshal, J. Swiatkiewicz, and P. N. Prasad, *J. Chem. Phys.* **84**, 7049 (1986a).

Rao, D. N., J. Swiatkiewicz, P. Chopra, S. K. Ghoshal, and P. N. Prasad, *Appl. Phys. Lett.* **48**, 1187 (1986b).

Rothberg, L., T. M. Jedju, S. Etemad, and G. L. Baker, *Phys. Rev. Lett.* **57**, 3229 (1986).

Rustagi, K. C., and J. Ducuing, *Opt. Commun.* **10**, 258 (1974).

Samoc, A., M. Samoc, P. N. Prasad, C. S. Willand, and D. J. Williams, unpublished work, 1989.

Sauteret, C., J.-P. Hermann, R. Frey, F. Pradere, J. Ducuing, R. H. Baughman, and R. R. Chance, *Phys. Rev. Lett.* **36**, 956 (1976).

Schott, M., and G. Wegner, in D. S. Chemla and J. Zyss (Eds.), *Nonlinear Optical Properties of Organic Molecules and Crystals*, Vol. 2, Academic, New York, p. 3.

Shank, C. V., R. Yen, R. L. Fork, J. Orenstein, and G. L. Baker, *Phys. Rev. Lett.* **49**, 1660 (1982).

Shirk, J. S., J. R. Lindle, F. J. Bartoli, C. A. Hoffman, Z. H. Kafafi, and A. W. Snow, *Appl. Phys. Lett.* **55**, 1287 (1989).

Sinclair, M., D. Moses, K. Akagi, and A. J. Heeger, in A. J. Heeger, J. Orenstein, and D. R. Ulrich (Eds.), *Nonlinear Optical Properties of Polymers, Materials Research*

Society Symposium Proceedings, Vol. 109, Materials Research Society, Pittsburgh, 1988, p. 205.

Singh, B. P., P. N. Prasad, and F. E. Karasz, *Polymer* **29**, 1940 (1988).

Singh, B. P., M. Samoc, H. S. Nalwa, and P. N. Prasad, *J. Chem. Phys.* **92**, 2756 (1990).

Skotheim, T. A. (Ed.), *Handbook of Conducting Polymers*, Vols. 1 and 2, Dekker, New York, 1986.

Stamm, U., T. Kobayashi, M. Taiji, M. Yoshizawa, and K. Yoshino, CLEO/QELS Conference Abstract, paper FHH3, 1989.

Stamm, U., M. Taiji, M. Yoshizawa, K. Yoshino, and T. Kobayashi, *Mol. Cryst. Liq. Cryst.* **7**, in press.

Stevenson, S. H., D. S. Donald, and G. R. Meredith, in A. J. Heeger, J. Orenstein, and D. R. Ulrich (Eds.), *Nonlinear Optical Properties of Polymers, Materials Research Society Symposium Proceedings*, Vol. 109, Materials Research Society, Pittsburgh, 1988, p. 103.

Su, W. P., J. R. Schrieffer, and A. J. Heeger, *Phys. Rev. Lett.* **42**, 1698 (1979).

Swiatkiewicz, J., P. N. Prasad, F. E. Karasz, M. Druy, and P. Glatoski, *Appl. Phys. Lett.* **56**, 892 (1990).

Swiatkiewicz, J., P. N. Prasad, and C. Lee, unpublished work, 1990.

Tomaru, S., K. Kubodera, S. Zembutsu, K. Takeda, and M. Hasgawa, *Electron. Lett.* **23**, 595 (1987).

Vardeny, Z., J. Strait, D. Moses, T. C. Cheung, and A. J. Heeger, *Phys. Rev. Lett.* **49**, 1657 (1982).

Vardeny, Z., E. Fhrenfreund, O. Brafman, M. Nowak, H. Schaffer, A. J. Heeger, and F. Wudl, *Phys. Rev. Lett.* **56**, 671 (1986).

Vardeny, Z., H. T. Grahn, A. J. Heeger, and F. Wudl, *Synth. Met.* **28**, C299 (1989).

Winter, C. S., S. N. Oliver, and J. D. Rush, *Opt. Commun.* **69**, 45 (1988).

Worland, R., S. D. Phillips, W. C. Walker, and A. J. Heeger, *Synth. Met.* **28**, D663 (1989).

Wu, J. W., J. R. Heflin, R. A. Norwood, K. Y. Wong, O. Zamani-Khamiri, A. F. Garito, P. Kalyanaraman, and J. Sounik, *J. Opt. Soc. Am. B* **6**, 707 (1989).

Yang, L., Q. Z. Wang, P. P. Ho, R. Dorsinville, R. R. Alfano, W. K. Zou, and N. L. Yang, *Appl. Phys. Lett.* **53**, 1245 (1988).

Yang, L., R. Dorsinville, Q. Z. Wang, W. K. Zou, P. P. Ho, N. L. Yang, R. R. Alfano, R. Zamboni, R. Danieli, G. Ruani, and C. Taliani, *J. Opt. Soc. Am. B* **6**, 753 (1989).

Yu, L., and L. R. Dalton, *J. Am. Chem. Soc.* **111**, 8699 (1989).

Zhao, M. T., B. P. Singh, and P. N. Prasad, *J. Chem. Phys.* **89**, 5535 (1988).

Zhao, M. T., M. Samoc, B. P. Singh, and P. N. Prasad, *J. Phys. Chem.* **93**, 7916 (1989).

11

NONLINEAR OPTICS IN OPTICAL WAVEGUIDES AND FIBERS

11.1 BASIC CONCEPTS OF GUIDED WAVES

An optical waveguide is a medium in which the propagation of an optical wave is confined in one or two-dimensions, the confinement dimension being comparable to the wavelength of light (Kogelnik 1979, Stegeman et al. 1986, Stegeman and Stolen 1989). Typical examples of optical waveguides are shown in Figure 11.1. A fiber is a waveguide in which the propagation of an optical wave is confined in two dimensions (both cross-sectional dimensions). A planar waveguide is an example of a medium that provides a one-dimensional confinement to yield a one-dimensional guided wave. A waveguide is formed by a region with a refractive index larger than that of the surrounding media. For an asymmetric waveguide for which the media on the two sides are different in refractive indices, there is a minimum film thickness for waveguiding to occur. For example, a planar waveguide can be formed by a film of thickness $\sim 1\,\mu$m deposited on a substrate with a lower refractive index and a surrounding cladding medium also with a lower index, provided the refractive index differences are not too small.

Figure 11.2 shows the features of a planar waveguide which consists of a film of thickness h in the z direction (direction of beam confinement) and refractive index n_f deposited on a substrate of refractive index n_s. The top cladding medium (which can be air itself) has the refractive index n_c. Conceptually, the propagation of a beam in the waveguide can be described by total internal reflection of the ray at both interfaces, as shown in Figure 11.2. The condition for total internal reflections at each interface requires n_f to be larger than n_c and n_s. If an integral number of wavelengths of light are traversed

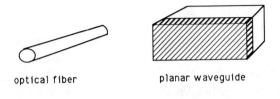

optical fiber   planar waveguide

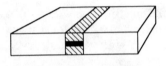

buried channel waveguide

Figure 11.1   Typical examples of optical waveguides.

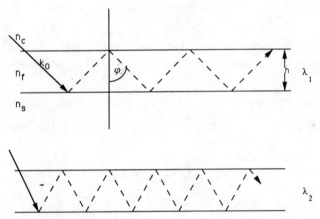

Figure 11.2   Ray propagation in a planar (slab) waveguide represented for two wavelengths, λ_1 and λ_2.

before the next interface, a standing wave pattern will develop and constructive interference will occur. Power can be transmitted along the x direction in the waveguide under these circumstances. Changing the wavelength to λ_2 means that either φ or the thickness of the guide must change to sustain the constructive interference condition. Changing the thickness of the guide without changing the wavelength implies that φ must change. The ray traveling in the thicker guide is required to make more reflections to traverse the same length in the x direction, which implies that the effective velocity of light in the thicker waveguide must be slower. This gives rise to a thickness-dependent effective refractive index in the guide. This is extremely useful for phase matching and will be discussed later. Now we provide a more quantitative description of the modes of an optical waveguide.

A wave propagating in the x direction must correspond to specific eigenmodes of the waveguide resulting from the resonance condition permitted by beam confinement (analogous to the resonance of a cavity). The general form for the field associated with propagation of a wave along the x direction can be written as

$$\mathbf{E} = \mathbf{f}(y, z)a(x)e^{i(\omega t - \beta x)} \tag{11.1}$$

where $\mathbf{f}(y, z)$ is the spatial field distribution within the waveguide. For a channel waveguide f is a function of y and z, but for a planar waveguide it is just a function of z. The term $|a(x)|^2$ gives the guided wave power in watts for a channel waveguide and watts per meter for a slab waveguide. The term β describes the propagation constant along the x direction and is equivalent to the propagation constant k_0 for the plane wave. It relates to the effective index N of the waveguide by an equation similar to that for a plane wave:

$$\beta = Nk_0 \tag{11.2}$$

The following discussion pertains to the simple case of a planar/slab waveguide. The effective index N is the analog of the bulk refractive index for a plane wave.

$$N = n_{\mathrm{f}} \sin \varphi \tag{11.3}$$

Perhaps it is appropriate to mention that in literature, β has sometimes been used to represent the effective index N itself. In such a case, the propagation constant has been represented by βk_0. The effective index N relates to the resonance conditions for the phase shifts occurring at both boundaries of the waveguide (Kogelnik 1979):

$$k_z = k_0\sqrt{n_{\mathrm{f}}^2 - N^2} \tag{11.4}$$

$$2k_z h + 2\phi_{\mathrm{fc}} + 2\phi_{\mathrm{fs}} = 2m\pi \tag{11.5}$$

In equation 11.5, ϕ_{fc} and ϕ_{fs} are the phase shifts suffered by a plane wave on reflection from the two interfaces: film–cladding and film–substrate. Here k_z is the wave-vector component along the z direction. The term m labels the different eigenmodes. The number of eigenmodes and their propagation constants is determined by (11.4) and (11.5) and is dependent on the refractive indices of the three media and the thickness of the waveguide. One can derive the field distribution for each eigenmode by detailed solutions of the Maxwell equation (Kogelnik 1979). Here only qualitative features will be discussed. The modes also possess well-defined polarizations. For an isotropic film slab waveguide, the modes are either TE or TM. In the TE mode, the polarization of electric

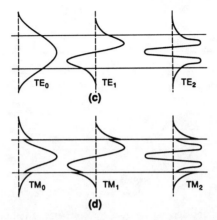

Figure 11.3 The electric field distribution for TE and TM modes for $m = 0, 1, 2$.

field distribution is only in the plane of the film. The TM mode is defined by the polarization in which only the magnetic field is in this plane. The unusual feature of the TM mode is that it also provides an electric field component (E_x) in the direction of propagation (x). In contrast, a plane wave propagating in a bulk medium has only transverse electric field components.

The electric field distribution for TE and TM modes of some low m values are shown in Figure 11.3. For the TE mode, only the E_y component of the electric field exists, which can be written for the mth order guided mode as (Stegeman et al. 1986)

$$E_y^{(m)} = \tfrac{1}{2}[f_y^{(m)}(z)a^m(x)e^{i(\omega t - \beta^{(m)}x)} + \text{c.c}] \tag{11.6}$$

The term $f_y^{(m)}(z)$ describes the details of the field distribution as a function of depth, as represented in Figure 11.3. The term $|a^m(x)|^2$ gives the guided wave power in watts per meter in the y axis as a function of propagation distance. For the TM modes, both E_z and E_x components of the electric field exist. Figure 11.3 describes the field distribution for E_z as a function of depth. For both TM and TE modes, the wave field extends into both the cladding and the substrate where it decays exponentially as a function of distance from the interface. Hence, a waveguide geometry permits interaction of the optical field with matter in all the three media (the guiding film, the substrate, and the cladding). Of course, the wave field is maximized in the film and exists only as an evanescent field in the cladding and substrate media.

Three important techniques for coupling radiation into a waveguide are represented in Figure 11.4. The grating and prism coupling methods are more commonly used for coupling into a planar waveguide (Kogelnik 1979, Stegeman et al. 1986, Stegeman and Stolen 1989). In both cases, the incident angle is adjusted so that the wave-vector component of the incident field parallel to the surface matches the guided wave vector $\beta^{(m)}$ for a guided mode m. For grating

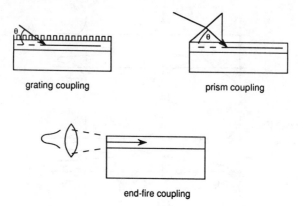

grating coupling   prism coupling

end-fire coupling

Figure 11.4   Three important techniques for coupling radiation into a waveguide.

coupling, the wave-vector component parallel to the surface has two contributions, one from the projection of the incident field wave vector ($k_0 n_c \cos \theta$) and the other from the grating ($2\pi/\Lambda$ in the lowest order, where Λ is the grating spacing). Therefore,

$$k_0 n_c \cos \theta \pm \frac{2\pi}{\Lambda} = \beta^{(m)} \tag{11.7}$$

In the case of prism coupling, the projection of the wave vector of the incident field at the base of the prism must be equal to the guided wave vector. Hence,

$$k_0 n_p \cos \theta = \beta^{(m)} = N^{(m)} k_0 \tag{11.8}$$

To meet the conditions set by equation 11.8, one must use a prism of high refractive index (n_p) with the condition $n_p > N$.

The end fire coupling matches the incident field spatial distribution to that of the guided wave. This technique is preferable when coupling into a two-dimensional waveguide such as an optical fiber or a channel waveguide.

The number of waveguide modes that can be propagated depends on the film index and the film thickness (or fiber core diameter). However, for a waveguide to remain a single mode, there is a maximum film thickness (or core diameter in the case of a fiber). For thicker films, the waveguide becomes multimode. Even in a multimode waveguide one can select the incident beam angle to launch almost all the power into a specific waveguide mode.

11.2   NONLINEAR OPTICS WITH GUIDED WAVES

For nonlinear optics, guided waves provide several advantages over plane waves in bulk media (Shen 1984, Stegeman and Liao 1983, Stegeman and Seaton 1985,

Stegeman and Stolen 1989). Nonlinear optics requires high light intensities (power/area). The strong beam confinement in one or two dimensions to the order of the wavelength (λ) of light leads to reduced area ($\pi\lambda^2$ in two-dimensional confinement) and, hence, increased intensity. For example, if a beam of 1 W power is confined in a waveguide of 1 μm^2 cross-sectional area (fiber or channel waveguide), the resulting intensity is 100 MW/cm^2.

To obtain high intensity in a bulk medium, one needs to focus the laser beam tightly to the smallest beam size determined by the diffraction limit. However, the tighter the focus, the shorter is the distance over which high power density can be maintained. Since most nonlinear optical processes depend quadratically on the interaction length, naturally one sacrifices the interaction length while trying to achieve high power density by a tighter focus. In contrast, a guided wave propagates without changing its profile, the propagation distances being solely determined by the propagation losses due to absorption and scattering of the guiding medium. Therefore, for a low-loss waveguide, these propagation distances can be mm $\rightarrow$ cm (intergrated optics waveguide) to m $\rightarrow$ km (in fibers). This long propagation distance provides a tremendous gain in the interaction length in the media.

As discussed above in Section 11.1, the electric field distribution extends into the bounding media (cladding and substrate). Therefore, any of the three media can be nonlinear, providing flexibility to design numerous nonlinear device structures. However, since the field distribution is maximum within the waveguide (the core), maximum nonlinear interaction will be achieved only by using the core as the nonlinear medium.

For nonlinear optical processes that need phase matching (harmonic generations), one utilizes the natural birefringence of the medium in a bulk and achieves phase matching by adjusting the angle. For nonlinear processes involving more than one beam, the phase matching in the bulk is achieved by using a noncollinear geometry with appropriate crossing angles. Of course, the noncollinear geometry in the bulk also reduces the interaction length between the beams. In a waveguide, one can take advantage of the possibility of different discrete modes which would have different effective index $N^{(m)}$. Therefore, one may be able to achieve phase matching (wave-vector conservation) by selecting one mode for one wave and another mode for a different wave. In addition, one may be able to use the core for guiding one wave and the cladding medium to guide another wave (for example, the second harmonic). With these phase-matching tricks unique to waveguides, one can still achieve long interaction lengths.

Finally, the biggest advantage of using nonlinear optics with guided waves is that it conveniently lends itself to integrated optics device structures compatible with fiber optics links. One can also take advantage of three dimensionality through channels and fibers to implement the concepts of optical neural network signal processing. Some of the specific examples of device structures are discussed in a later chapter.

11.3 SECOND-ORDER NONLINEAR OPTICAL PROCESSES

11.3.1 Second-Harmonic Generation in a Planar Waveguide

Second-harmonic generation (SHG) in a waveguide is a nonlinear process that has been relatively more extensively studied (Stegeman and Stolen 1989). The driving force behind the interest in this process is the realization that it would have immediate technological application in doubling of the diode laser frequencies. Other related processes that have also been investigated in a waveguide configuration are difference frequency generation, optical parametric oscillation, and amplification. However, for optical waveguides made of organic materials, only the SHG-type frequency mixing process has been investigated. In this section, the SHG process in a planar waveguide is discussed. The simple case of a codirectional SHG is considered where both the fundamental and the second harmonic propagate along the x direction, with the fundamental wave of frequency ω coupled to the waveguide at $x = 0$. Both the fundamental and the second harmonic signal are decoupled from the waveguide at $x = L$. The second-harmonic power $P(2\omega, L)$ obtained at the exit point $x = L$ is then given by the absolute square of the amplitude function $a^{(m, 2\omega)}(L)$ discussed above (Yariv 1973, Shen 1984, Stegeman et al. 1986):

$$|a^{(m, 2\omega)}(L)|^2 = (k_0 L)^2 \frac{d_{\text{eff}}^2}{N^{3(m', \omega)}} \frac{\sin^2(\Delta k(L/2))}{(\Delta k L/2)^2} |K|^2 |a^{(m', \omega)}(0)|^2 \qquad (11.9)$$

In the above equation, Δk describes the phase mismatch as

$$\Delta k = [\beta^{(m, 2\omega)} - 2\beta^{(m', \omega)}] = 2[N^{(m, 2\omega)} - N^{(m', \omega)}]k_0 \qquad (11.10)$$

The term d_{eff} is the second-harmonic coefficient, so that $d_{\text{eff}}^2/(N^{(m', \omega)})^3$ describes the waveguide figure of merit. The term K is an overlap integral describing the overlap of the electric field distribution functions for the fundamental and the second harmonic. It is a concept unique to the waveguide and for a planar waveguide is given as

$$K = \int_0^\infty dz \frac{d_{ijk}}{d_{\text{eff}}} f_i^{(m', \omega)}(z) f_j^{(m', \omega)}(z) f_k^{*(m, 2\omega)}(z) \qquad (11.11)$$

The indices i, j, and k are x, y, and z in general. For efficient codirectional SHG in a waveguide, two conditions have to be optimized: (1) the phase mismatch $\Delta k = 0$ (in other words, the process is phase-matched), and (2) the overlap integral K does not contain a product of field distributions that changes sign across the waveguide.

For phase matching to occur, the effective induces N must be matched, that is, $N^{(m, 2\omega)} = N^{(m', \omega)}$. For the same mode and polarization, for example, TE, $N(\text{TE}_m, 2\omega) > N(\text{TE}_m, \omega)$. Therefore, phase matching between the fundamental

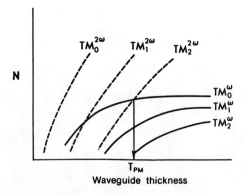

Figure 11.5 The dependence of the effective index N on the waveguide thickness for both the fundamental and the second harmonic for TM polarization. T_{PM} represents the thickness for phase matching, which also yields a favorable overlap of field distributions of the second harmonic and the fundamental.

and the second harmonic of identical mode and polarization is not possible. Since the effective index N is dependent on the mode and polarization, it is possible to phase match in a multimode waveguide by selecting the fundamental of one mode and polarization and matching its effective index with that of the second harmonic of a different mode and even different polarization.

Figure 11.5 exhibits the dependence of the effective index N on the waveguide thickness for both the fundamental and the second harmonic for TM polarization. The crossing points between the two curves represent the phase-matching condition and yield the optimum waveguide thickness needed.

As stated above, one also has to insure that the overlap of the field distribution functions is favorable, so that the overlap integral K is maximized. Figure 11.6 illustrates examples of good and bad overlaps for the fundamental (solid curves) and the second harmonic (dashed curve) for TE polarization. Meeting

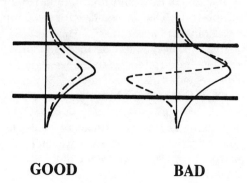

GOOD **BAD**

Figure 11.6 Examples of good and bad overlaps for the fundamental (—) and the second harmonic (- - -) for TE polarization.

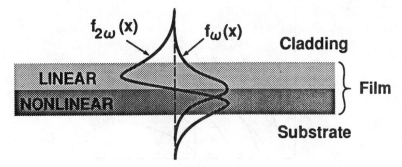

Figure 11.7 A two-layer waveguide designed to optimize the overlap integral.

simultaneously the two requirements discussed above is not an easy task. In the illustrative example presented in Figure 11.5, even though the phase matching is satisfied for two sets $TM_1^{(2\omega)}$, $TM_0^{(\omega)}$ and $TM_2^{(2\omega)}$, $TM_0^{(\omega)}$, only the latter set yields a more favorable overlap. Sometimes both conditions can also be met by using waveguides of special geometries. Tricks can be used to reduce the efficiency loss when the interacting modes are not all of the same order (same m value). An example is shown in Figure 11.7, where the waveguide film consists of two layers, only one of which is optically nonlinear. In this example, a favorable (constructive) overlap between the fundamentals and the second harmonic occurs in the nonlinear regime.

From the above discussion, it can be concluded that waveguide harmonic generation is an efficient process if phase matching between two modes with large electric field overlaps and large d_{eff} can be accomplished. Walkoff is not a problem in waveguides and does not limit the interaction length. On the other hand, thickness must be highly controlled as well as nonuniformities in the refractive index.

An early report of SHG in an organic waveguide is by Hewig and Jain (1983). They used an evaporated film of *para*-chlorophenylurea (PCPU) $\sim 0.9\,\mu$m thick deposited on a glass substrate. The waveguide supported multiple modes and the phase matching was obtained between the TM_0 and TM_2 modes. However, the reported conversion efficiency was extremely low, due both to the low overlap integral K and high optical losses in the waveguide.

Later reports of waveguide SHG have focused on the organic material, 2-methyl-4-nitroaniline (MNA) for which the largest coefficient d_{11} is 2.5×10^{-10} mV. In terms of figure of merit, this value is 2000 times larger than that of $LiNbO_3$. Stegeman and Liao (1983) had predicted by their theoretical calculation a higly efficient SHG in an MNA waveguide of thickness $0.43\,\mu$m. Their calculation yielded a conversion efficiency of 8.6% in a guiding length of 10 mm with a 100-MW input power. Sasaki et al. (1984) reported SHG in the 2-methyl-4-nitroaniline (MNA) crystal in a waveguide geometry. They vapor deposited a ~ 50-μm-thick MNA crystal on a tapered glass slab waveguide. The

SHG was achieved via the larger d_{11} coefficient by using different modes of the same polarization.

More recently, Itoh et al. (1986a, b) reported a more efficient SHG in a waveguide geometry by growing a thin tapered crystal of MNA. They used two different waveguide configurations. In the earlier configuration, they grew the tapered MNA crystal of thickness in the range of 0.5–5.00 μm from melt on the top of a planar glass waveguide (thickness 0.89–1.027 μm) that had been deposited on a fused quartz substrate by rf sputtering. Both the MNA and the glass waveguide acted as the guiding regions for the fundamental and the second harmonic. Then phase matching between TE modes of different order (m values) was obtained by translating the waveguide to vary the thickness. Again the orientation of the crystal was such that for the TE modes, the SHG utilized the d_{11} coefficient.

In a later experiment, Itoh et al. (1986b) grew from melt a tapered MNA crystal of thickness 4.8–6.5 μm to act as the sole waveguiding region. The fundamental from a Nd:YAG laser was coupled to this waveguide as a TE mode using a TiO_2 prism. The fundamental wave propagated along the x axis, which was perpendicular to the direction of d_{11} of the MNA film. Then the film was translated to adjust the thickness for the highest SHG signal corresponding to phase-matched conditions at appropriate thickness. Two peaks in the SHG signal were observed at two different thicknesses: $\sim 5.3\,\mu$m, corresponding to the phase-matching condition $TE_9^{(2\omega)} \to TE_{31}^{(\omega)}$, and $\sim 5.9\,\mu$m, corresponding to the phase-matching condition $TE_{10}^{(2\omega)} \to TE_{35}^{(\omega)}$. The observed thicknesses for the phase-matched SHG were found to be in agreement with the theoretical estimates.

A more promising group of materials for second-order nonlinear processes in a waveguide configuration is electrically poled polymeric systems. These polymeric systems can be a guest–host system in which a guest molecule (such as MNA) with a large β is dispersed in a polymeric host and is aligned by electric poling (discussed in Chapter 4) of this guest–host system near the glass-transition temperature of the host. Alternatively, the nonlinear molecular unit (with large β) can be chemically bonded as a side-chain group to the polymer backbone. The latter approach minimizes the problem of phase separation between the guest and the host at higher concentrations of the nonlinear guest group and it also imparts additional thermal stability to electric field-induced alignment. Much of the early device work concentrated on making large bandwidth electrooptic modulators, mostly using a Mach–Zehnder interferometer arrangement. This topic is discussed in more detail in the next chapter under device concepts.

11.3.2 Phase Matching in a Periodically Poled Waveguide

From the above discussion it is apparent that even though a waveguide configuration permits phase matching by selecting the fundamental of one waveguide mode and the second harmonic of a different mode, such conditions lead to a reduced overlap integral, given by equation (11.11). Somekh and Yariv (1972) proposed a clever scheme for phase matching in a waveguide, which does not depend on the matching of the guided wave indices of the fundamental and the second harmonic. Therefore, one can even use zero-order modes for both the fundamental and the second harmonic, which leads to a large value for the overlap integral. This type of phase matching is sometimes also referred to as quasi phase matching. It involves a periodic modulation of the nonlinear optical properties of the propagating medium. For a quantitative description of the principle of quasi phase matching interested readers are referred to the paper by Somekh and Yariv (1972). Here only a qualitative conceptual description is presented.

In the case of a finite phase mismatch Δk, the observed second-harmonic power shows an oscillatory behavior as a function of the propagation length through the nonlinear medium. This feature is discussed in detail in Chapter 5 and is shown again by curve c in Figure 11.8. The power flows from the fundamental to the second harmonic, which builds up as the length l increases to l_c, the coherence length of the nonlinear medium. When $l > l_c$, the power flows back from the second harmonic to the fundamental. Consequently, the second-harmonic intensity will continue to diminish and will become zero at $l = 2l_c$. This type of cycle then repeats. The reversal of the power flow from the second harmonic to the fundamental can be prevented in a peridically poled structure in which the nonlinear d coefficient of the medium is spatially modulated with a period of $2l_c$. One possible way to modulate the d coefficient is to reverse the direction of bulk polarization every coherence length. This changes the sign of the nonlinear coefficient. The resulting second-harmonic intensity in such a bulk is displayed by curve B in Figure 11.8. Nagel et al.

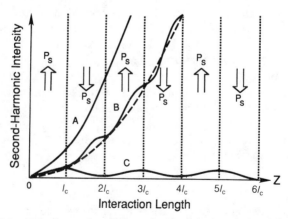

Figure 11.8 Representations of variation of second-harmonic intensity as a function of interaction length for three cases: (A) phase-matched process; (B) process in a periodically poled system for which the sign of the nonlinear coefficient changes every coherence length; (C) non-phase-matched process.

(1989) demonstrated this type of periodic poling in a lithium niobate crystal by producing oppositely polarized ferroelectric domains. They used a laser-heated pedestal growth technique with a rotating temperature gradient transverse to the growth direction. Since the crystal c axis is transverse to the growth direction, the temperature gradient, therefore, periodically poles the growing crystal along this axis. Lim et al. (1989) used such a periodically poled lithium niobate waveguide to demonstrate second-harmonic generation at 532 nm from a 2.6-mW cw Yag fundamental at 1.06 μm. They reported a conversion efficiency of $\sim 2\%$ W cm^2.

An alternative scheme of periodic poling is to turn off the nonlinearity at every coherence length, so that the nonlinearity vanishes between the region in which the power is supposed to flow back from the second harmonic to the fundamental. Khanarian et al. (1989) recently produced such a structure in a polymeric waveguide by periodically poling it with a periodic dc electric poling field. They reported a quasi-phase-matched second-harmonic signal with a polarization TM$_0$ produced from the fundamental input beam excited in the waveguide also as a TM$_0$ mode. The overlap integral given by equation 11.11 was therefore maximized.

11.3.3   Second-Harmonic Generation in Organic Crystal Cored Fibers

As an alternative to the planar waveguide structures discussed above, SHG in organic crystal cored fibers has also been investigated. One possible advantage of the crystal cored fibers can be in achieving uniform guiding dimensions over long lengths. The organic crystal cored fibers have been fabricated by growing the crystal from melt in a glass capillary using a modified Bridgemann technique (Nayar 1983, 1988, Vidakovic et al. 1987). The application of this technique requires that the material be chemically stable upon melting and does not undergo phase transitions, which can create defects. Using this method, defect- and void-free crystal cored fibers of several materials were grown up to 5 cm in length, having diameters in the range of 2–10 μm. These dimensions allowed for single-mode operation.

The theory of waveguiding in a fiber is similar to the one for a planar waveguide except that two-dimensional confinement now leads to mode structures labeled by two indices, m and n. Therefore, a specific transverse electric mode will be labeled as TE$_{mn}$. Again, for an efficient SHG in a fiber, one needs to fulfill the phase-matching condition and also maximize the spatial overlap integral for the electric field distribution. In addition, the direction of crystal growth should have components of the SH tensor coefficient in a direction transverse to the fiber axis. In many of the crystal cored fibers (e.g., m-nitroaniline; Nayar 1983) the SH tensor coefficient had the largest component along the fiber axis and, because the electric fields in the fiber are transverse, such crystal cored fibers could not be used for SHG.

SHG in a crystal cored fiber of benzil was achieved with a 6-μm core diameter using Schott glass SKN 18 as the cladding medium (capillary) (Nayar 1983).

The fundamental wavelength used was 1.06 μm from a Nd:YAG laser. However, in this waveguide phase matching could not be achieved for both the fundamental and the SH as guided waves. The coupling occurred with the fundamental as the guided wave and the SH as the radiation field. In other words, for the SH, the cladding medium (glass capillary) had a refractive index higher than that of the crystal core fiber and acted to guide the SH. Therefore, the second-harmonic signal was trapped in the cladding ring and exited in the form of a ring. Thus, this is not a phase-matched second-harmonic generation. The theoretical analysis shows that in such a case, the SHG efficiency is proportional to the interaction length rather than to the square of the interaction length observed when both the fundamental and the SH are waves guided in the core for a phase-matched process. A low conversion efficiency of $< 10^{-3}\%$ was estimated.

When the fundamental is a guided wave but the SH is a radiation field the interaction also leads to a very small overlap integral, which is also a reason for low conversion efficiency.

Recent works on crystal cored fibers include MNA (Nayar 1988), N-(4-nitrophenyl)-L-prolinal (NPP) and N-(4-nitrophenyl)-N-methylamino-acetonitrile (NPAN) (Vidokovic et al. 1987). No efficient SHG in a crystal cored fiber has been reported.

11.4 THIRD-ORDER NONLINEAR OPTICAL PROCESSES

A variety of third-order nonlinear optical processes have been investigated in a waveguide geometry (Stegemen and Liao 1983, Stegeman et al. 1986, Stegeman and Stolen 1989). Recently interest has focused heavily on the study of third-order guided-wave processes because they allow for the fabrication of all optical devices. Degenerate four-wave mixing, coherent anti-Stokes–Raman scattering (CARS), and stimulated Raman scattering and amplification have been demonstrated with guided waves. The most important process from the point of view of all optical signal processing is the intensity-dependent phase shift in a waveguide. One can achieve all optical modulation and switching in a waveguide geometry using the intensity-dependent phase shift because in a waveguide both the local intensity and the interaction length are maximized to significantly enhance the intensity-dependent phase shift derived at a given input power.

The intensity-dependent phase shift results from the intensity-dependent refractive index $n_2 I$ (discussed in earlier chapters). As is the case for the second-order processes discussed above, various device structures can be envisioned in which any of the three media can be nonlinear. Of course, maximum nonlinear ineraction can be obtained only when the guiding film is the optically nonlinear medium. This is the case that will be discussed here. The waveguiding condition requires that $n_f > n_c, n_s$. If the core (the film) is nonlinear, for waveguiding at all intensities the following condition must be met:

$$n_f = n_0^f + n_2^f I > n_c, n_s \tag{11.12}$$

In the low-intensity limit, $n_f \simeq n_0^f$. Based on a waveguide condition formulated for the low-intensity limit, one can describe the two limits of intensity-dependent perturbation. In the strong perturbation limit, the intensity-dependent part of the refractive index $|n_2^f I| > (n_0^f - n_c)$ or $(n_0^f - n_s)$. This is the case of strong optical nonlinearity (large n_2^f); both the electric field distribution and propagation wavevectors undergo large changes. One can see it more obviously, when n_2^f is negative. In such a case, at a certain power level n_f may become reduced to a value smaller than n_c or n_s and at that point the field distribution would change drastically since the film would no more guide the wave. The weak perturbation limit results for small optical nonlinearity where the intensity-induced refractive index change in the film is much smaller than the difference in the refractive indices $(n_0^f - n_c)$ or $(n_0^f - n_s)$. In this case, the field distribution of the guided mode remains essentially unchanged and only an intensity-dependent phase shift results. For organic systems, only the weak perturbation limit is applicable in view of their small $\chi^{(3)}$ values.

In a waveguide, the analog of the intensity dependent refractive index is the guided power dependent effective index N or propagation constant β:

$$\beta = \beta_0 + \Delta\beta_0 P_{g\omega} \tag{11.13}$$

where $P_{g\omega}$ is the guided wave power. A complicating feature in the waveguide is that due to long interaction length provided by the waveguide geometry, even a very small residual absorption leads to appreciable thermal effect. Most organic systems have weak absorption even in the region of their transparency due to three sources: (1) impurities and defects. (2) distribution of conjugation lengths in conjugated polymers (therefore, a polymer chain with a long effective conjugation would absorp at a much longer wavelength), and (3) presence of vibrational overtone absorption. For example, $-C-H$ overtone absorption creates weak absorption in the near IR. Weak absorption leads to thermal nonlinearity due to the dn/dT term. This thermal nonlinearity is slowly relaxing (typically in microseconds to milliseconds) and, therefore, is nonlocal and integrating in nature for optical pulses shorter than the relaxation time. The contribution to n_2 (intensity dependent refractive index) is given by (Assanto et al. 1988)

$$n_{2t} = \frac{dn}{dT} \frac{\alpha\tau_t}{\rho C_p} \tag{11.14}$$

where α is the absorption coefficient, ρ is the density, C_p is the specific heat at constant pressure, dn/dT describes the temperature dependence of the refractive index, and τ_t describes the thermal relaxation time. The intensity-dependent phase shift in a waveguide in the presence of both electronic and thermal nonlinearities has been theoretically analyzed by Stegeman and coworkers (Assanto et al. 1988). In the case of a transverse field (TE mode) propagating along the x axis, the guided wave nonlinear phase shift, ϕ^{NL}, is given by

$$\frac{\partial}{\partial x}\phi^{NL}(x,t) = A_e k_0 |a(x,t)|^2 + \frac{A_t k_0}{\tau_t} \int_{-\infty}^{t} dt' |a(x,t')|^2 \qquad (11.15)$$

The term a is the same as in (11.6). Except for a pulse experiment, it is also a function of time. $a(x,t)$ describes the guided-wave amplitude at point x and time t. The above equation assumes that the laser pulse width Δt is much shorter than the thermal relaxation time (τ_t). The first term on the right represents the purely electronic contribution; the second term is the contribution due to the integrating thermal nonlinearity. The terms A for electronic (A_e) and thermal (A_t) are given as

$$A_{e,t} = \int_{-\infty}^{\infty} dz\, n_{2e,t}(z)|f(z)|^4 \qquad (11.16)$$

For the case where the thermal relaxation time τ_t is much longer than the pulse width Δt, the effective integrating nonlinearity is reduced from the n_{2t} value of equation 11.14 by a factor of $\Delta t/\tau_t$. Therefore, to reduce the effective thermal contribution to the phase shift, one must use much smaller laser pulse width. Based on typical values of n_{2e} and n_{2t} for organics, the minimum laser pulse width to observe unambiguously electronic nonlinearity is predicted to be less than 100 ps. Even with the absorption coefficient $\alpha \approx 0.02\,cm^{-1}$, but with pulses longer than 10 ns, the thermal nonlinearity makes a significant contribution.

Most of the reported results of intensity-dependent phase shifts are very likely due to the dominant role of thermal effect, since the pulses used were of nanoseconds time scale. The first clear demonstration of the intensity-dependent phase shift derived from electronic nonlinearity is in the work reported by Burzynski et al. (1988).

In a grating or prism coupling experiment, the intensity-dependent phase shift in the waveguide changes the coupling angle θ through equations 11.7 and 11.8. Therefore, the optimum coupling angles at low and high input powers are different. If the coupling angle is optimized for the low power, then the guided-wave power measured by the output at the exit port will show a limiter action behavior as the input power is increased. Consequently, the coupling efficiency (defined by the ratio of output to input intensities) will drop as the input power is increased.

In the study reported by Burzynski et al. (1988), the waveguide was fabricated from spin-coating a 1.2-μm-thick film of polyamic acid. The absorption coefficient was estimated to be $< 0.25\,cm^{-1}$ at $\lambda = 633$ nm. The waveguiding loss, which has combined contribution due to losses by absorption and scattering, was estimated to be $1.2\,cm^{-1}$. Burzynski et al. performed experiments with laser pulses of three different pulse widths: (1) 400-fs pulses of 100-μJ/pulse energy at 602 nm, (2) 80-ps pulses of 10-μJ/pulse energy at $\lambda = 575$ nm, and (3) 10-ns pulses of 10-mJ/pulse energy at $\lambda = 633$ nm. From the theoretical analysis of their result, they concluded that the intensity-dependent phase shift observed with the nanosecond pulses were due to thermal nonlinearity, but the results of the 80-ps and 400-fs pulses reflected dominant contribution due to electronic nonlinearity.

Figure 11.9 shows the change of coupling angle θ for the TE_1 mode between

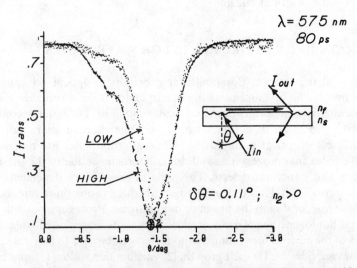

Figure 11.9 Change of the coupling angle θ for the TE_1 mode in a waveguide grating coupler experiment when going from low to high intensity. Pulses used are 80 ps wide and at 575 nm. The inset shows the grating coupling into the waveguide.

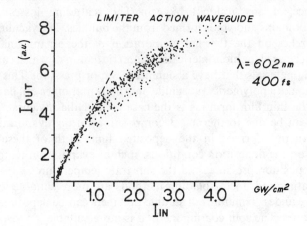

Figure 11.10 The observed limiter action behavior for the polyamic acid waveguide at 602 nm using 400-fs pulses.

the high and low intensities using 80-ps pulses at 575 nm. This result is also of fundamental importance because it provides the sign of n_{2e} (and hence $\chi^{(3)}$). At higher intensity, the coupling maximum shifts to higher angles, yielding a positive sign for n_{2e} and $\chi^{(3)}$. From the shift in the coupling angle, n_{2e}, and hence $\chi^{(3)}$, can be calculated. The calculated value of n_{2e} was found to be in reasonable agreement with that obtained by the degenerate four-wave mixing study which gave a value of $n_{2e} \simeq 3 \times 10^{-18} \, m^2/W$. Figure 11.10 shows the limiter action behavior observed in the I_{out} versus I_{input} curve obtained with pulses of 400-fs width at 602 nm.

11.5 MATERIAL REQUIREMENTS FOR WAVEGUIDES

The material requirements for nonlinear processes in optical waveguides are very stringent. The first and most important factor is the waveguide loss. This loss is due to both absorption and scattering. For both the second- and third-order processes, these losses would limit the effective interaction length one can use for optical nonlinearity. The scattering losses are derived from refractive index inhomogeneities and domain structures due to the presence of impurities and structural defects. This type of loss is very dependent on the processing condition. The absorption losses are due to absorption centers, which can be intrinsic, or due to the presence of impurities. Processing conditions and structural homogeneities of a material determine its waveguide losses.

Some of the various processing methods that can be used to fabricate organic optical waveguides are (1) melt growth, (2) vacuum deposition, (3) spin coating, and (4) doctor blading. The melt process is more convenient for materials that melt without decomposition and has been used for the fabrication of crystal cored fibers, as discussed above. In this case, the impurities are primarily due to thermal decomposition or oxidation of organic materials because of the presence of any residual oxygen. The scattering losses are due to crystalline domains and voids formed from the bubbles. The vacuum deposition technique relies on the thermal evaporation of the organic materials. This method can be used for monomeric or oligomeric materials which can be vacuum evaporated (materials that have a significant vapor pressure). This method can be used to make a polymeric waveguide only if the monomer can be polymerized in situ in the thin film form (as is the case for polydiacetylenes). In this case, the losses can be due to thermal decomposition during evaporation and any domain structure formed in the deposited film. Both of these defects are dependent on the deposition conditions, such as evaporation temperature and rate, the deposition pressure, and the substrate temperature.

Spin-coating and doctor-blading methods are more suitable for polymeric planar waveguides. Thin films of soluble polymers can be deposited using these two techniques. The spin-coating method is more suitable if a viscous solution can be formed to yield a slowly spreading film to produce a thickness of $\sim 1 \, \mu m$, required for optical waveguiding. The doctor-blading method yields a film of desired thickness ($\geq 1 \, \mu$m), when the solution is not as viscous. In this method, the film is spread using a knife. The method can also be used to fabricate waveguides of polymers that are not soluble by themselves, but can be processed through a soluble precursor. In this case, a waveguide of the precursor is fabricated and then converted into the nonlinear polymer. These methods create scattering centers due to contamination from the solvent and the airborne particles. Therefore, the use of ultrapure solvent and clean atmosphere (clean room) is important. Also, the spreading conditions (such as the nature of solvent and spinning speed) determine the refractive index homogeneity. These parameters have to be optimized by trial. Another important consideration in the choice of the method is the uniformity of the film, which plays a very important role in determining the waveguide performance.

For the fabrication of polymeric waveguides in the form of fibers, a number of techniques, such as gel spinning, can be used. For gel spinning, the polymer can be heated to the softening point or a thick viscous solution can be used. Then fibers are spun from the viscous material.

In addition to the requirements of low waveguide losses, other material requirements are (1) strong mechanical strength so that the waveguide is not easily damaged, (2) excellent environment stability, and (3) high optical damage threshold. The latter condition is especially important for optical waveguides in which the guided power density, because of the beam confinement, can be very high.

For third-order optical nonlinearity where the device may be based on the intensity-dependent phase shift, one can define a device figure of merit w (Stegeman and Stolen 1989) as follows:

$$w = \Delta n^{\max} \frac{L}{\lambda} \qquad (11.17)$$

In the above equation, $\Delta n^{\max}$ is the maximum refractive index change that can be accomplished by the increase of intensity. For many materials, especially with resonant optical nonlinearity, this value corresponds to the saturation value of Δn. For most organic materials, when nonresonant nonlinearity is used, this value corresponds to that derived from the intensity limit just below the optical damage threshold. In the case where α is the total waveguide loss (it includes contributions due to both adsorption and scattering) and L is the interaction length, the condition for high throughput is $L\alpha \ll 1$. Therefore, taking the limiting value $L\alpha \simeq 1$, one can define an effective W, a materials figure of merit, as

$$W = \Delta n^{\max} \frac{L}{\lambda} \frac{1}{L\alpha} = \frac{\Delta n^{\max}}{\lambda \alpha} \qquad (11.18)$$

The W value must be maximized. Many devices discussed in the later chapter require a value of $W > 2$.

For the second-order nonlinear optical processes in organic waveguides, electrically poled polymers have more desirable materials quality. They provide a medium with low loss, high mechanical strength and environmental stability, and high optical damage threshold. However, the question concerning their application focuses on the thermodynamic instability of the poled systems. The electric field aligned polar groups in the poled polymers represent a thermodynamically unstable ensemble. This alignment will undergo relaxation toward the thermodynamically stable structure. The detailed kinetic behavior of this relaxation over a long time has not been well studied.

For third-order nonlinear optical processes, waveguides formed by conjugated polymeric structures are more suitable. The problems with conjugated polymeric structures for waveguide applications are twofold:

1. Most conjugated polymers are insoluble and infusable. The two viable approaches used to solve this problem rest on chemical processing. In one approach, the conjugated polymeric structure is chemically modified by attaching a long flexible side chain (e.g., alkyl chain), which makes the polymer soluble. However, this may decrease the effective $\chi^{(3)}$ by reducing the packing density due to the presence of bulky side groups. This approach has been taken in the case of polydiacetylenes and polythiophenes, as discussed in Chapter 10. In the other approach, one uses a soluble precursor to form the waveguide and then convert it to the final polymer. This approach has been used for poly-p-phenylene vinylene and polyimides, also discussed in chapter 10.

2. A conjugated polymeric material is generally structurally inhomogeneous. It contains a distribution of molecular weights (chain distribution) and conjugation lengths (hence, $\chi^{(3)}$), the two not being related. The distribution of chain length may give rise to structural inhomogeneity, whereas the distribution of conjugation length creates difficulty in finding a truly transpaernt wavelength. Polymer purification, separation, processing, and characterization are, therefore, very important in the fabrication of good optical waveguides.

REFERENCES

Assanto, G., R. M. Fortenberry, C. T. Seaton, and G. I. Stegeman, *J. Opt. Soc. Am. B* **5**, 432 (1988).

Burzynski, R., B. P. Singh, P. N. Prasad, R. Zanoni, and G. I. Stegeman, *Appl. Phys. Lett.* **53**, 2011 (1988).

Hewig, G. H., and K. Jain, *Opt. Commun.* **47**, 347 (1983).

Itoh, H., K. Hotta, H. Takara, and K. Sasaki, *Appl. Opt.* **25**, 1491 (1986a).

Itoh, H., K. Hotta, H. Takara, and K. Sasaki, *Opt. Commun.* **59**, 299 (1986b).

Khanarian, G., D. Haas, R. Keosian, D. Karim, and P. Landi, CLEO abstract, paper THB1 (1989).

Kogelnik, H., in T. Tamir (Ed.), *Topics in Applied Physics*, Vol. 7, *Integrated Optics* Springer-Verlag, 1979. Berlin.

Lim E. J., M. M. Fejer, and R. L. Byer, CLEO abstract, paper THQ4 (1989).

Nagel, G. A., M. M. Fejer, and R. L. Byer, CLEO abstract, paper THQ3 (1989).

Nayar, B. K., in P. N. Prasad and D. R. Ulrich (Eds.), *Nonlinear Optical and Electroactive Polymers*, Plenum New York, 1988, p. 427.

Nayar, B. K., in D. J. Williams (Ed.), *Nonlinear Optical Properties of Organic and Polymeric Materials*, ACS Symposium Series, No. 233, 1983, p. 153.

Sasaki, K., T. Kinoshita, and N. Krasawa, *Appl. Phys. Lett.* **45**, 333 (1984).

Shen, Y. R., *The Principles of Nonlinear Optics*, Wiley-Interscience, New York, 1984, p. 505.

Somekh, S., and A. Yariv, *Opt. Commun.* **6**, 301 (1972).

Stegeman, G. I., and C. Liao, *Appl. Opt.* **22**, 2518 (1983).

Stegeman, G. I., and C. T. Seaton, *J. Appl. Phys.* **58**, R57 (1985).

Stegeman, G. I., and R. H. Stolen, *J. Opt. Soc. Am. B* **6**, 652 (1989).

Stegeman, G. I., C. T. Seaton, and R. Zanoni, *Thin Solid Films* **152**, 231 (1986).

Vidakovic, P. V., M. Coquillary, and F. Salin, *J. Opt. Soc. Am. B* **4**, 998 (1987).

Yariv, A., *IEEE J. Quantum Electron.* **QE-9**, 919 (1973).

12

DEVICE CONCEPTS

12.1 INTRODUCTION

Nonlinear optics is one of the few modern scientific frontiers where the huge surge in recent interest is not only owing to the quest for understanding of new physical phenomena occurring under intense laser fields, but also because of potential technological applications. The newly emerging technology of photonics utilizes photons instead of electrons to acquire, store, transmit, and process information. Concepts of optical computing, optical signal processing, and image analysis have been developed which utilize nonlinear optical processes to perform functions of frequency conversion, light modulation, optical switching, optical logic, optical memory storage, and optical limiter functions. To highlight these new applications, terms such as light wave technology and optical circuitry are also used.

Some of the applications, such as second-harmonic generation, image analysis, high-density data storage, as well as electrooptic spatial light modulation, can be expected to be realized in the not too distant future. The origins of the nonlinear processes are well understood and progress now depends on the development of a materials technology compatible with various device embodiments. A strong technology-driven push in this area comes from the prospect of utilization of third-order nonlinear optical effects which permit all-optical switching, an essential element for the future development of optical information processing and applications in broad-band communications. The two main advantages of using all-optical processing are the gain of speed, since optical switching in the nonresonant case can be in subpicoseconds, and gain in connectivity, because optical channeling through fibers or channeled waveguides permits three-dimensional connectivity. Based on the realization of the latter advantage, concepts of optical neural network for signal processing have been developed. However, a discussion of the concepts of optical computing architectures is beyond the scope of this book. Readers interested in learning more about optical computing are referred to the SPIE Proceeding on Optical Computing, Vol. 963 (1988). Compared to second-order devices, the devices utilizing third-order nonlinear optical processes are in the stage of infancy. The technology utilizing third-order optical nonlinearity can appropriately be described as high risk but high payoff and will take a long time to be developed.

TABLE 12.1

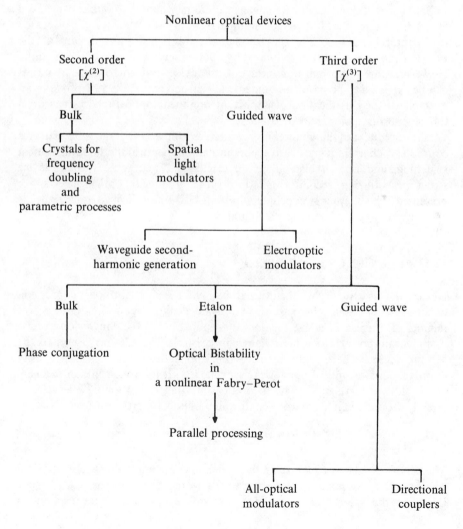

Devices based on nonlinear-optical effects utilize two important manifestations: frequency conversion and refractive index modulation. The latter effect provides a mechanism for light control and hence optical switching and optical bistability. In the case of the second-order effect, the refractive index modulation is produced by an applied ac field. Therefore, light is effectively controlled by an electric field. For the third-order effect, the refractive index modulation is induced by a change of intensity of a controlling optical field. In this case, light is controlled by light that is inherently much faster. Further classification of devices comes from the geometry of specific device structures in which the nonlinear optical material is utilized. The material can be used in a bulk form, such as crystals, slabs, and liquid cells. Most laboratory applications requiring second-harmonic generation of high-power laser sources utilize bulk crystals. Another device structure utilizing third-order nonlinearities is based on a Fabry–Perot etalon (or interferometer) geometry where the nonlinear medium forms the Fabry-Perot cavity. A nonlinear Fabry–Perot etalon is discussed in Chapter 9. In this case the resonance (and hence transmission) of the Fabry–Perot cavity can be modulated by the nonlinear response of the medium. However, more important nonlinear optical devices utilizing organic materials can be expected to be based on guided-wave structures. Table 12.1 provides an example of the classification of important nonlinear optical devices based on the general discussion presented above.

The remainder of this chapter is organized as follows: First, the general device concepts, such as frequency conversion and mixing, optical modulation, optical switching, and optical bistability, are discussed. Then some specific device structures utilizing second- and third-order nonlinear optical effects are presented. The chapter is conclude with a discussion of device structure needs and current status of materials availability.

12.2   FREQUENCY CONVERSION

Devices in this category will be utilized mainly for second-harmonic conversion of coherent radiation. There are two main wavelength regions and laser power regimes of interest. The first category of applications is the conversion of low- or moderate-power diode laser light sources from the 830-nm region to the 415-nm region. The driving force for this application is the packing density improvements in optical memories associated with improvements in focusing at the shorter wavelength. The theoretical limiting resolution of a lens (R) is directly proportional to the wavelength and numerical aperture of a lens (N.A.):

$$R = 0.61\lambda/\text{N.A.} \tag{12.1}$$

where N.A. is related to the refractive index and the angle that a cone of light coming into a lens makes with respect to the normal to the lens as N.A. $= n \sin i$. Doubling the value of R by halving the wavelength of light means that, in principle, four times the amount of information can be stored or detected from memory.

Devices for waveguide second-harmonic generation, fabricated from LiNbO$_3$ substrates, have been recently reported (*Electronics* 1988). In this device, channel waveguides are formed by proton exchange with lithium in the surface region leading to a slightly higher refractive index in the exchanged regions. The principle of operation of these devices is phase matching of the fundamental waveguided mode to a nonguided or radiation mode of the waveguide. Second harmonic is continuously coupled to the radiation mode and directed out of the waveguide at an angle of about 12.5° relative to the direction of propagation of the guided mode. In contrast to conversion between phase-matched guided modes where the interacting waves are confined over long distances, the process described above is inefficient and leads to a large (mm scale) anamorphic beam which may require sophisticated geometrical optics to render it useful for optical memory applications.

Organic crystals having values of $\chi^{(2)}$ over an order of magnitude larger than that for LiNbO$_3$ have been reported. It is also reasonable to expect that poled polymer films with $\chi^{(2)}$ values at least that of LiNbO$_3$ and perhaps significantly larger will also be developed. The advantages of these larger nonlinear coefficients for waveguide second-harmonic generation provide significant motivation for solving the materials problems associated with high-quality waveguide design and fabrication. To illustrate the point, the second-harmonic conversion efficiency is plotted as a function of the phase matched interaction length for a fundamental beam power of 10 mW and a waveguide cross-sectional

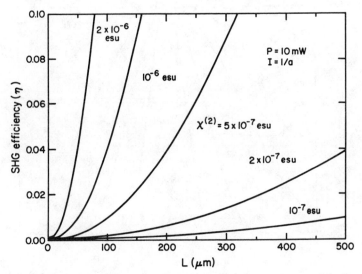

Figure 12.1 The second-harmonic generation efficiency (η) as a function of the phase-matched interaction length L.

area of $1\,\mu m^2$ in Figure 12.1. The drastic reduction in the interaction length (L) required to achieve 10% conversion efficiency, for example, as a function of $\chi^{(2)}$ is evident from the plots. The plot was made using equation 11.9, where the efficiency (η) is the amplitude of the harmonic signal divided by that for the fundamental beam, and the overlap integral between the modes is assumed to be of the order of the cross-sectional area of the waveguide. To illustrate the importance of achieving the desired degree of conversion in the shortest possible interaction length, the phase-mismatch function $\sin^2(\Delta kL/2)$ from equation 11.9 is plotted against Δk in Figure 12.2, where Δk is defined in equation 11.10. Since waveguide mode propagation constant β is dependent on the thickness and refractive index of the waveguide, then variation in thickness and refractive index inhomogeneities lead to phase mismatch and limit the interaction length. As the figure shows, the ability to achieve the desired degree of conversion in the shortest practical interaction length can have significant impact on waveguide tolerances.

Detailed issues of how to obtain organic waveguides with the required optical and geometrical quality and precision, optical transparency, phase matching, and mode overlap are the subject of considerable research interest. If a material and processing technology does emerge for fabricating such devices it could have considerable commercial importance.

The second category of applications is frequency doubling (and mixing) of high-power laser light sources into the mid and deep ultraviolet regions of the spectrum. The discovery that crystals of urea were uniquely suited for this

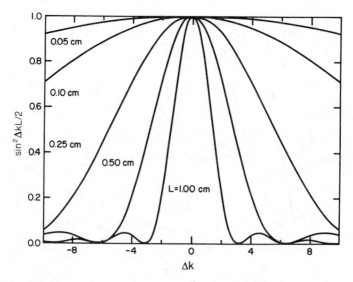

Figure 12.2 The phase-mismatch function $\sin^2(\Delta kL/2)$ plotted as a function of Δk for different interaction lengths L.

application (Halbout et al. 1979) led the way to a useful and convenient set of materials that enabled efficient phase-matched second-harmonic generation down to 244.2 nm in the UV. This was found to be a considerable advantage relative to potassium dihydrogen phosphate and similar crystals where special effects, such as cooling to $-125\,°C$, were required to exhibit phase-matched SHG to that wavelength.

The use of organic materials for other types of frequency conversion processes has been almost nonexistent. Hulin et al. (1986) reported the use of N-(4-nitrophenyl)-L-prolinol for parametric amplification of low-intensity, near-IR photons in the 1- to 1.6-μm spectral region. The new signal is readily distinguished in wavelength from the other signals with a monochromator and photodiode. Gains as high as 10^4 have been achieved, making this a promising technique for near-IR spectroscopy.

Another potentially promising area for organic-based devices is the parametric oscillator. A device of this type might be utilized to generate tunable near-IR radiation with high efficiency. The principle of this device is discussed in some detail in Chapter 5. The nonlinear material is part of a cavity formed by mirrors of appropriate reflectivity and geometrical arrangement for the desired wavelength of operation. The potential advantage of organic materials for this type of device is best illustrated by comparing it to $LiNbO_3$, which has been used for this application. By virtue of its low nonlinear coefficient $LiNbO_3$ requires about $5.7 \times 10^3\,W/cm^2$ to begin oscillation in a doubly resonant device. The threshold for oscillation is proportional to d^2 (Yariv 1975), so that we might except that organic crystals with d values 10–100 times larger should oscillate in the W/cm^2 range.

12.3   LIGHT MODULATION

Light modulation is derived from a change in refractive index of a nonlinear material. This refractive index change can be created by an external control, such as an electric field for second order or an intense optical field for third order. Light modulators find important applications both in laboratory signal processing as fast optical gates or modulators, and in optical signal processing. In relation to optical signal processing and computing, optical modulation devices can be used for beam control and for photo addressing. Light modulators utilizing second-order nonlinearities (electrooptic effect) are more widely used ($LiNbO_3$, KTP, KDP) because the large second-order effect creates a relatively large Δn compared to that derived from the third-order optical nonlineary. The principle for optical modulation is the phase shift induced in the propagation of an input optical signal by electric or optical fields, which leads to the modulation of the output optical signal. Numerous device configurations are possible for utilizing this phenomenon to change light intensity, polarization, frequency and so on.

For a linearly polarized light beam propagating along the optic axis in an anisotropic crystal, the phase shift ϕ can be written as

$$\phi = \frac{2\pi \Delta n L}{\lambda} = \frac{\pi n_0^3 r E L}{\lambda} \tag{12.2}$$

where r is appropriate component of the electrooptic constant tensor, E is the applied field, L is the propagation length in the nonlinear medium, n_0 is the linear refractive index, and λ is the wavelength of light. To introduce a desired phase shift, the voltage required depends on the value of the electrooptic constant r. Many light-modulation schemes require a phase shift of π. To compare, then, the relative strength of nonlinearity (and hence the performance of a modulator), one may specify the voltage required to create a phase shift of π. This voltage, V_π, characterizes the strength of nonlinearity. The higher the second-order optical nonlinearity, the lower is the π voltage required, and the easier is the light modulator to operate. The best second-order nonlinear optical organic materials have $\chi^{(2)}$ significantly larger than that for the best inorganic $\chi^{(2)}$ materials. Hence, organic materials should be quite attractive for electrooptic light modulation. Organic materials also offer another advantage in that their low-frequency dielectric constant is quite low leading to a small RC time constant, thus permitting a higher bandwidth for light modulation compared to that achievable using inorganic materials. Let us compare one specific example: LiNbO$_3$ versus 2-methyl-4-nitroaniline (MNA). The largest components of the electrooptic constant tensor are $r_{33} \approx 31 \times 10^{-12}$ m/V for LiNbO$_3$ and $r_{11} \approx 67 \times 10^{-12}$ m/V for MNA. The corresponding components of the dielectric constant tensors are $\varepsilon_{33} = 28$ for LiNbO$_3$ and $\varepsilon_{11} = 4$ for MNA, and the refractive indices are $n_3 = 2.1874$ ($\lambda = 0.7\,\mu$m) for LiNbO$_3$ and $n_1 = 2.0$ for MNA (Singer et al. 1987).

For all-optical light modulation using third-order nonlinear optical effects, the phase-shift produced by an optical field (the control beam) on itself is

$$\phi = \frac{2\pi \Delta n L}{\lambda} = \frac{2\pi n_2 I L}{\lambda} \tag{12.3}$$

where n_2 is the intensity-dependent refractive index, λ the wave length of light and, L the interaction length of the nonlinear medium. For conjugated polymers, which have large nonresonant $\chi^{(3)}$, we can take $n_2 = 10^{-7}$ cm^2/MW. Therefore, for $I = 100$ MW/cm^2 and $\lambda = 1\,\mu$m, the length required to produce a phase shift of π would be more than a centimeter. It would be difficult to make a bulk all-optical light modulator requiring a phase shift of π because diffraction effects on plane waves would not allow this high-power density to be maintained over a path length of $\geqslant 1$ cm. Consequently, to construct an all-optical light modulator utilizing organic $\chi^{(3)}$ materials, one will have to use optical waveguide structures. The advantage of the $\chi^{(3)}$ light modulator will, of course, be the speed (bandwidth) because modulation can be achieved at rates faster than terahertz.

In the following subsections we give some specific examples of light modulators.

12.3.1 Spatial Light Modulators

A spatial light modulator is a device for transferring the information content of one beam of light to another beam. In this regard they are a very important class of devices with a variety of applications, including optical processing, computing and beam steering, and control functions. An example of such a device is shown in Figure 12.3 (Lytel et al. 1987). In this device light impinges on a photoconductor, in this case amorphous Si, which is reverse biased with an applied voltage. Charge carriers are formed in the photoreceptor as a result of light directed onto it, in an image-wise fashion. The pattern of charge then drifts under the influence of the field to the interface between the light blocking layer and the photoreceptor, where it is trapped. Electric field gradients are generated in the adjacent electrooptic film, which, in turn, generate pattern-wise variations in the electric field-dependent refractive index in the electrooptic layer. A second beam impinging on the other side of the film can be used to read the phase information stored in the electrooptic film.

Depending on the nature of the electrooptic tensor, information content stored in the device can be read out in one of two ways. In the first way, modulation of the electrooptic layer is due to the electric field differences in the electrooptic layer derived from the charges associated with the image pattern and the ground plane. Here, the readout beam acquires a phase shift that is proportional to the electrooptic modulation and the device operates as an intensity-to-phase converter. A device of this sort can be an optical correlator, a convoluter of information, or a phase conjugator. A correlator compares this information in the input image with a reference image and produces spatial patterns containing information on how the two images correlate. The phase

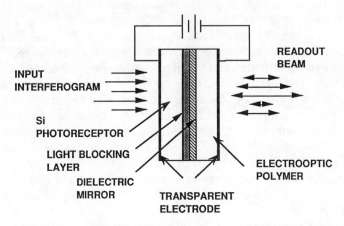

Figure 12.3   The design schematics of a spatial light modulator.

conjugator converts input light on the readout side to its complex conjugate. Physically, this means that the reflected beam will travel back along the same path as the incident beam but with its phase reversed. Any information content lost due to aberrations in the input path is restored as the reflected beam retraces itself. This is in contrast to a normal mirror where the angle of incidence and reflection are equal and phase reversal is not obtained. Devices of this sort are potentially important in laser beam-tracking functions and phase-locking functions from otherwise independent coherent light sources. This phenomenon is explained more fully in Section 12.7.

In the second method for transferring information, the modulation pattern in the electrooptic layer results from transverse field gradients associated with the charge pattern at the interface. Where a field gradient exists a phase difference will appear in the readout beam. A device of this type can act as an edge detector or as a method for encoding position information associated with a light-intensity pattern.

A number of different types of electrooptic materials have been used in spatial light modulators, each having certain advantages and disadvantages. Potassium dihydrogen phosphate (KDP) has been used in conjunction with high-resistivity Si to make a device with spatial resolution approaching 10 line pairs/mm and frame grabbing rates of approximately 1 kHz (Armitage et al. 1985). The low resolution of this device is related to the thickness of the electrooptic crystal, which is about $100\,\mu\text{m}$. The defocusing of electric field lines in the electrooptic material is thickness dependent. Using a similar architecture, nematic liquid-crystalline devices were also shown to operate both as transverse and longitudinal devices (Armitage et al. 1987). The liquid-crystalline based devices exhibited higher-resolution performance than the KDP-based devices due to the fact that thinner films could be used. The speed of these devices was considerably slower than the KDP devices due to the inherent molecular reorientation mechanism associated with the electrooptic effect in these materials.

Use of organic polymeric materials in such devices has not yet been reported. However, an analysis of the potential advantages was given by Lytel et al. (1987). Since organic films can be readily fabricated in the 1- to $10\text{-}\mu\text{m}$ range, resolution comparable to liquid-crystalline materials can be obtained. Since the electrooptic coefficients are expected to be considerably higher than KDP, with response times related to electronic motions, one would expect that the electro-optic coefficient would not constitute a fundamental limitation on the speed of the device. In addition, polymers tend to have dielectric constants around 4 at room temperature. A KDP-based device is typically operated near its ferroelectric Curie point ($\sim 57\,^{\circ}\text{C}$) where the dielectric constant is many times that of an organic polymer. The electrooptic coefficient gets considerably larger due to the electron–phonon coupling mechanism, which is responsible for the effect in that class of materials. The cycle time of the device or rate at which it can transfer images is limited by heat dissipation in the electrooptic material. Organic materials with their lower dielectric constants and, hence, capacitance should have a considerable advantage in this regard. Since the thermal conductivity of organic materials tends to be low, it will be important to design devices with careful attention to heat sinking and dissipation to take full advantage of their potential benefits.

12.3.2   Mach–Zehnder Interferometer

A Mach–Zehnder interferometer provides a convenient device structure for light modulation. This structure can also use nonlinear optical materials in form of optical waveguides. Light modulation using the Mach–Zehnder configuration can be fabricated from both $\chi^{(2)}$ and $\chi^{(3)}$ materials. The basic design of a Mach–Zehnder interferometer is shown in Figure 12.4. An input optical beam I_{in} is split in two parts, which travel through two arms A and B of a Y-type interferometer. In a guided-wave structure these two arms would be the two optical waveguide channels. The phase-shift introduced in the two arms is different so that when the two beams are recombined, the output intensity I_{out} is modulated due to the interference between them. If the relative phase shift between arms A and B is $\Delta\phi$, the output intensity is given as

$$I_{out} = I_{in}(1 + M \cos \Delta\phi) \tag{12.4}$$

where M is the modulation factor. In a Mach–Zehnder interferometer based on the electrooptic effect, the relative phase shift $\Delta\phi$ can be created simply by applying different electric fields (voltages) to the two different arms, assuming both arms (channels) are made of the same material. If the two arms are composed of different materials, the phase shift $\Delta\phi$ is given as

$$\Delta\phi = \pi \frac{V_B}{V_{\pi B}} - \pi \frac{V_A + V_0}{V_{\pi A}} \tag{12.5}$$

where V_A and V_B denote voltages being applied to arms A and B, $V_{\pi A}$ and $V_{\pi B}$

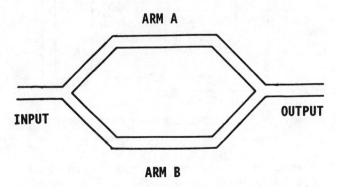

Figure 12.4   The basic design of a Mach–Zehnder interferometer.

are voltages needed to create a phase shift of π in arms A and B, and $V_0/V_{\pi A}$ represents the initial phase shift between two arms (when no voltage is applied).

For a Mach–Zehnder interferometer utilizing a $\chi^{(3)}$ material, one of the arms (B) is simply made of the $\chi^{(3)}$ material, while the other arm (A) is made of a material that behaves linearly in the intensity/range used for I_{in}. The modulation is created by changing the input intensity I_{in}, which causes a change Δn in the nonlinear arm B and, therefore, a corresponding relative phase-shift $\Delta\phi$. Use of spatially selective electrically poled polymers permits a very convenient method to fabricate a Mach–Zehnder waveguide electrooptic light modulators.

12.4 OPTICAL SWITCHING

Conceptually, a nonlinear relation, as shown in figure 12.5, between an optical input and the corresponding output of a device defines optical switching. The switching of the output power from a low to a high value occurs at a critical intensity I_c. Utilizing this optical switching behavior, one can design devices for differential gain analogous to that in electronic transistors. A differential gain is defined by the condition $dI_{out}/dI_{in} > 0$. Also, several logic operations for optical computing can be derived from optical switching. Figure 12.5 also illustrates how optical switching can be utilized for two specific logic functions: AND gate and OR gate operations. An AND gate operation requires the input intensity to be held at the input optical level I_h whereby two optical pulses (of appropriate intensity) would simultaneously be required to switch the device from the 0 (low output level) to the 1 (high output level) state. For an OR gate operation, the hold level is at a higher input intensity I_h. The device can then be switched from the 0 to 1 state by either of the two optical pulses.

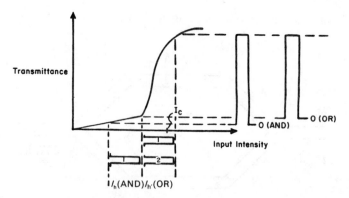

Figure 12.5 A typical nonlinear optical input–output relation exhibiting optical swtiching at a critical intensity I_c. The figure also illustrates the logical functions of AND and OR gate operations with two different input hold levels I_h and $I_{h'}$.

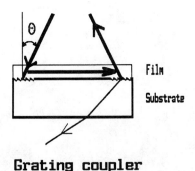

Grating coupler

Figure 12.6   A grating coupled optical waveguide.

Several schematic configurations have been proposed for optical switching. Here, two simple examples are illustrated. The first example is based on a grating (or prism) coupled optical waveguide, as shown in Figure 12.6. Maximum coupling into a specific waveguide mode is characterized by a specific coupling angle, as discussed in Chapter 11. The output power of the beam decoupled through another grating (or prism) is also maximum at this coupling angle. Now, the angle is detuned slightly from this optimum coupling value so that at low intensity, the coupling is not efficient. If the direction of angle detuning is matched properly with the sign of the optical nonlinearity ($\chi^{(3)}$ and hence n_2), then as the intensity increases, the refractive index changes in the proper direction to yield efficient coupling. At this point, optical switching from a low output to a high output level will take place. Optical switching can be seen from an analysis of the input and output pulse shapes. In the switched state a narrower output pulse shape relative to the input pulse shape is obtained. This is due to the fact that intensity in the wings of the pulse will be below the switching threshold and thus will not be transmitted by the device.

Optical switching behavior can also be achieved by using a nonlinear directional coupler, as shown in Figure 12.7, which was first proposed by Jensen (1982). A directional coupler is formed by two parallel single-mode waveguides that are sufficiently close to couple by means of evanescent wave overlap. The directional coupler can be formed by fabricating two-channel waveguides or by twisting together a pair of optical fibers. The coupling is analogous to two coupled oscillators. When light is launched into one channel, the overlap of the evanescent guided-wave field with the adjacent channel leads to a power transfer into the second waveguide. Power will transfer back and forth between the two channels with a periodicity determined by the coupling constant between the two channels, which is refractive index dependent. If one of the channels is made of a large $\chi^{(3)}$ material, then due to the difference in the intensity dependence of the refractive index of the two arms, the cross-coupling conditions become power dependent. In the case where the two channels are made of identical material, with $a_1(x)$ and $a_2(x)$ describing the respective guided-field

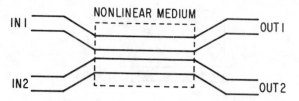

Figure 12.7   A nonlinear directional coupler.

amplitudes (see Chapter 11), the interaction between the two channels can be described by the following coupled wave equations (Jensen 1982):

$$-i\frac{d}{d\chi}a_1(x) = \Gamma a_2(x) + [\Delta\beta_0 a_1(x)^2 + 2\Delta\beta_0' a_2(x)^2]a_1(x)$$

$$-i\frac{d}{d\chi}a_2(x) = \Gamma a_1(x) + [\Delta\beta_0 a_2(x)^2 + 2\Delta\beta_0' a_1(x)^2]a_2(x) \qquad (12.6)$$

where Γ describes the coupling strength between the two channels and is strongly dependent on the channel spacing. The parameter $\Delta\beta_0$ is the power-dependent propagation constant of each waveguide channel, which is obtained by using the intensity-dependent refractive index averaged over the spatial distribution of the fields. This power-dependent propagation constant is discussed in Chapter 11. The parameter $\Delta\beta_0'$ describes the nonlinear effect of a strong field in one channel on the propagation characteristics of the other channel. In the above equation the waveguide attenuation has been neglected. The solution of the above equation have been described by Jensen. One specific case is when the device is designed to transfer all the power to the second channel at low input intensity (power P_i). This situation occurs when the length of the overlapping region is half of a characteristic length, called the beat length, which, in turn, is determined by the coupling coefficient Γ. In this case, as the power is increased the output power switches back from the second channel to the initial channel above a certain input power level. Although this type of optical switching has not been demonstrated for organic systems, it has been shown to take place in channel waveguides of GaAs–GaAlAs multiple-quantum wells (Likamwa et al. 1985) and in dual-core glass optical fibers (Gusovskii et al. 1985, 1987). More recently, Friberg et al. (1987) observed an ultrafast optical switching, using 100-fs pulses, between the two output ports in a 0.5-cm-length, fused-quartz, dual-core-fiber directional coupler in which the quartz itself was the optically nonlinear medium. Organic polymeric systems have values of n_2 many orders of magnitude larger than that of glass and can also be cast into device structures such as fibers. However, problems with processing high-$\chi^{(3)}$ polymers and fabrication difficulties have hampered the study of optical switching in organic systems.

12.5 OPTICAL BISTABILITY

Optical bistability is described by the response of a device that yields two output states I_{out} for the same input intensity I_{in} over some range of input values. Figure 12.8 typifies the input–output relation in the case of optical bistability. The system is defined to be bistable between the input intensities $I^{\downarrow}$ and $I_{\uparrow}$. Between these two intensities two output values are permitted. In the device operation, the optical switching occurs at two input levels. While increasing the input level, a switching from the low to high state of output occurs at $I_{in} = I_{\uparrow}$. However, when the input level is reduced from $I_{in} > I_{\uparrow}$, the device does not switch back to the low output state at $I_{\uparrow}$. The input intensity has to be reduced to a lower value $I_{\downarrow}$ before the device can now switch back to the low output level. The reasons for this behavior are explained below. This type of operation can conveniently be used for memory or latching functions in optical computing. The principle of optical memory operation can be explained by assuming the I_{in} level to be held at I_h, which is in between $I_{\downarrow}$ and $I_{\uparrow}$. Then with an encoding (writing) pulse of properly chosen intensity, the device switches from the 0 to 1 state and stays there even when the encoding pulse is gone. To erase the memory, the I_{in} level has to be reduced from I_h to $I_{\downarrow}$.

An optically nonlinear medium by itself cannot exhibit a bistable behavior. An external feedback is required to obtain optical bistability. The nature of this feedback depends on the device structure. Two sources of general reference on this topic are a special issue of *IEEE Journal of Quantum Electronics* (1981) and a book by Gibbs (1987). Use has been made of both dispersive and absorptive nonlinearities. In the absorptive nonlinearity case, a medium with intensity-dependent absorption (for example, a saturable absorber) is used; one labels this case as absorptive optical bistability. In the case of dispersive optical bistability, the intensity dependence of the real part of the refractive index is used. Optical bistability has been reported in many different device structures.

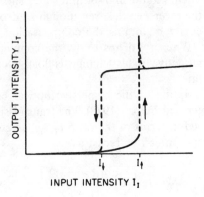

Figure 12.8 Representation of the optical input–output relation exhibiting optical bistability.

Here only two specific cases will be discussed. The first device structure is a nonlinear Fabry–Perot etalon which has been widely used to demonstrate optical bistability. The second structure discussed here utilizes a waveguide Bragg reflector. The behavior of a nonlinear Fabry–Perot etalon will be discussed in detail to explain the origin of optical bistability. First, a conceptual description of both the absorptive and dispersive optical bistability behaviors will be discussed. Then a brief quantitative description of dispersive optical bistability in a Fabry–Perot interferometer will be given. The reason for this ordering is that the dispersive effects are more important in relation to the nonresonant $\chi^{(3)}$ behavior of organic materials.

As discussed in Chapter 9, a Fabry–Perot etalon or interferometer is an optical resonator formed by two parallel mirrors of high reflectivity. The Fabry–Perot resonator is filled or partially filled with a nonlinear medium. In the case of absorptive bistability, this medium is a saturable absorber. When the input intensity is low (less than $I_\uparrow$), the beam is heavily attenuated by absorption inside the cavity. As a result, the output level is low. As the input intensity increases, at certain level $I_\uparrow$, the medium bleaches out due to saturation of absorption. The cavity then switches from the low- to the high-output state. In the reverse cycle, even when the input level is decreased below $I_\uparrow$, the local intensity (field) within the cavity remains high to keep the cavity in the high-output state well below $I_{\text{in}} = I_\uparrow$. A switching back to the low-output state occurs only when I_{in} is reduced now to below $I_\downarrow$. In this case the Fabry–Perot cavity mirrors provide the feedback needed for optical bistability.

For dispersive optical bistability, one uses a medium with an intensity-dependent refractive index in the Fabry–Perot cavity. Here, depending on the sign of optical nonlineary (n_2), the cavity is detuned in a certain direction and by a certain amount, the latter determined by the cavity parameters. At low input levels ($I_{\text{in}} < I_\uparrow$), the Fabry–Perot transmission is low. However, as I_{in} increases the refractive index changes in the appropriate direction, which compensates for detuning. At $I_{\text{in}} = I_\uparrow$, this compensation now leads to cavity tuning whereby I_{out} switches to the high-output state. The intensity (field) within the cavity is high. In the reverse cycle, even though I_{in} drops below $I_\uparrow$, the high intensity within the cavity maintains the compensation of detuning to keep I_{out} in the high state. When I_{in} drops to $I_\downarrow$, the compensation derived from intensity-dependent refractive index change is lost. As a result the output switches to the low state.

A quantitative description of the dispersive optical bistability follows the treatment of Marburger and Felber (1978). The transmission of a Fabry–Perot interferometer is defined in terms of the ratio I_T/I_{in}, as given by equation 9.18 of Chapter 9. This equation can be recast as

$$I_\text{T} = I_{\text{in}} \left[1 + \frac{4R \sin^2 \delta/2}{(1-R)^2} \right]^{-1} \tag{12.7}$$

The different terms in equation 12.7 are defined in Chapter 9 in the discussion of equation 9.18. Now the phase shift δ is assumed to be linearly dependent on the intensity within the cavity, which, in turn, is proportional to I_T. Hence,

$$\delta = \delta_0 + \delta^{(2)} I_T \qquad (12.8)$$

In the above equation δ_0 is the phase shift independent of the intensity, and, therefore, contains the value due to cavity detuning under low input intensity. The term $\delta^{(2)} I_T$ represents the intensity dependent phase-shift derived from the $n_2 I$ contribution. For maximum transmission, δ of the Fabry–Perot cavity must vanish. For transmission to be low at low intensities an initial detuning larger than $\delta_0 = 2(I_T/I_{in}\sqrt{R}) = 2\pi/F$ is required. This detuning is called the cavity instrument width, with F defined as the finesse of the cavity. To observe optical bistability an initial phase shift (δ_0) or detuning large enough to produce a valley in the I_{in} versus I_T plot is needed. Mathematically, it corresponds to the condition for which $dI_{in}/dI_T < 0$. This requirement leads to the condition that for optical bistability

$$|\delta_0| > \frac{\sqrt{3\pi}}{F} \qquad (12.9)$$

This value is fairly close to $2\pi/F$, the cavity instrument width. Therefore, to observe optical bistability, the cavity has to be detuned by a value larger than the cavity instrument width.

Earlier work on dispersive bistability in organic systems used liquid crystals and organic liquids such as nitrobenzene (Bischofberger and Shen 1978, 1979a, b). More recently, Singh and Prasad (1988) used a quasiwaveguide interferometer arrangement incorporating the film of a conjugated polymer, poly-4-BCMU, to observe optical bistability. This arrangement is equivalent to that of a Fabry–Perot interferometer with an oblique angle of incidence.

In a Fabry–Perot interferometer, the feedback mechanism is provided by reflections from the mirrors at the ends of the cavity. Winful et al. (1979) proposed a bistable device in which the feedback mechanism is distributed throughout the nonlinear medium as a periodic variation of the linear refractive index. They suggested that this nonlinear distributed feedback structure has a threshold for optical bistability that is comparable to that for a Fabry–Perot interferometer but with many distinct features. Some of these distinct features are (1) a truly bistable device as opposed to a Fabry–Perot device, which in strict sense is multistable, (2) a decrease in optical hysteresis width with increasing input intensity, (3) characteristic monochromatic spectral response, and (4) compatibility with integrated optics since a waveguide configuration is used. A schematic of a distributed feedback grating device is shown in Figure 12.9 (Stegeman et al. 1985, 1987). The device is fabricated on a nonlinear guiding film grating structure. The incident power in the guided mode P_i is reflected

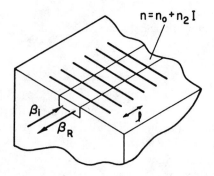

Figure 12.9   The schematic of a distributed feedback grating optical waveguide device.

by the grating. The transmitted power P_T is then

$$P_T = P_i - P_R \qquad (12.10)$$

where P_R is the reflected power.

A sinusoidal grating has a wave vector $k_g = 2\pi/l$ associated with it where l is the periodicity of the grating. The wavevector of the Bragg-reflected beam from this grating is β_R while that of the incident beam is β_i. The Bragg condition for reflection requires that

$$\beta_R = \beta_i + k_g$$

$$= \beta_i + \frac{2\pi}{l} \qquad (12.11)$$

For a geometry in which the reflected and incident beams are counter-propagating (normal incidence), $\beta_R = -\beta_i$ and we have

$$-\beta_i + \beta_R = 2\beta_R = \frac{2\pi}{l} \quad \text{or} \quad \beta_R = \frac{\pi}{l} \qquad (12.12)$$

Since the effective index of the guided wave is power-dependent because the waveguide is a nonlinear medium, the Bragg condition can be tuned optically. Qualitatively, the occurence of bistability can be understood as follows. At low intensity, the incident beam satisfies the Bragg condition given by equation 12.11. Consequently, most of the light is reflected, resulting in very low transmission. As the intensity is increased, the incident guided-wave power increases, leading to a change in the effective index of the waveguide, which now detunes the grating from the Bragg condition. The result is a further increase of the guided-wave power through distributed structures. The device eventually switches from a low- to a high-transmission state. In the reverse cycle, the intensity inside the distributed feedback structure is still high enough to maintain the Bragg detuning condition, even though the input intensity is reduced to below the value at which switching from the low- to high-transmission state had occurred. Consequently, the switching from the high- to low-transmission state occurs at a lower input intensity in the reverse cycle giving rise to optical bistability.

Mathematically, the behaviour of the incident and the reflected beams can be described by the following coupled equations (Winful et al. 1979, Stegeman et al. 1987):

$$i\frac{d}{dx}a_R(x) = \Gamma e^{-i\Delta\beta x}a_i(x) + 2\Delta\beta_0[a_R^2(x) + 2a_i^2(x)]a_R(x)$$

$$-i\frac{d}{dx}a_i(x) = \Gamma e^{i\Delta\beta x}a_R(x) + 2\Delta\beta_0[a_i^2(x) + 2a_R^2(x)]a_i(x) \tag{12.13}$$

where $\Delta\beta$ is equal to $\beta - \pi/l$. Therefore, $\Delta\beta$ is the initial wave-vector (propagtion-constant) mismatch at low intensity. The term $\Delta\beta_0$ is the intensity-dependent change in propogation constant of the waveguide, as discussed in Chapter 11. The terms a_R and a_i are the amplitudes of the reflected and incident waves (the respective power is proportional to $|a|^2$). The solution of equation 12.13 gives the variation of the transmitted versus the incident intensities as shown in Figure 12.10. This plot clearly exhibits a bistable behavior, since there are two output states for the input intensity between $I_\downarrow$ and $I_\uparrow$.

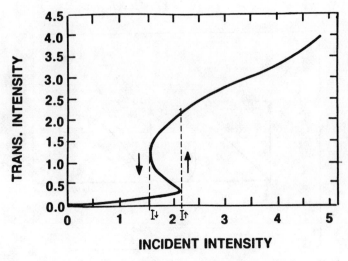

Figure 12.10 Theoretically computed variations of the transmitted versus the incident intensities for a distributed feedback grating Bragg reflector.

Sasaki et al. (1988) reported a study of bistability in an optical waveguide of a polydiacetylene with input and output grating couplers incorporating a distributed feedback grating between them. In their work, 100 layers of Langmuir–Blodgett films of the blue form of fatty acid polydiacetylene were deposited on a substrate. By using the interference pattern at the second harmonic of a Nd:Yag laser, periodic color changes from the blue to the red form of the polydiacetylene film were accomplished. This provided the refractive index variation for the distributed grating structure. The observation of an optically bistable behavior was reported. However, it is not clear whether the observed bistable behavior results from electronic or thermal nonlinearity.

12.6   SENSOR PROTECTION

Another important application of nonlinear optics is in the area of protection of sensors or human eyes. This application has received a great deal of attention in recent years due to very real possibility of eye damage or damage of optical sensors from exposure to laser pulses, especially ultrashort laser pulses, when the input optical intensity approaches a critical level. One can utilize nonlinear optical effects to produce desired optical switching (Figure 12.11A) or optical limiter (Figure 12.11B) response for preventing the exposure to intensity above a critical value (I_c). Figure 12.11A describes the optical switching that occurs at the critical input intensity I_c at which the output switches from a high value to a low value well below the optical damage threshold. Figure 12.11B describes the optical limiter configuration in which the output levels off to a safe limiting value beyond the input value I_c. Devices utilizing either a tunable filter or a

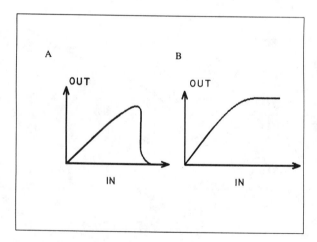

Figure 12.11   The optical switching (A) and optical limiter (B) behaviors of the input–output relation useful for sensor protection.

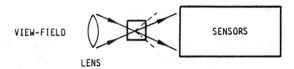

Figure 12.12   Use of nonlinear ($\chi^{(3)}$) focusing for sensor protection.

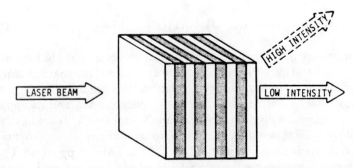

Figure 12.13   The schematics of a distributed feedback Bragg reflector for sensor protection.

fixed-notch filter have also been proposed (Kushner and Neff 1988). The advantage of using a nonlinear process as opposed to processes utilizing photochemistry is the broad-band operation and gain in the response speed. However, not all sensor protection would require ultrafast speed. It is unlikely that one device would meet all the specifications against laser threats. Based on the response speed, intensity level, and wave-length requirements, specific devices utilizing appropriate nonlinearity, nonlinear material, and device structure would be required.

To demonstrate some device concepts, two examples are presented here. One simple approach can be to place a third-order nonlinear material in the focus of an imaging lens, as shown in Figure 12.12. In the low-intensity range, the beams passing through the $\chi^{(3)}$ material go to the sensor with the normal image fields unchanged. As the intensity of the beam increases, the refractive index of the $\chi^{(3)}$ material changes, causing significant self-focusing or defocusing at high input intensity. The net effect is effectively diffusing the dangerous high-intensity input.

A second type of device utilizes the distributed feedback Bragg reflector concept discussed above. To adapt it for the application in sensor protection, a slightly modified structure with periodic variation of the refractive index can be utilized. Such a structure is shown in Figure 12.13 (Foster-Miller Inc., private communication). It is fabricated using alternating layers of optically linear and nonlinear materials, which provide distributed Bragg reflector due to the difference in the linear refractive indices of the two materials. This layered structure provides a three-dimensional Bragg reflector as opposed to the guided-wave distributed feedback grating reflector described above. At low intensity, the conditions are set for a high transmission (detuned from Bragg reflection condition). As the input intensity increases, the refractive index change in the nonlinear medium causes the Bragg condition to change, reducing the transmission. The same configuration can also be used for an optically tunable filter at lower input powers.

12.7   OPTICAL PHASE CONJUGATION

An important application of nonlinear optical processes is in the area of image reconstruction, and the correction of optical distortions for dynamic holography. The principle used is optical phase conjugation, achieved by using the nonlinear optical response. A simple discussion of this phenomenon and its application to real time holography is presented in an article in *Scientific American* (Shkunov and Zel'dovich 1985). A more detailed discussion of the application of nonlinear optical phase conjugation is provided by O'Meara and Pepper (1983).

One convenient arrangement to obtain phase conjugation is degenerate four-wave mixing in a counterpropagating geometry, as discussed in Chapter 8 (Figure 8.5). In this geometry, the signal beam I_4 (Section 8.4) is generated counterpropagating to beam I_3. In other words, if I_3 is the probe beam carrying the information (image), I_4 is generated as a phase conjugate of I_3, which means that it has the same wavelength as I_3 and at each point in the space it is the conjugate of I_3. The conjugate term simply implies the retracing (reversal) of the overall phase factor. In the mathematical form, if the wave I_3 is described as a plane wave

$$E_3(r, t) = \varepsilon(r)e^{i(\omega t - kz)} \tag{12.14}$$

then the phase conjugate wave I_4 is given as

$$E_4(r, t) = \varepsilon^*(r)e^{i(-\omega t + kz)} \tag{12.15}$$

The field amplitude $\varepsilon^*(r)$ for the wave I_4 is the complex conjugate of the field amplitude $\varepsilon(r)$ for the wave I_3. Furthermore, the phase term kz for I_4 is opposite in sign to that for the wave I_3. This effect has important consequences in cases in which beam I_3 has undergone any phase abberation (distortion). Because of the phase reversal, beam I_4 reverses the phase abberation and thus reconstructs the original phase. In terms of holography, this implies that if beam I_3 undergoes any phase distortion created by objects (optical elements) in its path, the distortion is corrected by beam I_4 to produce a high-quality image. Thus, four-wave mixing describes a dynamic (real-time) hologram in which all three processes—recording, developing, and read-out—occur simultaneously. In addition to its application in dynamic holography, there are many other possible applications of optical phase conjugation. Two of these are (1) production of highly directed laser beams, and (2) the self-targeting of radiation. The production of highly directed beams can be useful in correcting for phase distortions in laser amplification.

For most applications using phase conjugation, the time response does not have to be ultrafast (picosecond or shorter). For this reason, it is more important to have a nonlinear optical material, which would produce phase conjugation at a relatively low laser power. Inorganic photorefractive materials, such as barium titanate, produce phase conjugation even with cw lasers. The reason is that these materials are, strictly speaking, $\chi^{(2)}$ materials and the photoinduced refractive index grating is formed due to a space charge field created by photogeneration of charge carriers and their subsequent separation. Phase conjugation, therefore, involves the Pockel's (electrooptic) effect to produce the refractive index variation and can consequently be generated at much lower peak power. It appears unlikely at this stage that the organic $\chi^{(3)}$ materials would compete with inorganic photorefractors for phase-conjugation applications. Specially designed second-order nonlinear materials might offer considerable advantage relative to inorganic materials due to their low dielectric constants and high electrooptic coefficients. Lower dielectric constants would result in a larger internal electric field associated with the space charge distribution.

REFERENCES

Armitage, D., W. W. Anderson, and T. J. Karr, *IEEE J. Quant. Elec.* **QE-21**, 1241 (1985).

Armitage, D., J. I. Thackara, W. Eades, M. Stiller, and W. W. Anderson, *SPIE Proc.* **824**, 34 (1987).

Bischofberger, T., and Y. R. Shen, *Appl. Phys. Lett.* **32**, 156 (1978).

Bischofberger, T., and Y. R. Shen, *Opt. Lett.* **4**, 40 (1979a).

Bischofberger, T., and Y. R. Shen, *Phys. Rev. A* **19**, 1169 (1979b). *Electronics*, p. 36 (July 10, 1988).

Friberg, S. R., Y. Silberberg, M. K. Oliver, M. J. Anderjco, M. A. Saifi, and P. W. Smith, *Appl. Phys. Lett.* **51**, 1135 (1987).

Gibbs, H., *Optical Bistability*, Academic, New York, 1987

Gusovskii, D. D., E. M. Dianov, A. A. Maier, V. B. Neustruev, E. I. Shklovskii, and I. A. Shcherbakov, *Sov. J. Quantum Electron.* **15**, 1523 (1985).

Gusovskii, D. D., E. M. Dianov, A. A. Maier, V. B. Neustruev, V. V. Osiko, A. M. Prokhorov, K. Yu. Sitarkii, and I. A. Shcherbakov, *Sov. J. Quantum Electron.* **17**, 724 (1987).

Halbout, J. M., S. Beit, W. Donaldson, and C. L. Tang, *IEEE J. Quantum Electron.* **QE-15**, 1176 (1979).

Hulin, D., A. Migus, A. Antonette, S. Tedoux, J. Bodon, J. L. Ouder, and J. Zyss, *Appl. Phys. Lett.* **49**, 761 (1986).

IEEE J. Quant. Electron. **QE-17** (1981).

Jensen, S. M., *IEEE J. Quant. Electron.* **QE-18**, 1580 (1982).

Kushner, B. G., and J. A. Neff, in A. J. Heeger, J. Orenstein, and D. R. Ulrich (Eds.), *Nonlinear Optical Properties of Polymers, Materials Research Society Symposium Proceedings*, Vol. 109, Materials Research Society, Pittsburgh, 1988, p. 3.

Lattas, A., H. A. Haus, F. J. Leonberger, and E. P. Ippen, *IEEE J. Quantum Electron.* **QE-19**, 1718 (1983).

Likamwa P., J. E. Sitch, N. J. Mason, R. S. Roberts, and P. N. Robson, *Electron. Lett.* **21**, 26 (1985).

Lytel, R., G. F. Lipscomb, J. Thackara, J. Altman, P. Elizondo, M. Stiller, and B. Sullivan, in, P. N. Prasad and D. R. Ulrich (Eds.), *Nonlinear Optical and Electroactive Polymers*, Plenum, New York, 1987, p. 415.

Marburger, J. H., and F. S. Felber, *Phys. Rev. A* **17**, 335 (1978).

O'Meara, T. R., and D. M. Pepper, in R. A. Fisher (Ed.), *Optical Phase Conjugation*, Academic, New York, 1983, p. 537.

Optical Computing, SPIE Proc. **963** (1988).

Saifi, M. A., and P. W. Smith, *Appl. Phys. Lett.* **51**, 1135 (1987).

Sasaki, K., K. Fujii, T. Tomioka, and T. Kinoshita, *J. Opt. Soc. Am.* **B5**, 457 (1988).

Shkunov, V., and B. Zel'dovich, *Scientific American*, **253**, 54 (1985).

Singer, K. D., S. L. Lalama, J. E. Sohn, and R. D. Small, in D. S. Chemla and J. Zyss (Eds.), *Nonlinear Optical Properties of Organic Molecules and Crystals*, Vol. 1, Academic, Orlando, FL, 1987, p. 437.

Singh, B. P., and P. N. Prasad, *J. Opt. Soc. Am. B* **5**, 453 (1988).

Stegeman, G. I., and C. T. Seaton, *J. Appl. Phys.* **58**, R57 (1985).

Stegeman, G. I., C. T. Seaton, and R. Zanoni, *Thin Solid Films* **152**, 231 (1987).

Winful, H. G., J. H. Marburger, and E. Garmire, *Appl. Phys. Lett.* **35**, 379 (1979).

Yariv, A., *Quantum Electronics*, Wiley, New York, 1975, p. 450.

13

ISSUES AND FUTURE DIRECTIONS

Many of the scientific and technological issues relating to the development of the field of nonlinear optics in organic and polymeric systems have been discussed throughout this book. In this final chapter a brief overview of some of the outstanding issues and future directions that face this field is presented. The issues for second- and third-order nonlinear optical effects and the development and utilization of materials exhibiting them have similarities and important differences. For both second- and third-order nonlinear optical processes the understanding of the basic physics of the processes is relatively complete. Waveguide-based technologies utilizing second- and third-order processes face similar issues to greater or lesser degrees: processability, damage resistance, mechanical and thermal stability, compatibility with dissimilar materials, and patternability. On the other hand, the ability to describe and account for the excited-state charge distributions and dynamics of conjugated polymeric structures is at a more primitive level. Since third- and higher-order nonlinear optical processes are often mitigated by resonant interactions with single and multiphoton excited states, predictive capabilities and overall understanding are also at a much lower level relative to second-order processes. Because of the complexity and richness of the higher-order nonlinear optical processes one might anticipate that investigations into the fundamental physics of electromagnetic interactions in various media will continue to reveal new physical insights and perhaps new effects of interest.

For second-order nonlinear optics the issues to be addressed in the future are divided into three areas: (1) those issues related to the design and synthesis of structures with optimized molecular nonlinearities; (2) local field effects, which relate the molecular nonlinearity to the macroscopically observed value; and

(3) the achievement of optimized symmetry, orientation, and format for various applications. Research resources required for successfully addressing these areas are quite different. The first two areas require sophisticated quantum chemical, synthetic, and laser-based characterization capabilities. The third area may require research capabilities in crystal growth and structure determination, polymer synthesis and characterization, thin-film fabrication and processing, and other capabilities and techniques of materials science.

Many of these same issues pose major challenges for the development of materials for third-order nonlinear optics. The great number of electronic and other contributions to the third-order nonlinear response make experimentation in this field extremely difficult and complex. Ultrafast laser tools are often needed to perform the various experiments required to sort out various contributions to the nonlinear response in the time domain. Sophisticated theoretical methods will also need to be developed to better account for the multitude of processes that occur and to provide a better basis for the rationale design of new materials.

13.1 MICROSCOPIC PROCESSES

Although the understanding of microscopic processes that lead to second-order nonlinear optical effects is at a high level, several challenges remain, particularly in the area of predictive capabilities. Semiempirical quantum-chemical calculations are currently quite useful in establishing trends in molecular hyperpolarizabilities by finite field and sum-over-states methods. The level of confidence in predicted absolute values for the calculated coefficients is at a much lower level. This points to the need for computational research at the ab initio level to obtain descriptions of electronic structures that more accurately account for the influence of the electric field on electronic distributions in molecules of interest. This is particularly important for heteroatom and organometallic systems where the influence of the heavy atom could dominate the electronic properties of the molecule.

The state of understanding of third-order nonlinear optical responses, particularly at the microscopic level, is considerably more primitive than for second-order processes. Quantum-chemical approaches for calculating γ do not, in general, exhibit very good agreement with experimentally determined values for conjugated systems. However, the relationship between π-electron conjugation and γ is satisfactory for small molecules. Ab initio methods have a sound theoretical foundation but these methods, which are limited in use to small molecules, cannot be applied successfully for large molecules. This is partially due to computational inefficiencies, and also to the inadequacy of these methods to account for the dynamics of charge separation and chain or lattice deformations. Currently, one has to rely on semiempirical methods, but these methods are not yet at a stage at which they can be used with any degree of confidence to predict third-order nonlinearities.

13.2   LOCAL FIELDS

Related to the issue of predicting β and γ of an isolated molecule is that of accounting for the influence of neighboring molecules on their values in condensed phases. This is usually taken into account as a correction through the use of a local field factor. The problem of accurately determining the local field is a complex and formidable one, since the field appearing within the microscopic volume element containing the molecule of interest is determined by the external field along with the contributions from charges on neighboring molecules and currents associated with their motion. A complete solution to this problem would involve incorporating these sources of electric fields into Maxwell's equations. The result would be dipolar, quadrupolar, and higher-order electrical, and magnetic contributions to the local field. The use of the dipolar assumption is the basis for the simplified approaches for arriving at local field factors. In this approximation, the contributions from higher-order multipoles are relatively insignificant relative to dipolar contributions form neighboring molecules. The Lorentz model of local field factors is discussed in Chapter 4. It assumes that the molecule resides in a spherical cavity surrounded by a dielectric continuum and that the field inside the cavity can be arrived at by an appropriate integral over the surface of the cavity. While this model is convenient due to its simplicity, its assumption of spherical symmetry for the polarizable molecule may be grossly inadequate for this type of molecule where the charge distribution can be highly asymmetric. For third-order effects derived from electronic excitations such as excitons and polaritons the assumptions may also be inadequate. Clearly, a major theoretical challenge remains in arriving at the proper description of the local field.

13.3   MATERIALS FOR SECOND-ORDER NONLINEAR OPTICS

The development of crystalline materials for second-order nonlinear optics has seen considerable effort. Much of the effort has been based on powder SHG assessments of candidates for crystal growth. A number of useful candidates have been identified by this technique and crystals with large nonlinear coefficients identified and characterized. At this time these materials are not widely used due to the quality of the crystals that have been obtained and their environmental stability. Crystalline surfaces are composed of small molecules with relatively high vapor pressure and tend to deteriorate rapidly. An additional set of issues relates to transparency of these materials. Much of the motivation for their development relates to the efficient doubling of diode lasers in the 830-nm region of the spectrum. Very few, if any, aromatic chromophores with strong donor–acceptor substituents exhibit sufficient transparency at the second-harmonic frequency to be of use for this application. Additional concern surrounds the optical absorption associated with overtone bands due to C–H stretches. At the power densities required to produce efficient SHG, residual absorption could lead to heating and material or device degradation. An investigation of alternatives to π-electronic systems could yield significant benefits in this area. Recently, a class of materials using crystalline complexes of iodoform has been reported (Samoc et al. 1987, 1990) which show very efficient second-order processes without relying on π-electronic structures. Some of these structures, for example, the crystalline complex between iodoform and sulfur, are a hybrid organic–inorganic system. Clearly, considerable work remains to be done on molecular and crystal design as well as packaging and protection for this type of application.

There has also been considerable interest in adapting known organic crystals for use in planer waveguides and fiber-optic cores. The large nonlinear coefficients combined with the power confinement over long interaction lengths in waveguides offers the promise of extremely efficient devices. Extreme focusing requirements can be traded off against long interaction lengths to, perhaps, reduce susceptibility to laser damage. While some progress has been made in growing crystalline substrates in these formats, the general observation has been that control of the proper crystal orientation in the waveguide format has to date been unachievable. Based on the paucity of publications in this area, one might conclude that efforts to accomplish this are being pursued by relatively few groups. It might be expected, however, that these problems will eventually yield to the sophisticated techniques of materials science and perhaps lead to significant classes of applications.

At a more fundamental level, an important set of issues for the development of materials for second-order nonlinear optics is the control of macroscopic symmetry and order. For crystals, one would like to incorporate chromophores with large values of β into appropriate space groups, depending on the application, with optimized geometrical arrangement within the unit cell. Apart from empirical guidance based on molecular topology and nonbonded as well as weak bonding interactions, such as hydrogen bonding, predictive capabilities for crystal structures are virtually nonexistent. Progress in this area might greatly facilitate the development of optimized organic materials for second-harmonic generation and other applications.

The most promising approach to utilizing organic materials for electrooptic effects through the second-order nonlinearity is that of poled polymers discovered by Meredith et al. (1982). This approach relies on the electric field-induced orientation of a nonlinear chromophore, either as a dopant or a side- or main-chain component of the polymer. In any case, the principle is the same. The chromophores must be aligned in a softened state of the host medium and the alignment fixed through physical or chemical hardening of the system. Physical hardening occurs when the system is lowered in temperature significantly below its glass transition temperature and chemical hardening involves the establishment of chemical crosslinks, preferebly involving the nonlinear chromophore itself. The main concern with this approach has been stability of the induced metastable aligned state. Hampsh et al. (1988) have extensively characterized the relaxation of poling induced alignment in glassy polymer systems and various factors that influence it. A number of approaches have been explored for overcoming the loss of alignment associated with the thermodynamic properties of glasses. Ye et al. (1988) showed that hydrogen bonding can lead to considerable increase in stability by establishing a weak crosslinked network. Recently, Robello et al. (1989) and Eich et al. (1989) showed that photoinduced crosslinks in multifunctional acrylic systems or diepoxide–diamine condensation reactions, respectively, can lead to considerable increases in the stability of poling. Much work remains to be done to achieve the wide range of materials properties required for successful device applications. In addition to the issues relating to stability, entire sets of application specific properties undoubtedly exist. Among the more important and difficult properties to control are the electric and dielectric properties of poled films and associated materials in waveguide structures. These properties are crucial to device fabrication and performance, as well as to overall waveguide optical quality and thickness control.

Poled polymers have sufficiently promising properties that work has progressed on the exploration of prototype devices for electrooptic and integrated optic applications. Electrooptic coefficients similar to those obtained for $LiNbO_3$ have now been obtained in a number of polymeric structures. Lytel et al. (1988) demonstrated a number of fundamental device functions, including waveguiding and electrooptic phase retardation, from poled polymer structures. The study of prototype devices and progressive degrees of integration of various passive and active optical functions and components remains a major challenge and will undoubtedly uncover a variety of problems that remain to be solved.

Beyond crystals and poled polymers, a very attractive and potentially useful approach to second-order nonlinear materials is the use of molecular assemblies. The Langmuir–Blodgett (LB) and self-assembled monolayer (SAM) techniques discussed in Chapter 4 offer approaches that are conceptually very attractive for controlling symmetry and order. Since noncentrosymmetry is not a requirement for third-order nonlinear optics, a compelling reason for persuing these classes of materials for third-order processes does not exist. A possible exception may be the exploration of superlattices where semiconducting and insulating structures are fabricated in a periodic array with well-defined geometrical and energetic parameters. In the case of II–VI semiconductors, this type of structure has been shown to exhibit extremely large dynamic third-order nonlinearities.

Although the LB approach has been known for many decades, only very recently has a significant body of chemical understanding of the significant molecular design issues that control film quality and stability begun to emerge. Device concepts based on LB films can be conceptually divided into two catagories: those requiring only a few layers and those requiring hundreds of layers. Examples of devices based on the first catagory are surface plasmon electrooptic modulators and optical sensors. Nonlinear optical waveguides are examples of the latter category. The long-standing problem with LB films has been the formation of domain-like structures due to weak structural correlations and propagation of defects in the films. These domains can lead to severe light-scattering problems, limiting their usefulness for waveguide structures. A challenge for the future is to gain control over the chemical and structural factors that lead to domain formation and eliminate them as a source of concern. SAM methods rely on chemical functionality for inter- and intralayer stability, which may provide for additional structural control and perhaps provide a useful route to materials for nonlinear optic applications. Theoretical and experimental studies in these areas are leading to detailed understanding of intralayer packing and surface requirements for high-quality multilayer films (Tillman et al. 1989).

13.4   MATERIALS FOR THIRD-ORDER NONLINEAR OPTICS

The process of optimizing materials for third-order nonlinear optical applications is a complex one and no single approach is likely to emerge as dominant, due to the range of potential applications in photonics technologies and their particular requirements. It is clear, however, that electronic design strategies will have to emerge for the various categories of applications ranging from the femtosecond response times, perceived to be important for optical computation and switching technologies in communications, to the slower (microsecond) but highly light-sensitive materials required for photorefractive applications. For all anticipated applications the optimization of the nonlinear coefficient is perceived to be of extreme importance and its magnitude is a vital component of any device figure of merit. The requirements for molecular and condensed phase properties can be vastly different for the range of applications anticipated for these materials.

While crystals, polymers, and molecular assemblies are anticipated to be of importance for the development of second-order nonlinear optical materials, polymers and composite materials are anticipated to dominate the progression of organic materials for third-order nonlinear optics. There are several reasons for making this projection. First of all, the largest nonlinear responses observed to date have been in conjugated organic polymers. The delocalization of electrons in these systems appears to be an inherent requirement for large nonlinear responses. Electronic delocalization in these systems is also highly one dimensional relative to bulk semiconductors, which also enhances the nonlinearity through confinement of charge displacement to one dimension. Polymers may also potentially offer simplified routes to fabrication and processing of thin films. Although the stringent symmetry requirements of second-order nonlinear optics do not apply here, there are still advantages for having axially aligned polymeric systems. Since one component of the $\chi^{(3)}$ tensor is usually dominant it is desirable to express this component, in a macroscopic sense, by aligning the material so that its interaction with the optical fields can be maximized.

While it is impossible to anticipate all of the requirements that may be required of materials for the variety of devices and integrated optical circuits based on third-order nonlinear optical applications, some general needs can be anticipated. First, optical damage thresholds will need to be high, since enormous peak powers can be generated in ultrashort laser pulses. The prospect for achieving this is good since optical damage thresholds in excess of $10\,GW/cm^2$ have already been observed. In addition mechanical, thermal, and environmental stability will certainly be needed. It is highly desirable to have a soluble polymer for processing purposes since waveguide optical quality has been achieved through the spin-coating process in polymer films. A number of routes are under exploration for simultaneously achieving excellent optical quality and the large nonlinear response of extended conjugated systems, which are often highly insoluble. Chemical derivatization with electronically inert solubilizing groups has proved useful in achieving solubility in highly specialized organic structures. The concept of using a soluble precursor route (prepolymer) to achieve solubility and then promoting conversion to the polymer has also indicated promise.

Some of the areas where progress in science and technology is critical to the development of third-order nonlinear optical materials are summarized below:

1. *Theoretical Modeling.* To increase the magnitude, especially for third-order nonlinearities, predictive capabilities need to be improved. Progress here might enable the prediction of polymer structural and other functionalization requirements that might be needed.

2. *Molecular Engineering through Polymer Design and Chemical Synthesis.* Progress in synthetic and polymer chemistry and chemical engineering is important for sustained progress in producing systematically varied structures where the impact of important electronic or structural features can be studied. Synthetic strategies must be based on the rigorous processing requirements associated with waveguide fabrication and performance issues. Examples of such issues are control of domain structures that can lead to refractive index inhomogeneities, environmental stability, photostability, mechanical compatibility with substrates, and adhesion.

3. *Material Processing.* Processing conditions are extremely important in determining the ultimate quality of fibers and films of organic and polymeric materials. One promising avenue for optimization of electronic properties, as well as the range of needed ancillary properties, is through the use of polymer blends and composites. An example of this approach is the recent development of a composite between sol–gel processed silica and polyparaphenylene vinylene (Wung et al., in press). This silica gel is an excellent linear optical medium with extremely low optical losses, but it has a very low third-order nonlinear coefficient. Special processing techniques were developed to overcome the general tendency for phase

separation to occur at higher concentrations of polymer. The resulting films have excellent optical quality and a nonlinear coefficient consistent with the concentration of the polymer in the glassy medium.

4. *Device Structures.* Considerable opportunities exist in the development and fabrication of device structures. A number of preliminary device structures have been explored, but as properties of materials are improved and their intrinsic advantages and limitations realized, refinements in devices that can benefit from their advantages will undoubtedly be pursued.

In summary, the field of organic nonlinear optics offers many exciting opportunities for both fundamental research and technological application. As in other high-tech areas, such as microelectronics and genetic engineering, science and technology can be expected to share a vital interplay where advances on one front enable advances and present new challenges on the other. The involvement of a variety of scientific and engineering disciplines including chemistry, polymer science, physics, optics, and device engineering in both academia and industry will be essential to address the challenges that lie before this field. We hope that this monograph will contribute in some way to the progress in this frontier area of science and technology.

REFERENCES

Eich, M., B. Reck, D. Y. Yoon, C. G. Willson, and G. C. Bjorklund, *J. Appl. Phys.* **66**, 3241 (1989).

Hampsch, H. L., J. Yang, G. K. Wong, and J. M. Torkelson, *Macromolecules* **21**, 526 (1988).

Lytel, R., G. F. Lipscomb, M. Stiller, J. I. Thackara, and A. J. Ticknor, in J. Messier, F. Kajzar, P. Prasad, and D. Ulrich (Eds.), *Nonlinear Optical Effects in Organic Polymers*, Nato ASI Series, Vol. 162, Kluwer, Dordrecht, 1988, p. 277.

Meredith, G. R., J. G. Vandusen, and D. J. Williams, *Macromolecules* **15**, 1385 (1982).

Robello, D. R., A Ulman, C. S. Willand, U. S. Patent 4,796,971 (1989).

Samoc, A., M. Samoc, J. Sworakowski, J. Funfschilling, M. Staehelin, and I. Zschokke-Granacher, Mat. *Science (Wroclaw)* **13**, 224 (1987).

Samoc, A., M. Samoc, P. N. Prasad, C. S. Willand, and D. J. Williams, submitted to J. Phys. Chem.

Tillman, N., A. Ulman, and T. L. Penner, *Langmuir* **5**, 101 (1989).

Wung, C. J., Y. Pang, P. N. Prasad, F. E. Karasz, Polymer (in Press).

Ye, C., N. Minami, T. J. Marks, Y. Yang, and G. K. Wong, *Macromolecules* **21**, 2910 (1988).

APPENDIX

UNITS

In the following table, the units for the quantities, $\alpha, \beta, \gamma, \chi^{(1)}, \chi^{(2)}$, and $\chi^{(3)}$ in the international system (SI) and electrostatic system (esu) have been tabulated. A commonly used unit for $\chi^{(2)}$ and $\chi^{(3)}$ is based on the MKS system of units and has the dimensions of $(pm/V)^{n-1}$ where n is the order of the nonlinearity. These units and conversion factors are included for $\chi^{(2)}$ and $\chi^{(3)}$. The definition of the conversion factor, Q, is located at the appropriate column heading and N is the physical quantity of interest. In the table, C = coulomb, sC = statcoulomb (sometimes referred to as esu), J = joule, m = meter, pm = picometers, and V = volt. A thorough discussion of units and the relationships among the various systems is given by J. D. Jackson, *Classical Electrodynamics*, Wiley, New York, 1975, p. 811ff.

Physical Quantity	MKS	SI	esu*	$N_{SI} Q I = N_{esu}$ Q	$N_{MKS} Q = N_{esu}$ Q
α		$C^2 m^2/J$	cm^3	8.988×10^{15}	
β		$C^3 m^3/J^2$	cm^5/sC	2.693×10^{20}	
γ		$C^4 m^4/J^3$	cm^7/sC^2	8.068×10^{24}	
$\chi^{(1)}$		$C^2/J\,m$	1	8.988×10^9	7.960×10^{-2}
$\chi^{(2)}$	pm/V	C^3/J^2	cm^2/sC	2.693×10^{14}	2.387×10^{-9}
$\chi^{(3)}$	pm^2/V^2	$C^4 m/J^3$	cm^4/sC^2	8.068×10^{18}	7.112×10^{-17}

*These units are often quoted as esu.

INDEX